HELPING CHILDREN

A Social History

Murray Levine and
Adeline Levine

State University of New York at Buffalo

New York Oxford
OXFORD UNIVERSITY PRESS
1992

Oxford University Press

Oxford New York Toronto
Delhi Bombay Calcutta Madras Karachi
Petaling Jaya Singapore Hong Kong Tokyo
Nairobi Dar es Salaam Cape Town
Melbourne Auckland

and associated companies in
Berlin Ibadan

Published by Oxford University Press, Inc.,
200 Madison Avenue, New York, New York 10016

Library of Congress Cataloging-in-Publication Data
Levine, Murray, 1928–
Helping children : a social history / Murray Levine and Adeline Levine.
p. cm. Includes bibliographical references and index.
ISBN 0-19-506699-5
1. Child welfare—United States—History 2. Community mental health services—United States—History. I. Levine, Adeline.
II. Title.
HV741.L445 1992
362.7'0973—dc20 91-11192

An earlier edition of this book was published in 1970 as *A Social History of Helping Services*.

2 4 6 8 9 7 5 3 1

Printed in the United States of America
on acid-free paper

The first edition of this work was lovingly dedicated to David and Zachary and to our other colleagues. This new edition is dedicated not only to them but also with much joy, to our new family additions — Joanna and Arielle.

FOREWORD

Over the years, I got into the habit of asking people: "Is there a book that had an impact on you but seemed to have no impact on your field?" With few exceptions, they could come up with, and quickly, at least one book. When I was a graduate student (after the Civil War) I read J. F. Brown's *Psychology and the Social Order*. Appearing as it did in the middle of the Great Depression, that book appealed to those relatively few people with an inclination to political radicalism. It had no impact on mainstream social psychology. It has not been in print for decades. And yet, the contents of that book were prodromal of much that would occupy several fields of psychology after World War II. So if a book has gone into oblivion, it may or may not be deserved; if a book has a status akin to a best seller, it does not follow that it will pass the test of time.

The first edition of this book by Murray and Adeline Levine went out of print after its first printing was exhausted. That, perhaps, was predictable because much education in psychology, but not only psychology, is ahistorical in content and thrust. It came out at a time when the social sciences were still optimistic that they could fathom the workings of the human mind and provide secure bases for societal and institutional efforts to help those whose life was marred by disorder, unhappiness, defects, or self-defeating dependence. One can characterize the post–World War II era as one in which rendering help to others became an ever-growing, ever more costly business and industry. The helping services, the so-called human services, represent a dramatically significant portion of the gross national product.

Coming out, as this book did, in an era of optimism (albeit near its end) people were future, not past, oriented. The past had to be overcome, not understood. We had little or nothing to learn from the feckless past efforts of the human services. Even so, the book was a critical success which meant that few people read it, aside from people like me, and aficionados of the historical perspective who sometimes put the book on lists to be studied by graduate students for comprehensive examinations, or who kept it alive by photocopying library copies after their personal copies were lost or stolen.

The above is prologue to my "finding" that the first edition of this book was mentioned to me by a number of people as having altered their thinking. In addition, about five years ago I became aware that it was being cited in the literature. In accord with a major thesis of that book, its new life is not independent of the larger social scene. The era of optimism is over and the search has begun to understand why we are in the mess practically and conceptually speaking, we are in. Social history, heretofore a museum of relics, is taking on a compelling significance. Perhaps there will even be a new era of optimism in the not as yet foreseeable future. In marching to the future, we can no longer ignore drummers from the past. Not all drummers, of course, because not all of them were as insightful as the Levines.

This revised and enlarged edition is an instance of snatching victory from the jaws of defeat. We are indebted to the Levines (and their publisher) for not allowing a superb book to remain ignored. However often disconfirmed, the reappearance of this book does indicate that there is justice in this world. In my foreword to the first edition, written at a time when many were discovering the field of community mental health, I said: "What this book demonstrates is that this 'new' field may be new to us but it was truly old hat to a variety of individuals who lived decades ago. . . . What the reader will find in this book are the accomplishments of some heroic figures who both reflected and transcended their times. . . . In the end there is the word, whereas at the beginning there was the idea. The Levines help us to recapture the idea." The ideas are still worth recapturing.

Yale University
1991

Seymour B. Sarason

PREFACE

One day in 1963, we began to discuss the puzzling fact that many helping services for children originated at the turn of the nineteenth century. From that discussion grew our plan to write a social history, showing how services for children and modes of treatment were integral parts of the political economy, not simply the consequence of scientific developments within the fields. The book, *A Social History of Helping Services* was published in 1970. In 1990, we decided to prepare a new edition because we found that the *Social History*'s subject matter, and its concepts, were pertinent to problems of delivering services to children and families today. The new generation of people entering the helping professions were just entering elementary school at the time the *Social History* was published. The experiences of the 1960s were ancient history for them. Nonetheless, as we observed the development of the mental health fields we saw new services developing as outgrowths of 1960s programs, and as we entered the 1980s, we saw the renewal of orientations that emphasized individual psychopathology. The problems associated with poverty continued to be with us, but for a decade or more, efforts to deal with those problems were not front and center on the national agenda. With the growth of attention to the homeless population, and with the growth in our understanding of the need for preventive programs, the concepts and approaches of the past are worth renewed attention, if only to show us some of the paths that earlier workers traversed. If our attention to problems related to poverty is indeed cyclical, we believe that sometime within the next ten years, we may enter a period in which we will pay renewed attention to several important social problems. After the 1960s, welfare problems, child protection, and issues related to birth control and abortion became increasingly important to the mental health professions. The first edition of this book did not cover those problems. This book does because we believe that we understand more about where we are and where we might be going when we understand where we have been. It is in that spirit that we offer this new edition.

This edition differs from the first one in several respects in addition to its new title. First, we have added four new chapters which cover developments in child-related services from 1930 to 1980, welfare, child protection, and abortion and birth control. Second, we have eliminated one chapter that discussed the precursors of the community mental health movement of the 1960s, feeling that this material is now well known and has been amply covered elsewhere (Levine, 1981;

Levine & Perkins, 1987). Third, we have edited each remaining chapter to take into account more recent scholarship. Finally, we have rewritten the final chapter because it seemed to reflect too much the optimism of the 1960s. The basic ideas remain, but in condensed form. We have also added some comments to that chapter relevant to the additional topics covered in this edition.

We want to thank colleagues who expressed the wish that the first edition had not gone out of print. This new edition is a response to their encouragement. We also wish to thank Joan Bossert, our editor at Oxford, for her enthusiasm and her faith in the project, and Margaret Yamashita, our copy editor, for her diligence in helping to polish the manuscript.

Buffalo A.L.
April 1991 M.L.

ACKNOWLEDGMENTS

Through their support and interest, our colleagues at the Yale Psycho-Educational Clinic have helped us to write this book. We owe a special debt to Seymour B. Sarason, who first brought the necessity for historical studies to our attention, who shared ideas with us, read and commented on earlier drafts of all the chapters, and who allocated some of the resources of the clinic to this work. Without his interest and encouragement, this work would not have been undertaken, much less completed.

We are also very much indebted to a number of other people who graciously granted interviews, or who read and criticized various chapters. These people include E. K. Wickman, George S. Stevenson, Everett S. Rademacher, David Levy, and Milton Senn, who all contributed to the chapter on the child guidance clinics. Morris Viteles was helpful in shaping the discussion of Witmer. Rube Borough and Charles Larsen were among many who contributed to the chapter on Ben Lindsey. Staughton Lynd read much of the manuscript, ostensibly to check historical fidelity, but he also offered some rather keen psychological observatrions which were most helpful. Other friends and colleagues who read through and commented upon a goodly portion of earlier drafts include Barbara Frankel, Janet Rosenberg, [the late] Marcia Guttentag, and Paul Weiss.

Our sons, David and Zachary, deserve special mention and commendation for participating with their parents in this work for a good part of the last several years. ''When you can't lick 'em, join 'em'' must have been their motto, for they joined in making the book almost a family project. They also contributed a great many potential titles, all of which were wittier, if less descriptive, than the one we finally selected for the book.

The following publishers kindly granted us permission to quote from copyrighted works. The specific quotations are identified in the text.

American Journal of Orthopsychiatry
Appleton-Century-Crofts
Commonwealth Fund
Little, Brown Company
Horace Liveright Publishers
Macmillan Publishing Company
New York Public Education Association

The Psychological Clinic (Morris Viteles, Ed.)
Russell Sage Foundation

We also acknowledge our indebtedness to the original authors of material now in the public domain, upon which we relied heavily. This includes the anonymous letter to Jane Addams, first published by Christopher Lasch in his volume *The Social Thought of Jane Addams* (Bobbs-Merrill, 1965). The specific quotations are acknowledged in the text and in the bibliography.

We owe another special debt to Anita Miller, who in this instance was not only secretary, but also research assistant, private eye, locator of missing persons, and cheering section. That the manuscript was completed in good style is due in no small measure to her intelligence, her energy, her capacity for work, and her general good will.

New Haven M.L.
1970 A.L.

We want to thank all the colleagues who used the first edition of this book and urged us to update it. We also thank Bridget Winter, Lee Gordon, Nancy Komorowski, and Kavita Munjal for helping get the final manuscript printed and copied, with cheer and good humor.

Buffalo M.L.
1991 A.L.

CONTENTS

I

THE PROGRESSIVE ERA AND ITS AFTERMATH: RESPONSES TO SOCIAL PROBLEMS

In the mid-1960s, out of nothing more than idle curiosity, we began looking into the early history of clinical problems and clinical services for children. It quickly became apparent, however, that the early history of clinical services for children was important to the community mental health movement then getting under way. Now, more than twenty years later, that early history continues to be important to the development of services for children and adolescents. Although "everyone knows" that Lightner Witmer started the first psychological clinic in 1896 and that William Healy started the first "real" psychiatric clinic for children in 1909 as part of the juvenile court in Chicago, it was not until we started reading the original material that we could find references to what those early clinics actually did. It was all too easy to assume that early clinics were unsophisticated versions of contemporary clinics and to dismiss them as having no contemporary relevance. But as soon as we began our research, we knew we had tapped a rich vein, indeed.

Going back even further than the late nineteenth century, we learned that there has always been a need to care for those unable to care for themselves, whether they be widows and orphans, the sick, the old and the disabled, or the mentally ill and the mentally retarded. From colonial times until the early 1800s, the colonists followed the patterns set by the English poor laws of 1601. The predominant solutions were institutional: the undifferentiated almshouse. In the early nineteenth century, dependent youth became more of a problem because the market for apprentices was declining as the economy moved into a factory system of production. Prisons and asylums were built during that first quarter of the nineteenth century, and in the mid-1820s "houses of refuge" were built for boys who had committed less serious offenses, along with runaways and disobedient and vagrant children. In that early period, specialized institutions also were constructed for orphans and different types of disabled children.

Care in the community did not begin until the cities, especially those in the Northeast, experienced massive immigrations first from Ireland and later from Germany. Italian and eastern European immigrants came later, beginning in the 1880s. These waves of immigration were accompanied by poverty, slums, and family disorganization. In the 1840s and 1850s, some community services for children and youth were offered, often religiously based, and were provided by voluntary organizations. The immediate predecessors of the present-day clinical services did not emerge until between 1880 and 1930. Those services were part of the community, were concerned with education, and were oriented toward prevention. In short, the early models of clinical services demand close examination precisely because they began with the types of services that the community mental health movement was offering in its early years in the 1960s. It thus was quite a shock to us to discover that Yale's Psycho-Educational Clinic, established in 1963 as an innovative clinical service, was somewhat reminiscent of the 1890s!

We were not alone in our innocence. In the spring of 1966, one of us participated in a panel discussion on the relationship between day care centers specializing in educational programming for emotionally disturbed children and public school settings. The problems that the 1966 workers were discussing bore an amazing similarity to problems faced by workers several generations earlier, but they were being approached as if they were brand-new. Today, in 1990, those concerned with modernizing service systems for children and adolescents are still dealing with approaches akin to those adopted by the early workers described in this book. Because of the efforts of community health workers over the past twenty years, people now planning children's services are generally more aware of the need for approaches that do not rely totally on inpatient treatment, in a hospital or a residential treatment facility, or outpatient psychotherapy. But even with this greater awareness, the experiences of the early workers still are helpful.

1

Introduction

The services for children established at the turn of the century and the following two decades started out with what we call a community orientation, or a preventive orientation, but in a relatively few years, these services changed. Although Witmer's clinic was closely involved with the public schools, and the Commonwealth clinics of the 1920s made a conscious effort to relate to the public schools, it was commonplace by the mid-1960s for child guidance clinics to have either no contact or minimal contact with the teachers and principals of children they treated. Because of the influence of the 1960s community mental health movement, many clinics started consultation programs with schools, and so today there may be somewhat more communication with the schools about children under treatment, although the restriction of confidentiality may limit two-way communication. The core service in most child and adolescent clinics is still, however, some form of group or individual counseling. Some clinics working with children from poverty populations, and those working with abused and neglected children, offer a wider range of services, but such programs are still considered innovative.

The contemporary child guidance clinic, although no longer called that, is professionally staffed and offers individual, group, or family psychotherapy, as well as intake and diagnostic services. Most clinics have a rather small investment in consultation with community agencies or schools; a very small proportion of clinic time is spent in community education. In these budget-conscious days, agency executives must spend the bulk of their time pleading with local and state governments to fund the existing programs. Consequently, few have time to spend lobbying government agencies and officials to improve social conditions that may cause or exacerbate some mental health problems. Clinics do not ordinarily seek cases; they still operate on the medical model of treating those who voluntarily present themselves for treatment. Ironically, some thirty years after the community mental health movement began, with its emphasis on prevention, preventive programming is barely beginning to come into its own right, although our thinking about prevention has probably become more sophisticated in the intervening years.

From a historical perspective, individually oriented clinical services for children have been a failure. The fact that any treatment-oriented, in-clinic service would be immediately swamped was apparent from the earliest days, and so the early clinics consciously tried to avoid getting bogged down. Nonetheless, they quickly acquired discouragingly long waiting lists. Even though they intended to devise preventive methods of helping, here, too, they were unsuccessful. Witmer's clinic, Healy's clinic, the visiting teacher movement, the settlement houses, and the early Commonwealth Fund demonstration clinics were closely tied to a variety of community agencies; whereas contemporary clinics usually are not. The goal of influencing those at home or in school to care for mental health problems more effectively—a goal of the early clinics—has succeeded only to the extent that psychiatric teaching has influenced professional social work, but even this influence has been limited. The graduate schools of social work turn out relatively few trained workers each year, a great many of whom function as psychotherapists and have only a limited social influence.

In the sixty years between the 1840s and the turn of the century, many of the innovations in treating mental health problems were made by people outside the professional mental health fields. The early settlement house workers were responsible in part for the establishment of the juvenile court, with its system of probation workers, and the first probation worker was the model of the *indigenous nonprofessional,* a term that became prominent in the service vocabulary in the 1960s. The settlement house workers also originated the visiting teacher concept, the forerunner of the school social worker. The court had been operating with some effectiveness as a helping agency for ten years before the psychiatrist William Healy began working with it. And Judge Ben Lindsey, who created the juvenile court in Denver, regarded the court of law as an institute of human relations whose function was to help people and not to punish them. Lindsey, noted for his contributions to law and later notorious for his advocacy of companionate marriage, has been forgotten as a clinician, although his therapeutic achievements and his methods reflected a rare human artistry. Nor should we forget that Lightner Witmer—trained as an experimental and not as a clinical psychologist—frequently used schoolteachers in therapeutic roles. Indeed, in at least one of the cases he reports, the important therapeutic work was done, with his knowledge and encouragement, by a housemaid. The clinic associated with the Cincinnati court (circa 1914) extensively used volunteers from the Big Brothers, the Rotary Club, and other civic organizations to work with children needing help.

The child guidance movement itself owes its existence to Clifford Beers, a Yale graduate and a layperson in the field, who suffered a severe psychosis. After he recovered, Beers organized the National Committee for Mental Hygiene in 1909 (Beers, 1933), the group that pressed for opening child guidance clinics as preventive agents. Even earlier, Charles Loring Brace, a minister and a great-grandfather of the social work profession, offered services for the homeless and indigent population. Child protective services started in the 1870s, influenced by those concerned with protecting animals and separate from the Charity Organization Societies, which then were still made up of volunteers but were rapidly moving toward having a professional staff. As a case in point, although Margaret Sanger was trained as a nurse, her work in opening birth control clinics depended as much on her background as a political radical as on her experiences as a nurse.

Regrettably, mental health personnel tend to suffer from what Sarason and his colleagues (1966) have called "professional preciousness." Preciousness is the feeling that one's professional training has uniquely fitted one for offering mental health services and that anyone who does not have the requisite training not only will be unable to perform such services but will in fact do harm. As a result of experiences in the 1960s and a shortage of personnel, that attitude may be somewhat less prevalent today in regard to the delivery of public services. But preciousness continues to be revealed in professional turf battles, accreditation standards, and licensing requirements. This attitude resists innovation in practice and is highly sensitive to infringements on professional roles and functions in many clinical settings. In fact, one could argue that the current selection and training process may produce many professional workers who are ill prepared to work with the more serious problems in the community. In any case, the contemporary mental health professional is a Johnny-come-lately in a field largely developed by those who were in no way trained for the tasks they performed. Such knowledge, however, provides a basis for humility and may help us understand that various practices are not necessarily direct consequences of advances in science.

The fact that social services originated with nonprofessionals is another reason to study the social context of those services. They did not start out as conscious extensions of scientific hypotheses, and if they were a response to social need, then the services must indeed reflect their times.

A comprehensive and detailed history of mental health and social services is not our intention. Rather, our book considers those services for children that began roughly between 1890 and the mid-1920s. From about 1890 to World War I is the first well-demarcated period, and the decade from 1920 to 1930 is another (Phillips, 1990). They represent, respectively, a period of reform and a period of conservatism, and the change from the one to the other can be seen in the philosophy and implementation of services in the two periods. After the first edition of this book was published, we became interested in several other contemporary services whose origins we had not discussed before. These services became important in the 1960s and later but originated outside the period we first studied. These services are birth control, abortion, welfare, and child protection. Each of these was started in response to social changes in an earlier period of immigration and urbanization, or in response to economic distress. The first edition of this book ended in the 1930s. The succeeding years of the Great Depression, the post–World War II period, and the 1960s, however, also represent periods of social change followed by periods of conservatism.

We have focused on services for children because they are among our primary professional interests, but there is nothing in our thesis that restricts itself to specific time periods or to work with children. In fact, we suggest that those interested in sociohistorical research examine other periods of change and conservatism to determine whether the principles we have specified pertain to other kinds of helping agencies and to other ages. Grob's (1966) excellent study of the state mental hospital in the mid-nineteenth century and Rothman's (1971) study of the asylum are examples.

We have tried to be true to the data—human documents—but our approach may be more similar to that of the artist or the historical novelist than to the quantitative empiricist. We wanted to tell a coherent story even if we could not count the number of positive and negative instances found in some sample of documents in our time period of interest. Like Jules Henry (1965), we are as interested in events that occurred often enough to be noticed and recorded as we are in how often the events occurred. Our professional conscience, nonetheless, requires us to warn others about how we undertook this project. We have great admiration for historians such as Gordon, Horn, Katz, Lubove, Platt, and Ryerson, whose works we have consulted for this new edition. But we also feel our sociological and clinical training enables us to examine the historical material from a useful and different perspective.

Our purposes, then, in writing this book are several. First, we simply want to describe a variety of helping services and the conditions under which they arose. We want to call attention to certain modes of operation and the problems that earlier workers encountered, so that we will neither repeat the past nor lose valuable models. Second, we want to explore the thesis that social and economic conditions and the intellectual and political spirit of the times greatly influence the mental health problems that concern us and the forms of help that flourish. As a corollary of this thesis, we might argue that changes in the forms of help are shaped at least as much by the predominant social forces of the times as they are by thoroughly supported developments in the science of human behavior.

More specifically, our thesis states there are essentially two modes of help, the situational and the intrapsychic. The situational mode assumes that people are basically "good" but have been exposed to poor conditions and therefore have not reached their full potential. Improving their situation thus should improve their psychological states. Improving people's situations may mean creating new services or new community facilities or modifying existing community agencies to better serve human needs. On the other hand, the intrapsychic mode assumes that people are in difficulty not because of the situation but because of personal weaknesses and failings; it assumes that the situation is more or less irrelevant, that what must be changed is not the circumstance but the person.

We believe that situational modes of help, which demand that we question the social environment—and change the social environment—flourish during periods of political or social reform. At such times society is ready for change and so will accept change. Intrapsychic modes of help, because they focus on the inadequacies of the individual, assume the goodness of the environment. Intrapsychic modes of help therefore are prominent during periods of political or social conservatism, for such modes usually support the status quo by placing the onus of both the problem and the change on the mind and emotions of the individual. That is, a conservative ideology believes that the present is the best of all possible worlds. If a person is having difficulty in this best of all possible worlds, then clearly something is wrong with that person. On the other hand, during periods of reform or social change, we attribute responsibility for personal difficulties to the social environment; community agencies or community structures are not meeting people's needs. Such agencies thus need to be changed or created in order to help people. We cannot claim to have originated this thesis. Cohen (1958) mentions a similar viewpoint in his history of American social work, and Lubove (1965) also uses this theme. But the following elaboration of this thesis is ours.

No one period is exclusively dominated by one or the other form of help; instead, each period is marked by an intermingling of both forms of help. Each reform period has conservative critics, just as each conservative period has reform critics. Each period contains the seeds for the dominant form of the next period, so that we cannot expect unambiguous support for our thesis at any one time. The hypothesis itself is not readily amenable to empirical test; it serves only to direct attention to observations that themselves may become clear only in historical perspective. We stated the general hypothesis based on our observations of the programs and periods covered by the first edition, aided by our participation in some of the community mental health programs of the 1960s. We leave it to the reader to decide whether our thesis helps illuminate the approaches of the 1980s. The thesis should have continuing value because it directs attention to variables that mental health professionals ordinarily ignore. In 1990, those concerned with modernizing service systems for children, who have the advantage of observing the problems in implementing community mental health programs, undoubtedly already know how political and social realities limit and shape the potential for action. We stated this some twenty years ago, and we are restating it now because we believe it continues to be relevant to today's services.

2

An Era of Reform, 1890–1914

Almost all modern professional community-based services for children were established between 1890 and the beginning of World War I as part of a general concern about their welfare. Cohen (1958) lists the following agencies that pertained directly or indirectly to child welfare and to services to children: Family Service Association (1911), U.S. Children's Bureau (1912), National Committee for Mental Hygiene (1909), the national organization of the YWCA (Young Women's Christian Association) (1906), National Federation of Settlements (1911), Boy Scouts (1910), Camp Fire Girls (1910), Girl Scouts (1912), and Pathfinders of America (1914). To these we could add the Public Education Association of Philadelphia (1881) and of New York (1895) and Lightner Witmer's psychological clinic (1896). The juvenile courts in Denver and Chicago (1899), the visiting teacher movement (1906), and Healy's Juvenile Psychopathic Institute (1909) belong to the same period. These services also had precursors in even earlier efforts to deal with the problems associated with poverty and immigration. The Progressive-era programs, however, are more direct ancestors of present-day professional services, and so we shall emphasize these in Part I.

The turn of the century marked the start of the "century of the child" (Kanner, 1962; Key, 1909), and the field of child study began about this time in Europe as well as in the United States (Kessen, 1965). That professional clinical services for children were now being offered should not be a surprise. A renewed and deepened concern for people, which encompassed all areas of health, education, and welfare, was part of the zeitgeist of this period. In the United States, the concern with children grew out of the ferment accompanying great social and economic changes.

In the years after the American Revolution, with the important exception of the African slaves, the United States was a relatively homogeneous country, racially, ethnically, and religiously. Like the earliest British, Dutch, and French settlers, most of its citizens were white, from western and northern Europe; a British-derived culture predominated; and by 1820 the German-speaking and other language minorities were assimilating rapidly. Although cities were growing, the United States was largely an agrarian country and had not yet felt the full force of the Industrial Revolution. By the 1840s, however, a great deal had changed. Andrew Jackson had been elected president in 1828, marking the first time that a common man and not a landed gentleman had assumed that office. In addition, the principal source of wealth was moving from agriculture to commerce and manufacturing. Because manufacturing drew people from their homes into the factories, it changed the pattern of relationships within the home and between employer and employee (Johnson, 1978). Steamboats, canals, railroads, and the cities teeming with immigrants contributed to a new sense of vitality, and many basic institutions no longer seemed appropriate. The laws governing rights to land changed, accordingly, to those promoting its development (Horwitz, 1977). The apprenticeship system was breaking down, and religious revivals swept the land. When Catholic immigrants arrived, they wanted to have their own parochial schools, and this led to conflicts over the school system, which contributed to the passage of compulsory school attendance laws (Sarason & Doris, 1979). Reformers took as their targets education, labor, politics, debt, war, health, family life, church, prisons, asylums, the welfare system, and the care of both dependents and deviants (Degler, 1959).

Beginning in the 1840s, the stream of Irish immigrants grew into a torrent. Between 1846 and 1855, nearly 1.3 million Irish people arrived. At about the same time German immigration increased rapidly as well. Most of these immigrants settled in the northeastern cities; only relatively few went to the frontier (Seller, 1977). In contrast with an earlier time, by 1850, 15 percent of the population of the Northeast was foreign born, and by 1860, nearly 20 percent had been born abroad. The immigrants had a difficult life. Most arrived with little money, and employment was not easy to find, especially for the thousands of rural people suddenly planted in an urban center. The Irish "faced exhausting difficulties in making a place for themselves in the city's economic life. There was no one to help them; the hard-pressed Catholic priest and the overburdened benevolent and immigrant-aid societies could assist only a few" (Handlin, 1970, p. 77). Irish women often entered service as domestics, and many of the men took degrading and repugnant positions, for that was all that was available to them. From the time the Irish arrived in the United States until the Civil War, too many competed for too few jobs.

> Urban families worked together at the most menial and unpleasant tasks in order to survive. Impoverished Germans in New York City became scavengers. Men, women and children gathered discarded bones from slaughter houses and filthy rags from hospitals and gutters. In their tenement apartments, they boiled the rotting flesh off the bones, washed and dried the vile-smelling rags, bagged the products, and sold them to refuse dealers for a few cents a bag. (Seller, 1977, p. 77)

It was in this context in the 1850s that Charles Loring Brace began his work among New York's "dangerous classes."

The immigrants' increasing misery forced society's attention to the human consequences of the industrialization and urbanization following the Civil War when the railroad, mining, oil, steel, meat-packing, and other industries flourished. The administrative and organizational genius of the early economic barons combined science, invention, and machinery to devise techniques for large-scale production that both responded to and encouraged the demands of a swiftly growing nation. Industrial expansion was accompanied by still more rapid urbanization and a mushrooming of social problems. In 1860, some 80 percent of the people lived in rural or farming environments, but by 1920, nearly 60 percent lived in urban environments. Those coming to the cities were mainly immigrants drawn by the voracious needs of an expanding industry for a labor force. When the number of Irish and German immigrants started to fall, beginning in 1880, Italian and other southern Europeans, as well as Russians and other eastern Europeans, began arriving in unprecedented numbers.

By the 1870s and 1880s, farming had become less attractive as an enterprise, by the 1890s the costs of establishing a farm were becoming prohibitive, and opportunities on the western frontier were diminishing. Lower prices for farm products, rising costs of transportation for both farm products and manufactured goods, and financial policies favoring creditors helped depress farmers and accelerate the process of urbanization. At this time, then, the children of farmers started moving to the cities (Beard & Beard, 1927).

According to one view, what happened to the people in the period of progress following the Civil War was not important. Not only did the government follow a laissez-faire policy, but the intellectual spirit of the day also supported the position of the ''man of affairs,'' as Eric Goldman (1956) showed so vividly. Liberty meant the right to acquire and keep property without the interference of government. Competition was the correct means of distributing wealth. If the poor were poor, it was because they had too many children who ate up their incomes. According to this theory, the immigrant groups and blacks were born to live in squalor and poverty because they lacked an inherited instinct for liberty. God gave people their abilities, and any attempt to interfere with their use was a violation of His will.

Herbert Spencer, a brilliant and facile English sociologist, came to America and was lionized because he formulated social Darwinism, which held that society was an organism that evolved according to the laws of the survival of the fittest (Spencer, 1873). The businessman had evolved to survive. Any social change that benefited the poor and the halt could result only in social disaster because it interfered with ''nature, red in fang and claw.'' As expounded by William Graham Sumner at Yale, such interference would weaken the race, threaten the economic system by placing an extreme burden on ''the forgotten (middle-class) man,'' and fly in the face of American ideals of laissez-faire (David, 1940; Hofstadter, 1955; Sumner, 1883).

The concepts of social Darwinism provided a philosophy for scholars and a rationale for the dominant classes and also what Eric Goldman (1956) called ''conservatism's steel chain of ideas''; that is, there was actually no point in trying to change social institutions if people were born to live in a certain way.

A conservative philosophy maintains that the problem resides within the person and therefore that corrective measures, if any, should change the person. The alternative viewpoint, that people are essentially good but are shaped by their environments, is generally characteristic of a liberal or reform philosophy. As we shall argue, the very form of professional practice in the mental health professions is shaped by whether the nation happens to be in a conservative or a reform period.

Although some social scientists or clinicians might refute the idea that theories and methods of professional practice are molded by political and social philosophy, this is a concept that deserves attention. Pastore (1949), for example, showed an important connection between scientists' professional stands in the nature–nurture controversy—an issue on which liberals and conservatives general differ—and their public stands on political issues. Scientists who are political conservatives clearly favor the nature viewpoint. If one can discern such generalized attitudinal influences on scientific positions, it should also be possible that other aspects of scientific activity are shaped to some degree by the spirit of the times.

The Conditions for Change

After the Civil War, not all was well with the American system, and not all Americans accepted their inferior positions passively. The depression of 1873 was followed by disorder. On a political level, during this era the Populists, the Progressives, the Socialists, and the Anarchists rose to the fore. The labor movement took hold, and it was marked by strikes, violence, rioting, and the use of federal and state armed forces to maintain order. Turmoil and fear of social conflict, or worse, were widespread. Anthony Comstock's crusade against obscenity took place in this climate of concern about a loss of true American values. Just as the threats and reality of urban riots provided the background for reforms in the 1960s, so did violence and unrest create the climate for reform in the last part of the nineteenth century.

The intellectuals of that time, many of them moved by the wretched conditions of the poor, opposed the imperiousness of social Darwinism and emphasized an environmental determinant of human behavior. Such ideas permitted social reformers not only to examine society but also to demand concrete reforms (Goldman, 1956; Hays, 1957). Henry George (1879) decried the inevitable association of poverty with the increase in national wealth. Muckrakers drew attention to abuses in industry. Lincoln Steffens (1904) and Jacob Riis (1902, 1917) focused on political corruption and life in the tenements. Religious leaders began to seek a Christian approach to the problems of the unfortunate (Abell, 1943; Hopkins, 1940). Later, between 1900 and 1917, at least ten of the major philanthropic foundations were established, including the Russell Sage Foundation and the Commonwealth Fund, both important to social and clinical services for children.

The social justice movement, composed of lawyers, intellectuals, ministers, and upper-class women, sought changes from state and federal legislatures that would enable the urban poor to rise above their circumstances (Hays, 1957). No small part of their work, and a good part of the work of the early settlement house workers, was gathering data on the condition of women and children in the factories. Jane Addams (1910), for example, describes an incident in which she attempted to use an ergograph (an apparatus for measuring the work capacity of a muscle) to demonstrate that factory women were indeed tired after a long day. To some extent, surveys and other types of empirical studies of the social environment were undertaken because hardheaded legislators demanded proof that working conditions were really harming women and children. In fact, the science of sociology was seen as a means of promoting meaningful social reform (Abell, 1943; Hopkins, 1940).

The effects of industrialization and urbanization on the individual, the family, and the sense of community were profound. For numerous immigrants, these effects were compounded by the difficulties of having to change their entire ways of life (Handlin, 1951; 1959). Most of the immigrants were forced here by poor conditions in their own lands. They were largely unskilled, and one-third were illiterate. They came from rural environments and so had to adapt both to the city and to the factory. Industrialization, with its specialization of function, removed workers from their households, subjected them to the external discipline of the factory, made them dependent on the vagaries of the labor market for existence, and took away their sense of competence and pride in workmanship. A worker was no longer an individual in a community of individuals, with many diffuse ties to others. Rather, such ties became role specific, and the worker's place in the group became nebulous.

The urban situation also made relationships more impersonal, utilitarian, and transitory. Primary-group controls become less effective, and the social order was maintained by a variety of secondary institutional controls. The police, social agencies, correctional agencies, investigating commissions, and the like became more important to people's lives. These agencies, which followed explicit, impersonal rules and had powers of enforcement, took over bit by bit those control functions that previously had been maintained by internalized folkways and mores governed by personal relationships (Wilensky & Lebeaux, 1965).

The nuclear family became more important than the extended family. Migration compounded family problems, for when the men migrated first, as many did—sending for their wives and children after saving up money for their passage—the families often suffered severe internal disruption. The now-smaller family intensified the ties among its members. At the same time, the family members' need to support themselves and their families drew more and more lower-class women out of the home, thereby changing their roles to some extent. With women working long hours under poor conditions in factories or home sweatshops, the strain on small families became even more acute. Divorce rates increased drastically, and particularly in the lower classes, families broken by divorce, desertion, or the premature death of one parent became more common. In rural communities, in which there still were extended family ties, the broken family was not as severe a social problem, for the community took care of its own. But in the urban, industrial environment, specialized agencies became necessary.

Problems of youth were exacerbated. The place and value of children in the family had changed. As labor became more specialized and tasks in the home became more limited, children became less valuable economically and more dependent. For middle-class families, this meant that a smaller family was more desirable, and so interest in birth control and abortions grew. This does not mean, however, that working-class children did not work. They did, on their own or alongside their parents, in both factories and the tenements. Some of the greatest abuses of this period were connected with child labor, an evil not really controlled until federal and state legislation in the 1930s and 1940s. Poverty pulled children from the schools into the factories, but even the factories required minimally educated workers and citizens.

The schools, geared to the educational needs of the wealthier groups, proved to have curricula and teaching methods out of harmony with the immigrant children's backgrounds. The school system changed slowly and with difficulty. Reformers, intellectuals, and scientists recognized that irrelevant and inadequate school programs were forcing children out of school, contributing to delinquency, and wasting human resources. Jane Addams (1910) became a member of the Chicago School Board precisely because she hoped to be able to influence school programs to change, thus helping eliminate one cause of delinquency.

Immigrant children assimilated much faster than their parents did, which led to intergenerational conflict and a lessening of the authority and influence of the older generation. Delinquency and school problems were most likely to occur in the first generation of immigrant children born on American soil.

Social Problems Among the Jews

From about 1890 on, eastern European Jews came to the United States in large numbers, fleeing pogroms, restrictive legislation, and economic recession in their homelands. They came to the cities at the same time as did Italians and other south and central European peoples. These groups displaced the northern Europeans in the oldest and poorest sections of the cities.

Many people believe that in the "good old days" people were poor, but because they had close-knit families, worked hard, and sacrificed for their children to go to schools, they "pulled themselves up by their bootstraps." Today, some make invidious comparisons between the earlier immigrants and today's African-American and Latino urban poor. In general, the stereotyped view of the past is only partially correct, for what it denies is that there were social problems among all of the immigrant groups, even the Jews, who are often cited as an outstanding example of successful immigrant adaptation (Glazer & Moynihan, 1963).

In fact, a variety of social problems, like those of today, were endemic to all immigrant groups, including the Jews. The following centers on the Jews as a group only because we believe that revealing their severe problems will demonstrate that any group, white or black, Jew or Gentile, exposed to severe urban poverty will develop similar social problems. We feel that this evidence explains the reasons for offering helping services, and it also helps show current urban problems in a better perspective.

Family Disorganization

Jacob Riis (1917), the reformer concerned with New York City's urban problems, described New York's all-Jewish Tenth Ward, the Lower East Side, as having the greatest population density known in any country and the population as being housed in overpriced and overcrowded tenements. He told the health officers of the city that it was known as the "typhus ward" and told the official body that dealt with suicides that it had also been dubbed the "suicide ward." It was known "among the police as the 'crooked ward' on account of the number of 'crooks,' petty thieves and their allies, the 'fences,' receivers of stolen goods who find the dense crowds congenial" (Riis, 1917, p. 81; see also Goren, 1970).

Jewish family life in the miserably crowded and filthy cities was not as close or as free of problems as we have been taught to believe:

> It is doubtful that there were specific relief needs peculiar to the Jewish population in the United States aside from religious and cultural considerations, despite recurrent concern that this might be the case. Thus, desertion in the early 1900s was considered to be a "Jewish problem" and a major cause of dependency in Jewish families. However, subsequent investigations proved that desertion was at least as common, if not more so among non-Jewish families. (Stein, 1958, p. 201)

An example of family disorganization can be found in a description of a child treated in Lightner Witmer's clinic in Philadelphia. The child was an eight-year-old girl, of Russian-Jewish parentage, brought to the clinic because she had spent two years in first grade with no progress. She was brought not by her parents but by a visiting nurse. She was one of seven children, poorly nourished, dull, sullen, and unwilling or unable to answer simple questions. Her home was described as follows:

> The living room has one window, contains a table, a few chairs, a stove and lounge, no carpet, dirty clothes piled in one corner, many flies and a barking dog. The table is covered with a piece of black oil cloth on which there is usually to be found pieces of brown bread and glasses of tea. . . . The family never sits down to the table; no meals are prepared. . . . Bread is always on the table and the children take it when they feel like eating. . . . One hydrant at the entrance suffices for the different families; there is underground drainage, but an offensive odor comes from the water closets. (Witmer, 1908–9a, p. 143)

Confirmation of problems in family stability comes indirectly from an analysis of themes prevalent on the Yiddish stage in those days. The tremendously popular plays centered on the suicides of fallen women, illicit sexual relationships between wives or daughters and the ubiquitous boarder, and marital difficulties stemming from changes in life-style wrought by the new environment. A not-infrequent villain was the cruel husband (Hapgood, 1967). Unwed mothers were by no means rare. The Kimpetoran Society, formed to alleviate distressing conditions among indigent Jewish mothers on New York's Lower East Side, listed many single young mothers among the recipients of "a scuttle of coal, a clean bedsheet, a few diapers, and $5" (Lukas, 1967, p. 49).

Many stories and plays tell of the heartbreaking problems which arose when the first generation began to break away from old customs and parental authority. Lincoln Steffens (1931) describes, as more than occasional, fights in ghetto homes between fathers who attempted to control their children by physical punishment and young toughs who did not hesitate to strike back. The fights were severe enough to bring the police.

Delinquency

Delinquency was a common problem on the Lower East Side, and it was by no means always minor.

> The young people of Jewtown are inordinately fond of dancing, and after their hard day's work will flock to these "dancing schools" for a night's recreation. . . . [I]t happens that a school adjourns in a body to make a general raid on the rival establishment across the street, without the ceremony of paying the admission fee. Then the dance breaks up in a general fight, in which likely enough someone is badly hurt. The police come in, as usual, and ring down the curtain. (Riis, 1917, p. 110)

And the problem was not restricted to New York's Lower East Side. At least 7 percent of Healy's (1915) one thousand delinquents in Chicago were of Russian-Jewish parentage, and judging from the listing of the other eastern European birthplaces of his group, as many as 15 or 20 percent of the youths may have been Jewish. About 70 percent of all delinquents had foreign-born parents, whereas the native-born white population of Chicago produced only 1 to 5 percent of all delinquents. Healy deliberately avoided giving a breakdown by religious grouping, stating that it would do no good. He also pointed out that Jews were seen in unusual numbers in his clinic because of the efforts of Jewish fraternal and social agencies to look after their own delinquents. In any case, Jewish delinquents were not rare in Chicago around 1910.

In Philadelphia, about 20 percent of all delinquents coming before the juvenile court were Jews. The same source indicates that the juvenile court in Philadelphia specifically selected a Yiddish-speaking probation worker to help cope with the problems (International Prison Commission, 1904). In Denver, the reports of the juvenile court for the years from 1908 to 1910 list figures by religious group. Between 16 and 20 percent of all children who came before the juvenile court were Jewish (Report of the Juvenile Court, 1908–9, 1909–10). Several of the boys mentioned in one of Judge Lindsey's books had Jewish names, and it is specifically stated that half of one of the gangs that worked with Lindsey was composed of Jewish children. Moreover, some of Lindsey's strongest support as a political independent came from the families of Russian-Jewish boys whom he had befriended (Lindsey & O'Higgins, 1910). An undated pamphlet from the juvenile court in Denver, describing the state industrial school at Golden, states that teachers from Denver gave religious instruction to Jewish boys on Sunday afternoons.

Further evidence of delinquent and criminal activity among the Jews on the Lower East Side may be found in the novel *Jews Without Money* (Gold, 1930). Coulter (1913), a juvenile court clerk and the founder of the Big Brother movement in New York, implies but does not state that faginism (teaching young children to steal) and pickpocketing were largely Jewish enterprises on the Lower East

Side. Riis (1902) states that the fighting gangs among the Irish were matched by the thieving gangs among the Jews. As late as 1930, fully 20 percent of the children brought to the attention of the juvenile court in New York City were Jewish. Although it is true that more than a third of the Jewish offenses in 1930 were for peddling or begging without a license, a sizable proportion were for other, more antisocial acts. By 1952, Jewish children seen in the New York City courts had dropped to 3 percent, a tenth of the rate in proportion to their numbers in the community (Robison, 1958).

The available evidence shows that delinquency of all kinds was by no means uncommon among the Jews who came to the cities before World War I. Because there was a fairly high rate of delinquency among the Jews in cities as widespread as New York, Philadelphia, Chicago, and Denver, urban conditions—poverty, crowding, and unfamiliarity with the ways of the city—apparently produced similar effects. Our assumption is that crime and delinquency were at least as prevalent among other immigrant groups in the cities; the literature of the time presents ample evidence of rowdyism, thieving, fighting, gangsterism, attacks on police by crowds in the streets, and myriad other antisocial disorders among almost every one of the immigrant groups.

Prostitution

Evidence of Jewish delinquency often occasions surprise, but evidence of the involvement of Jews in prostitution at that time is still more surprising. The openness and prevalence of prostitution among the Jews are mentioned in the works of Jane Addams (1910), Lincoln Steffens (1931), and Michael Gold (1930). Steffens and Gold portray nearly identical scenes in which children, through the windows of their own tenements, view prostitutes entertaining their customers. The Lower East Side had a long-standing reputation as a red-light district, a reputation confirmed in the story about the first visitor to the College Settlement, which was located in a district settled largely by Russian and Polish Jews. That visitor was a policeman who unwittingly characterized the neighborhood and his relationship with it by asking for a bribe not to interfere with the business activities of seven young women (Davis, 1959). Lillian Wald (1915), the most famous of the visiting nurses, delicately refers to the few blocks of the red-light district near her Henry Street Settlement in New York, which for years she would not enter out of a sense of shame.

Prostitution was widespread in the country, and Jews were heavily involved, as both prostitutes and procurers. Maude Emma Miner, secretary of the New York Probation and Protective Association, cites a number of case histories of prostitutes in New York (Miner, 1916). The names she uses in perhaps a quarter of the cases are Jewish: Rose Schafer, Rachel Goldberg, Sophie Bergman, Yetta Rosen, Bertha Levy, and Jennie Rosenberg. When a Jewish name is mentioned, furthermore, there is no statement explaining that the Jewish prostitute was rare. On the contrary, Miner writers:

> It is a comparatively new phenomenon in the life of the Hebrew people to have any of its daughters in prostitution. The moral standards of Jewish women have

been high. Less than a generation ago it was rare in New York City for a Jewish girl to be living a life of prostitution. Now many are entering the life. (Miner, 1916, p. 35)

The presence of Jewish prostitutes was matched by the involvement of Jews in the organization of prostitution. The Report of the U.S. Immigration Commission on the Importation and Harboring of Women for Immoral Purposes (1909, cited in Miner, 1916) states that of a group of 218 procurers, 65 percent were Russian or Italian) (at that time there were few Russian immigrants other than Jews). The report goes on to record that "there were two organizations of importance, one French, the other Jewish." Further, "the activity of Jewish procurers in seducing young girls and turning them into prostitution was much greater than the French, whereas the French were more willing to import women who were already familiar with the life" (U.S. Immigration Commission, 1909, p. 23).

In discussing prosecutions of procurers, Miner states that Morris and Lena Cohen boasted of having the largest clearinghouse for women in New York City, and another procurer named Samuel Rubin was given a long-term sentence for that crime. Confirmation of the presence of Jews in prostitution also is found in the Report of the Moral Survey Committee of Syracuse, New York (1913):

Visited Madame X-9's parlor house, X-10. Met there a girl I know. She introduced me to the madam as one of the boys. I stated my errand to Syracuse, to open a house if possible, as New York is now so "tight." She discouraged me by informing me that "they" will not stand for Jews, that the minute a Jew opens up, they get to citizen X-11 who gets to official X-12 and has them driven from the town. I explained to her that my woman is a Gentile and that I wouldn't mix in.

The fact that the undercover investigator was Jewish and openly identified himself as such to the madam speaks for itself. In Goren's study of the development of communal life among Jewish immigrants, he states that "Even the highly praised penchant for establishing mutual aid societies produced its monstrous aberration in the New York Independent Benevolent Association. Created by brothel owners and procurers the Association provided its members, unwelcome in other Jewish societies, with traditional health and death benefits rendering 'assistance in case of necessity' " (Goren, 1970, p. 137).

We have found references to Jewish cadets (a term for pimp) and Jewish prostitutes in reports from Hartford, Connecticut (Report of the Hartford Vice Commission, 1913), and in a pamphlet entitled "Facts for Mothers and Fathers" issued in New York at about this same time. Many of the other reports of vice commissions from cities throughout the country contain similar data on prostitution, but the ethnic backgrounds of the people involved are not obvious because they are identified by case number only. The birthplace of the individual or the parents indicated as being in an east European country is common, however, and a substantial proportion of these persons may have been Jews.

A Jewish source, the Report of the International Society for the Rescue of Jewish Women and Children (1927), documents the problem of "white slavery" among the Jews. It does not give any figures, though, and the participants—

including Bertha Pappenheim, who was probably Sigmund Freud's and Josef Breuer's Anna O. (Oberndorf, 1953) and at this time a social worker—differ among themselves about the extent of the problem after World War I. But they do indicate that there was indeed a problem in the prewar period. This source describes in great detail—and in a form sufficiently melodramatic to rival the best productions of the Yiddish stage—the tragedy of young women who were forced into the life of a prostitute. Immigration had resulted in a shortage of young men in the small villages, or *shtetlach,* of eastern Europe. Operators would come to the villages and tell the young women they could have a husband in the New World. After a proxy ceremony, the women were loaded aboard ships and taken to Hong Kong, South America, or the United States where they were delivered to brothels and literally held as slaves. The rescue society described raids on ships in which the group would forcibly release women from the clutches of their captors.

Apparently it was not easy for young women to find husbands in America, for many were introduced into prostitution by cadets who falsely promised to marry them or who took advantage of their naïveté by entering into fraudulent marriages with them. In many instances, however, the motivation to enter prostitution was a desire to escape the hard, drab life of the sweatshops and tenements and to satisfy a yearning for the fineries that the uptown shops provided for wealthy women. The problem may have been more common among the first American-born generation than among its immigrant sisters, just as delinquency may have been in some part a function of the intergenerational gap. The existence of the problem in the early 1900s may give us a somewhat more sympathetic understanding of the ghetto youth of today who see hustling and prostitution as their way of adapting to and conquering the hard life of the city.

The Public Schools

The eastern European Jews did exceptionally well in the public schools. Lillian Wald (1915) is one among many who stated that Jewish parents urged education on their children, willingly making great sacrifices to support them. Jacob Riis (1917) also mentions that the eager Jewish students quickly became top scholars, taking all of the prizes in the schools they attended. Lincoln Steffens (1931), Hutchins Hapgood (1967), and others discuss the vigorous intellectual life among the young adults in the Jewish ghetto. Although the general picture is undoubtedly true, there is another side of the educational picture that has been neglected. When we see the nature of the educational problem among the educationally oriented immigrant Jews, we can assume that it was considerably worse among other groups. The schools themselves contributed heavily to the problem of educating immigrant children.

The generation preceding the Civil War had largely won the fight for free public schools and universal education (Sarason & Doris, 1979), but placing the schools under the aegis of the politicians did nothing to help promote rapid change when change became necessary. Normal schools to train teachers were opened before the Civil War, and education quickly fell victim to professionalization, bureaucratization, and a variety of corrupt political influences. By the 1870s the city schools began to feel the burden of a school population rapidly increasing as a result of immigration, urbanization, and the new compulsory attendance laws. The schools thus responded with rigid systems for grade level, pupil promotion, and supervision of instruction, as well as all the other institutional trappings essential to processing masses of people but antithetical to individual needs.

In these and subsequent years, the immigrants, along with migrants from the farms, continued to flood into the badly lighted, poorly heated, and unsanitary schools. Class sizes in excess of sixty were all too common. By 1909, more than 57 percent of the children in the thirty-seven largest school districts in the United States had foreign-born parents, and in New York and in Chicago, at least two-thirds of the children had foreign-born parents (Cremin, 1964).

One of the problems the schools faced was the compulsory school attendance laws. The fight for universal, free, tax-supported public schools had been more or less won earlier, but until 1852, no state required attendance at school (Sarason & Doris, 1979). Between 1870 and 1890, the bulk of the compulsory school attendance laws were passed, and by 1912, every state in the union had such a law. These laws were passed for a variety of reasons. The labor unions supported them as a means of curbing low-wage child labor, and so compulsory school attendance laws often were passed in tandem with child labor laws. The laws were favored by humanitarian reformers not only because they struck at child labor but also because they held promise of creating a literate populace, which was indispensable to a democratic society.

As with many good ideas, the implementation left much to be desired. School attendance laws, forcing an overwhelming population on the schools, indirectly provided the single most important force in the development of clinical services for children, and eventually for change in the school programs. When the laws were passed, the problems they created for the schools were immediate and immense. Between 1890 and 1920, the nation's population increased by 68 percent; the public school population increased by 70 percent; and average daily attendance increased almost 100 percent, reflecting the enforcement of the attendance laws.

More children were going to school and more children stayed in school to an older age, but the school system was not prepared for the onslaught. According to a prominent educator regarding the status of compulsory school attendance laws before the 1890s:

> From the start it has been found that the satisfactory execution of attendance laws requires special adjustments within the schools. Children forced to come to school against their will, or with little interest in school cannot be classified with children who attend regularly. . . . unless special schools are provided. The school authorities themselves will hardly cooperate in enforcing the law if such is not the case. New York State passed its first compulsory law in 1874. After fourteen years' trial, it was found that the law had not modified school attendance, had not secured the cooperation of school principals and was most substantially a dead letter. The machinery for its execution was inadequate, but at bottom it was a question of lack of accommodation for the difficult pupil. (Monroe, 1911, p. 294)

A similar circumstance in France led to the development of the intelligence test. France had passed a compulsory public education act in 1882, with provision for its enforcement. By 1904, the French Ministry of Public Instruction knew it had a problem. Children were not responding to the school as it existed, and so Alfred Binet and Theophile Simon were commissioned to devise methods for selecting children who could not adapt to the regular curriculum and who thereby reduced the efficiency of the teachers and the other children. In this way, the intelligence test became part of the planning for the introduction of special classes into the public schools (Peterson, 1926). (The social-class bias in intelligence tests and the consequences for defining intelligence are discussed by Eels et al., 1951).

In the 1890s, when southern and eastern Europeans began to arrive by the hundreds of thousands, the schools were faced with additional problems. The newcomers could not cope with the school's rigid curriculum, and one result was the retention of large numbers of pupils. When the term *retardation* was originally used, it did not necessarily mean retardation in mental and physical growth. Rather, it meant that students were too old for their grade. The extent of the problem in the urban schools of the time is revealed in a number of reports. In 1905–6, in the schools in Camden, New Jersey, fully 26 percent of the pupils were two or more years behind (Bryant, 1906–7). In New York, 30 percent were retarded; in Boston, 22 percent; in Philadelphia, 37 percent; and in Kansas City, almost 50 percent (Cornman, 1906–7; see also Heilman, 1906–7). The differences among the cities reflect administrative regulation as much as differences in the people who settled there. In later years, the problem was solved by the simple expedient of promoting almost everyone, by administrative fiat.

Successive issues of Witmer's *Psychological Clinic* reflect the deep consideration given to school retardation. Bryant attributed the problem to the "serious efforts to enforce provisions of compulsory education enactments [that] have led to the presence of large numbers of children who are unable to make steady progress in the school" (1906–7, p. 42). The efforts to enforce the laws reflected the interest of reformers who made up the social justice movement. Jane Addams (1929), for example, noted that one of her settlement workers made an intensive study of pupil attendance in Chicago schools. Abbott and Breckinridge (1917), two former Hull House workers, also reported on attendance in the schools in Chicago.

Although there were a few sporadic efforts in the United States to open special classes in the 1870s, the first special classes did not begin until 1896. By 1905, the idea of special classes, classes originally begun as disciplinary units and not to meet intellectual needs, had taken hold. Most of the larger cities had them, but in insufficient numbers to meet the need. The public at large had so little understanding of the problems the schools faced that the first such special class was greeted by an article in the local paper calling it the "fool class" (Kanner, 1964).

Lightner Witmer was one of the leaders in instituting special classes, helping the Philadelphia school systems organize them, and with others (Van Sickle, Witmer, & Ayres, 1911), he wrote an important monograph on special classes. Witmer promoted the special-class concept through his journal, *The Psychological Clinic,* seeing it as a means of treating children with special educational needs. He did not intend to remove problem children from the regular schools; on the contrary, he intended to use special classes to provide for individual differences within the framework of the regular public school. An editorial in an issue of *The Psychological Clinic* (Witmer, 1908–9c), featuring two articles on special classes, details his view. He emphasized that the introduction of special classes into the school system was part of the reform of the school system in response to the school's failure to offer adequate education to all children. This editorial also spells out Witmer's conception of the classroom teacher as an applied psychologist and the special classroom as the clinical psychologist's laboratory.

These classes and the problems that the immigrant children presented to the public schools are depicted in a report by Helena T. Devereux, one of the early special-class teachers. She later founded the Devereux Schools, currently the largest private group of residential treatment centers for retarded and emotionally disturbed children in the United States. Her report, quoted almost in full, with only extraneous details omitted, offers a view of the classroom from the teacher's vantage point:

> On January 15, 1908, a class for exceptional children was formed in a large school in the foreign slum district of Philadelphia. . . . These children were reported not only as being stupid, but "queer." It is hard to define exactly what was meant by this term but it did mean that the child to whom it was applied was singled out from the other children. Other children not placed in the class were dull and slow but still normal, while each of these children seemed to have a personality differing from the normal child and also from the purely incorrigible type.
>
> From January, 1908 to February, 1909, forty-three children have attended the class, some for one month and some for the year, some for both sessions and some for only one. Among these children thirty-four were of Russian Hebrew extraction and four of Italian. Of the twenty-six who first formed the class, six were respectably cared for. The others were much neglected, and as they lacked even the pride of the ordinary child, they were, indeed a forlorn looking little company, insufficiently fed and poorly clothed. They ranged in ages from eight to fifteen years, being from two to five years retarded as compared with the average child. . . .
>
> Before the class was started the idea was that the manual work (woodwork, basketry, paper, sloyd, and sewing) might be a means of awakening the child and might form a "peg" on which to hang the real mental training. My only idea was to immediately teach the child the rudiment of school work and so return it to the regular grade when it was sufficiently advanced. When I had been with the class for one week I abandoned my scheme of memory and sense perception, training the intellect only as a side issue, and I determined to base all my work on the development of the emotions. The great need of those children as I read it then was to make them less like little animals—to instill humanity into them. I could understand then wherein these children were queer. They were subnormal or rather freakish in disposition and temperament. They were selfwilled, passionate, malicious, and all their shortcomings were very glaring, largely because they had not the sense of their fellows to see when it was expedient to be "good." With each child I picked out the moral defect or defects which were most emphasized such as selfishness, untruthfulness, stubbornness and temper, and determined to overcome them. I tried to make them see in every way what was the

right thing to do, and not only to make them do it but to make them want to do it. I cared less that they should learn to write their names or finish an article in woodwork beautifully than they they should learn truthfulness, obedience and promptness, and truly they learned to watch, to listen, and to do. The bearing that this phase of the work, the training of the emotions, had upon the results cannot be overestimated. It laid the foundation for the other training—the mental side. I certainly agree with the educators who say that when once the real personality of the backward child is revealed it is more childlike in its trust, kindliness and simplicity than the normal child. When once they recognized that the teacher was their friend their attitude towards school life gradually changed. When definite instruction for the purpose of advancing them in their school work was given, the children were earnest little workers who wanted to learn. Their previous training had increased their power of attention so that the teaching was comparatively easy. I found each child had some serious defect, such as the inability to do arithmetic, written language, or reading. Even now when they are returned to the grade, some still have a persistent defect but it does not mar their entire standing. . . .

This class has in one year been a blessing to thirty three [sic] children, children who, though never doomed by nature to spend their lives in an institution, would surely have drifted to one, which in some cases might easily have been a prison, had not some interest been taken in them. This class was begun amid many difficulties, there being no public funds provided for the necessary materials. Then too, the work was marred by my inexperience. The success of the class, especially as far as the regeneration of the boys is concerned, has been achieved largely through the woodwork which was the basis of manual, emotional and mental training, and that was made possible by the generosity of a Philadelphia merchant. May my last word be a plea for more classes in our city, properly equipped, where the three-fold motto shall be interest, persistency and encouragement to aid in making useful men and women out of ''the least of God's little ones.'' (Devereux, 1909–10, pp. 45–48)

Educational Problems Among the Jews

The reader might already have noted that Helena Devereux (1909–10) reported that thirty-four of forty-three children in her special class in Philadelphia were Russian-Jewish. In an earlier article, Heilman (1906–7), in presenting the need for special classes in the Philadelphia schools, described twenty cases of retarded children in need of special schooling. Of these twenty children, twelve were Russian-Jewish, and four were Italian. The remainder were Irish, Swedish, German, and native born. Although we shall show that many Jewish children were not successful in the public schools, there is no evidence—and no one should assume— that retardation was specifically a Jewish problem. We do know that Jewish children generally did well in school. We do not know how many of the immigrant and first-generation children did poorly in school. We suspect that the numbers may have been large, but we do not know that the problem has been studied along religious or ethnic lines.

The visiting teachers, forerunners of the school social workers, began their work in the urban slums. In New York, Boston, and Hartford, Connecticut, some of the first visiting teachers were placed in school districts that were from 81 to 96 percent Jewish. The problems that they handled included deficient scholarship, truancy, incorrigibility, adverse home conditions, and neglect. There is no evidence to indicate that the problems were found any more frequently among Jews than among other groups, but clearly they were sufficiently common to warrant placing some of the visiting teachers in schools in predominantly Jewish neighborhoods (National Association of Visiting Teachers, 1921). Another indication of the problems in education may be the fact that in the early 1900s the type of school that would now be called a "600" school (a school for emotionally disturbed or difficult-to-manage youth) was established on the predominantly Jewish Lower East Side.

In general, the poor and dirty slum-dwelling Jews were not always received with open arms by the public schools. Jacob Riis pointed out that some of these children went to religious schools. "But the majority of the children seek the public schools, where they are received with some misgivings on the part of teachers, who find it necessary to inculcate lessons of cleanliness in the worst cases by practical demonstration with washbowl and soap" (Riis, 1917, p. 113).

There were important cultural differences between the middle-class teachers and the lower-class children. The following passage is from *Jews Without Money*, a novel by Michael Gold, who was born on the Lower East Side in 1894. Although the novel was popular in the 1930s and is considered a protest piece of the time, it is largely autobiographical. One can therefore assume that Gold is depicting his own public school experience in the early 1900s.

> I first admired Nigger in school, when I was new here. He banged the teacher on the nose. [Nigger was the name of a white Jewish schoolmate.]
>
> School is a jail for children. One's crime is youth and the jailers punish one for it. I hated school at first; I missed the street. It made me nervous to sit stiffly in a room while New York blazed with autumn. I was always in hot water. The fat old maid teacher (weight about 250 pounds), with a sniffle, and eyeglasses, and the waddle of a ruptured person was my enemy.
>
> She was shocked by the dirty word, I, a six-year-old villain, once used. She washed my mouth with yellow lye soap. I submitted. She stood me in the corner for the day to serve as an example of anarchy to a class of fifty scared kids.
>
> Soap eating is nasty. But my parents objected because soap is made of Christian fat, is not kosher. I was being forced into pork-eating, a crime against the Mosaic law. They complained to the principal.
>
> O irritable, starched old maid teacher, O stupid proper, unimaginative despot, O cow with no milk or calf or bull, it was torture to you, Ku Kluxer before your time, to teach in a Jewish neighborhood.
>
> I knew no English when handed to you. I was a little savage and lover of the street. I used no toothbrush. I slept in my underwear, I was lousy, maybe. To sit on a bench made me restless, my body hated coffins. But Teacher! O Teacher for little slaves, O ruptured American virgin of fifty-five, you should not have called me "LITTLE KIKE!"
>
> Nigger banged you on the nose for that. I should have been as brave. It was Justice. (Gold, 1930, pp. 22–23)

Some Speculations

Why has the evidence of social problems among Jews of delinquency, family instability, involvement in prostitution, and school problems been forgotten? The Jews, of course, had much to be proud of. There were many self-sacrificing mothers and fathers and many good students. The second generation did have less difficulty than did the immigrants or the first generation, and by the third generation, the antisocial problems had almost disappeared.

In addition, popular writers who were not Jewish, such as Hapgood, Steffens, and even Riis (he spoke of the stubborn people of "Jewtown" who refused to recognize Christ), were attracted to the artistic and intellectual life in the Jewish ghetto. These writers had Jewish friends and came to understand the culture in its many dimensions. They also were liberal reformers, eager to bring about social change in the corrupt municipalities of the time. Because immigration remained largely unchecked until after World War I, many saw the immigrants as a threat to the well-being of American society. Many liberal writers thus felt compelled to point out the immigrants' contribution to the American economy and the American culture by defending the presence of the immigrants and supporting efforts at reform. Taking account of the blatant anti-Semitism, reform-oriented writers may have felt it necessary to minimize reports of Jewish antisocial behavior. Because many of the social scientists of that time were also reformers, they may not have studied Jewish involvement in antisocial behavior. Healy (1915), for example, tells us that he deliberately suppressed data showing the numbers of delinquents in the different religious groups. Jews may also have received favorable attention because of their sobriety. The literature emphasizes that there were no Jewish alcoholics, in a time when prohibition was a favorite cause among reformers.

The Jews' adaptation to America is all the more remarkable considering their social problems, and their reduction in a relatively few years deserves attention. The contribution of Jewish welfare and social agencies and the tradition of self-help in the Jewish community is worthy of detailed study in light of our contemporary understanding of social process (Goren, 1970).

Summary

Our survey of the period has shown that the turn of the century was a reform period, that there was considerable social unrest at the time, and that various social institutions were plagued with troubles stemming from immigration, industrialization, urbanization, and poverty. The general reform movement was concerned about how people lived, and it was in this context that reformers and intellectuals took an interest in the schools, children, and their welfare.

3

Lightner Witmer and the First Psychological Clinic

To add to the troubles of the public school system—then as now—the schools received a great deal of attention from the reformers. The reason is not hard to find. John Dewey expressed the viewpoint of intellectuals and reformers when he said: "I believe that education is the fundamental method of social progress and reform" (1897, p. 437). Clearly, anyone who wanted society to be different saw that the schools were the first point of attack. Cremin (1964) documented the history of the progressive education movement and showed the variety of changes made by Dewey, his followers, and others. Sterile formal methodology, the irrelevance of the curriculum to the real world, the lack of manual training, and corrupt political influences were variously attacked by muckrakers and other intellectuals. The vulnerability of the schools caused by their myriad problems also was important to obtaining some degree of curricular change.

The attention that intellectuals, particularly academics, gave to education provided a significant background for Witmer's interest in the public schools and his decision to establish an applied service. John Dewey had gone to the University of Chicago in 1894 to head the departments of philosophy, psychology, and pedagogy. The University of Chicago at that time was the heartland of reform Darwinism. In 1896, the chairman of its sociology department, Albion Small, addressed the National Education Association and, in resounding terms, insisted the school was the prime medium for social change. Dewey, at that time a respected philosopher and psychologist, was on the Hull House board of directors and contributed to the development of the educational program at Hull House.

Earlier, G. Stanley Hall had given lectures to teachers on educational problems, and when he went to Clark University in Worcester, Massachusetts, he formed a department of pedagogy. In 1891 he founded the *Pedagogical Seminary,* a journal devoted to the new scientific approach to education (Hall, 1923). By 1895, courses were offered in the new pedagogy in at least five major universities. In 1899, Teachers College became a part of Columbia University, and E. L. Thorndike introduced educational psychology as a respectable field of study (Boring, 1957).

Among other American intellectuals who took an interest in the public schools was William James. In 1892, he delivered a series of lectures entitled "Talks to

Teachers'' to the public school teachers of Cambridge. James then gave these popular lectures throughout the country, and they finally appeared in book form in 1899 (James, 1899). In them, James tried to show what the new psychology had to say about the learning and teaching process.

The topic of education was in the air. The *Psychological Review* of 1897 records a controversy between Hugo Munsterberg, the founder of applied psychology, and James McKeen Cattell, who was Witmer's predecessor as director of the psychological laboratories at the University of Pennsylvania. Munsterberg argued with force and conviction that the new psychology, experimental laboratory psychology, had nothing whatever to say to the public school teacher, because it was irrelevant. Cattell, defending the experimental method, asserted that the findings of psychology were solidly established in laboratory research and that eventually the scientific soundness of experimental psychology would enable it to make a contribution. In the meantime Cattell was satisfied that the new psychology was scientific.

Given the problems the schools were experiencing and the interest of eminent psychologists, philosophers, and sociologists in the public school system, it is not surprising that in 1896 when Lightner Witmer opened the first psychological clinic in the United States—if not the world—at the University of Pennsylvania, it focused on educational issues. Witmer (1915) himself explained that his clinic was a product of broader social forces. In a public lecture he gave in 1913, Witmer discussed the history of services for the exceptional child, recognizing the work of Pereire, Seguin, Pinel, Itard, and other pioneers in the work with the mentally retarded. He referred to this earlier work as the first movement and went on to say:

> The second movement for the study and educational treatment of exceptional children was inspired, or at least made possible by the development of modern science, especially psychology and hygiene. It began about the year 1890 in this country, and was the result of the introduction of school medical inspection, the enforcement of compulsory education, the failure of society to provide for the care of the blind, deaf, feebleminded and delinquent children, the development of a type of applied psychology called clinical psychology, and lastly the increased efficiency of public school administration, which led to the discovery that millions of children in the United States were not obtaining the elementary education which they were supposed to be getting and which a democracy insists is necessary for self governing citizenship. The earlier movement for the training of exceptional children was not conceived as part of the problem of general education. It was in part a scientific movement and in part a humanitarian movement. To some extent also it was a blind effort of society at self preservation by removing from society (temporarily at least) certain unfit elements. The later movement for the educational treatment of exceptional children is a part of the public school problem. It represents a conscious effort to introduce scientific procedure into common school practice. It has had and will have far reaching consequences not only upon educational practice with respect to all children but also upon the development of the sciences of psychology and hygiene. It led directly to the founding of psychological clinics at the University of Pennsylvania and other institutions of learning, and to the establishment of similar clinics in connection

> with the public school system of many cities. I have taken part in this movement for the more scientific treatment of exceptional children from its inception in this country, and from the beginning I have been inspired by the belief that the study and training of the exceptional child constituted the most favorable point of approach for a psychology applied to educational practice. (Witmer, 1915, pp. 535–37)

Although Lightner Witmer was very much an academician and experimental scientist, in keeping with the spirit of the times he turned toward practical problems, aligning himself with Thomas Huxley in their belief in the continuity of pure and applied science (Witmer, 1906–7). The lecture just quoted was written nearly twenty years after the first clinic opened, but Witmer's earliest writings expressed the same ideas. In the first written description of the clinic, a paper read at a meeting of the American Psychological Association in 1896, Witmer (1897) talked about his interest in working out a relationship between the psychology department and the public schools. In the second volume of his journal, *The Psychological Clinic,* Witmer stated in an editorial that both the clinic and the journal were "propagandistic in spirit," that it was his purpose to urge the school to realize its possibilities as a social force, as an institution essential to "the development of the individual and the progress of the race" (Witmer, 1908–9a).

Witmer's clinic began in the spring of 1896 as part of the psychology department of the University of Pennsylvania, and there is still a bronze plaque over the entrance to the clinic in College Hall, marking both the event and the place. The clinic's 22,000 case records are now microfilmed (Levine & Wishner, 1977). Beyond that plaque and beyond Sarason's (1958) attention to Witmer as a clinician, Witmer's work tends to be noted on ceremonial occasions, but its substance has been forgotten. Yet the substance of Witmer's work is interesting, for it includes approaches similar to those used today.

Witmer's Influence

Lightner Witmer was born in Philadelphia, educated in private schools and the University of Pennsylvania, and was a law student at Penn for one year. In the philosophy department he met G. S. Fullerton and James McKeen Cattell, who encouraged him to study at Leipzig for his Ph.D. On his return to the United States Witmer was made director of the laboratory of psychology at the University of Pennsylvania, succeeding Cattell, who had moved to Columbia the previous year.

Evidently Witmer was a complex person. He was an intellectual, he was interested in dangerous sports, and he enlisted in the Spanish-American War to find out how he would behave in battle. Although he was an outspoken liberal in his political opinions, he lived in the manner of an aristocrat, becoming more and more selective of friends until some thought of him as a distant, solitary figure. In his later years, colleagues and students said he was disorganized and erratic. He was never accorded the professional recognition he deserved, but those who had known him in his earlier days spoke of him reverently as an individual, a clinician, and an intellectual (Brotemarkle, 1931).

It is difficult to understand why Witmer's contributions continue to receive so little attention by professional psychologists. The *Psychological Review* after 1896 never refers to Witmer's work and does not even take notice of the beginning of his journal. It is possible that Witmer's influence was not given more credit because of academic inbreeding. Many of his students—Twitmeyer, Miller, Viteles, Brotemarkle, Murphy, and Phillips—stayed on with the clinic or with the University of Pennsylvania psychology department, and many of them developed branching interests that Witmer encouraged. Morris Viteles went into vocational guidance and then industrial psychology; Edwin Twitmeyer specialized in speech problems; and Robert A. Brotemarkle concentrated on college personnel problems. Each established a separate branch of the clinic (Levine & Wishner, 1977), and some of Witmer's other students worked in the Philadelphia area. Therefore, many of his students simply may have not been able to carry the word far afield. In later years, some potential students may have been lost because Witmer did not validate his clinical methods.

Witmer's influence is undoubtedly underrated. Certainly after the publication of *The Psychological Clinic,* beginning in 1906, his influence must have been fairly widespread. But Witmer was aware of his lack of recognition, and in a paper written in 1925 he complained that he had been given that most sincere tribute, imitation, but too often without acknowledgment (Witmer, 1931). The Phipps Psychiatric Clinic at Johns Hopkins University, for example, offered a special class focusing on the educational process as the principal mode of therapy. C. MacFie Campbell (1919), a noted figure in American psychiatry, spoke in glowing terms of the effects of this special-class program in reducing the incidence of children's truancy and the number of their appearances before the juvenile court, changing the children's personal habits, and offering the children's parents additional help and education. The program is highly reminiscent of Witmer's special classes at the University of Pennsylvania (Witmer, 1911), but Campbell's article does not mention Witmer.

With the establishment of like clinics in subsequent years, Witmer's complaints were justified. By 1934 there were at least eighty-five psychoeducational clinics affiliated with universities, although they were likely to be located in schools of education and not in psychology departments. They were engaged in training and service, helping the educational, vocational, or social adjustment of children from preschoolers to high school students (Witty & Theman, 1934). Many psychological services connected with public schools, including special-class programs, were probably also influenced directly or indirectly by Witmer's clinic.

Shakow (1948) and Watson (1953) suggested that Witmer's narrow experimental orientation and his emphasis on the cognitive aspect of retardation, to the exclusion of psychodynamic considerations, led him away from what came to be the main clinical interest. It is true that Witmer did not use psychoanalytic concepts per se, but he did acknowledge some intellectual indebtedness to Freud (Witmer, 1931). However, the case records in Witmer's clinic showed very little change over the years in his approach. The Stanford–Binet intelligence test was introduced at a relatively early date, but the basic diagnostic battery did not seem to change very much until after World War II, long after Witmer's retirement (Levine & Wishner, 1977). Witmer's clinic clearly was not at the forefront of change in clinical approaches.

Witmer emphasized the experimental method, publishing *Analytical Psychology* (Witmer, 1902), a laboratory manual of standard psychological experiments, after he started the clinic. Even his clinical thinking is couched in experimental terms, as he intended his applied work to produce basic scientific knowledge. In 1925 Witmer first used the term *psychonomics* and defined it as conforming with a fundamental or universal law of thought (1931). The clinical method, in his view, was dedicated to scientifically explaining the full range of behavior, not only abnormal behavior. Witmer thought of the therapeutic method of diagnostic teaching as an experiment based on the development and testing of hypotheses about individual cases of educational retardation.

Witmer was a highly sophisticated clinician who employed methods that repay study, especially the use of the educational setting as a means of dealing with the problems of emotionally disturbed children. Witmer developed effective approaches for a variety of disturbed children, not only those with intellectual retardation. His case studies, his discussion of the clinical method, and his insistence on considering the qualitative aspects of performance, in addition to the quantitative distribution of test scores, in no way support the view that he was a narrow experimentalist. The neglect of his ideas is a real historical mystery.

A partial explanation is suggested by some evidence that Witmer's temperament and personal style contributed to the neglect of his work. For example, in 1908, he attacked both William James and Hugo Munsterberg for their lack of devotion to laboratory experimentation and scientific method (Boring, 1957; Roback, 1964). Indeed, Witmer's attack was so vicious that Munsterberg wanted to lodge a formal protest with the American Psychological Association (APA), (Roback, 1961), and finally James wrote Munsterberg in an attempt to dissuade him from responding to Witmer's attack (James, 1926, vol. 2, p. 320). Why Witmer should feel so vehemently about the two men is unclear. Viteles (personal communication), who recalled that Witmer also was at odds with Cattell, suggested that Witmer's attack on several past presidents of the APA, combined with his aloofness, may have encouraged further neglect of him (see also Roback 1961, 1964).

Witmer did serve on APA committees after 1908. But in a world so small that the entire membership of the organization could meet in a single lecture room in College Hall at the University of Pennsylvania, his personal style might well have had an important effect on the reception given his ideas. Someone once wisely said that a cause ought not to be judged by its first adherents.

Origin of the Clinic

As a university student, Witmer taught English in a college preparatory academy. There he tutored a young man and succeeded in helping him sufficiently so that he was able to enter the University of Pennsylvania. Witmer again encountered him, this time as a student in his college course, and the young man, albeit with great difficulty, finally did manage to graduate from a professional school. This experience had a strong impact on Witmer, who felt that the young man might have had to struggle less had he had special training in his earlier years.

The case that actually prompted Witmer to open his clinic was brought to him by an elementary school principal who was taking psychology courses at the University of Pennsylvania. She challenged Witmer to apply the new psychology to diagnose and cure a spelling deficiency. Witmer was appalled that the science of psychology had absolutely nothing to say about spelling, for spelling failure was, after all, a deficiency of memory, a subject about which psychology was supposed to have furnished authoritative knowledge. Witmer wrote: "It appeared to me that if psychology was worth anything to me or to others it should be able to assist the efforts of a teacher in a retarded case of this kind" (1906–7, p. 4). Beginning with that case, in the spring of 1896, Witmer started seeing a few children several hours each week to help them overcome specific disabilities in subject matter.

By December 1896, Witmer (1897) had developed a farsighted plan for "practical work" in psychology that investigated mental development in schoolchildren, as manifested in "mental and moral retardation," the investigation to be conducted by means of clinical and statistical methods. Witmer urged the opening of a clinic, supplemented by a hospital school to treat such children, and he proposed that the clinic be used to train teachers, social workers, and physicians. His final goal was to train students for a new profession, that of the psychological expert who would examine and treat mentally and morally retarded children and who would work with the school system. Although his clinic began in 1896, it had few cases until 1907, the year that Witmer began publishing *The Psychological Clinic*. Thereafter, the clinic operated continuously until 1961 (Levine & Wishner, 1977).

The Clinical Method and the Community

The term *clinical* characterizes Witmer's method. The clinical psychologist was interested in the individual child and examined and proposed treatment to achieve the next steps in the child's mental and physical development. Witmer realized that the clinical psychologist operated as a scientist attempting to discover cause-and-effect relationships in applying remedies to children suffering from some form of retardation. He did not restrict the term *clinical* to refer to the study of the abnormal only: "For the methods of clinical psychology are necessarily invoked wherever the status of an individual mind is determined by observation and experiment and pedagogical treatment applied to effect a change, i.e., the development of such individual mind" (Witmer 1906–7, p. 9).

Earlier, Witmer thought that clinical psychology would be closely allied with medicine, but by 1906 his position had changed considerably:

> Although clinical psychology is closely related to medicine, it is quite as closely related to sociology and to pedagogy. The school room, the Juvenile Court, and

the streets are a larger laboratory of psychology. An abundance of material for scientific study fails to be utilized, because the interest of psychologists is elsewhere engaged, and those in constant touch with the actual phenomena do not possess the training necessary to make their experience and observation of scientific value.

While the field of clinical psychology is to some extent occupied by the physician, especially the psychiatrist, and while I expect to rely in a great measure upon the educator and the social worker for the more important contributions to this branch of psychology, it is nevertheless true that none of these has quite the training necessary for this kind of work. For that matter, neither has the psychologist, unless he has acquired this training from other sources than the usual course of instruction in psychology. In fact, we must look forward to the training of men to a new profession which will be exercised more particularly in connection with educational problems, but for which the training of the psychologist will be prerequisite. (Witmer, 1906–7, pp. 7–8)

Witmer's statement is surprisingly relevant to the community mental health movement, and in its early days his clinic was not an institution separate from the schools but was closely tied to them. When Witmer read his 1896 paper announcing a program in practical work in psychology, he included the following two points: First, he wanted to institute special or upgraded training classes for children who were mentally or physically disabled. These classes were to be organized under the control of the city school authorities but to be in "harmonious and effective relationship" with the psychology department. Second, he insisted that the program include instruction in psychology for public school teachers, for he felt that they needed, above all else, practical studies of children (Witmer, 1897).

Witmer was close to, and widely respected by, educators in the Philadelphia area and elsewhere. He helped establish special classes in the public school and wrote an extensive monograph on special classes with J. H. Van Sickle and L. P. Ayres (1911). The Philadelphia system for many years used Witmer's term *orthogenic* for its classes for "backward" children. In addition, Witmer lists a number of school superintendents and supervisors who were associated with him in the clinic. For example, the principal who brought the second case to him was later one of the lecturers in a summer institute that Witmer held in 1907, along with O. P. Cornman, a principal, and J. E. Bryant and E. Twitmeyer, both superintendents of schools. Another teacher, Mary Marvin, was associated with Witmer from the beginning. Classes for observation and experimentation were also part of the clinic teaching and therapy program (Witmer, 1911). These were more than summer institutes condescendingly offered for teachers, for here clinical psychologists and teachers participated in one another's training, a situation rarely found today. This influence was continued through the many education students who took his courses.

Diagnosis

The literature does not contain a good description of the clinic's intake policy, referral sources, and other operational matters. During the first ten years, Witmer conducted almost all of the psychological examinations himself. Cases were referred from a variety of sources, including school personnel, the visiting nurse service, and private sources. His procedure included an examination by medical specialists for neurological disorders, possible defects of vision or audition, and even diseased tonsils and adenoids, conditions thought to be related to retardation. Apparently Witmer used an extensive battery of psychological tests, including tests of memory, attention, sensory acuity, and association ability as well as the Witmer Formboard and the Witmer Cylinders (the latter tests standardized by his students). In later years the Binet test was used. Although Witmer emphasized the individual, he was a pioneer in obtaining normative data for his tests.

Witmer emphasized that test scores revealed only minimal information, that a test score could be composed of various correct and incorrect performances, and that such a score hid the details of the individual's performance. He taught his students that each individual should be examined carefully and that the qualitative data were the most important to obtain. Where one person would see only failure in an incorrect response, Witmer constantly sought signs of intelligence in the way that the individual handled a particular task. Many of his case studies do not even report IQs.

Diagnostic Teaching—Therapeutic Method

Witmer's contributions to psychodiagnostics are probably of less interest than are his contributions to therapy. Witmer worked with a concept of diagnostic education, feeling that workers could not make a final diagnosis until they had attempted to teach a child something. For Witmer, diagnostic teaching was an experimental method for understanding the individual case by systematically devising educational techniques to test hypotheses about the factors that caused or maintained the condition. He stressed continuing diagnosis, based on a prolonged period of educational treatment, with every step determined by whatever was known about the child to that point, and modified as more was learned. Repeated measurement of actual achievement validated the correctness of the ongoing diagnosis and the effectiveness of the remedial treatment.

The essence of Witmer's method was to teach a child something beyond his or her known level of performance, an attempt carried out consistently several times a week for periods up to a year or more, to the point that the teaching could be continued at home or in school, under the clinic's supervision. In the early days teachers and sometimes the principal accompanied the child to observe the examination. Even in later years when the direct ties to the public schools were no longer strong, this two-way communication was maintained. Sometimes teachers were asked to come to the clinic to participate in examining the child. Students preparing for clinic teaching were sent to the schools to make classroom observations, and frequently they used the clinic teaching in schools or hospitals (Viteles, personal communication).

Clinic teaching was the primary mode of outpatient therapy. Witmer tried to determine which of a child's abilities could be employed to help him or her move toward a preferred pattern of adaptation. Witmer always asked what could be done next to help the child to take yet another step, to increase the child's ability to learn.

The Hospital School

The hospital school, a facility vital to Witmer's approach, was first established in 1901. It was a residential center providing medical care, nursing care, and education primarily to private patients on a fee basis, but exceptionally interesting cases were admitted with little or no fee. Children were admitted for diagnostic purposes, and an attempt was made to train them. They would stay from a few days to a year or more. The training included education and self-care; the hospital school was not just an inpatient institution. In some instances the children were taught at classes held at the University of Pennsylvania Clinic, and sometimes the schooling took place at the hospital school. In addition, the children were often enrolled in local public school classes (Witmer, 1908–9a), and the trainer (Witmer's term for the hospital school supervisor, originally a nurse) stayed in close touch with the public school teacher. Witmer was clearly aware of the problem of returning the child to the community, and he tried to bridge the gap between the institutional and the normal setting.

Therapeutic Methods

The trainers had fairly sophisticated psychotherapeutic techniques at their command, but they used them in natural settings as the problems arose. For example, in one case, outbursts of rage, largely related to one child's jealousy of and competitiveness with another, were treated by the trainer and other attendants, who discussed the problem with the boy over and over again (Witmer, 1908–9d). At one point, this boy was given the task of playing a game with another child without losing his temper. When he did so, he happily reported the fact to the trainer. The trainer taught him to label his undesirable behavior as a "mean mood," and to some extent the child seemed to try to control his "mean mood" to win the trainer's approval.

The assistant trainer interpreted this patient's behavior as self-defeating, discussing it with him when it happened and in the situation in which the behavior occurred. The assistant trainer pointed out how his "mean moods" caused him to be shunned by others, and the staff generally held out the expectation that he should and would begin to control himself better. This consistent treatment had an effect. The boy was ordered to leave the dinner table to wash his hands. He got up, pouted angrily, and said, "I won't come back!" But he washed his hands, and when he came back, he explained, "I nearly got mad, but I just said to myself 'I will control my temper.' " (The reader may be interested in comparing this technique of treating this boy with Redl's life space interview—(Redl, 1966)—and with contemporary cognitive approaches.)

Although Witmer was not much of a theoretician, an examination of his cases and those of his students suggests that he frequently assumed that a low level of frustration tolerance, accompanied by a lack of demand from a parental figure that a child learn to tolerate the tension of frustration, was a cause of retardation. In one article Witmer wrote:

> I believe that the home and school, chiefly through neglect of discipline, permit the minds of many children to remain undeveloped during the formative period. The discipline that is required is not merely that which makes for obedience. It is the discipline of work and strenuous effort, the discipline that trains the memory, the will, the attention and forms habits of work which permit children to assume progressively more difficult tasks. (Witmer, 1908–9b, p. 158)

Witmer seemed to place great reliance on the consistent, firm demand that individuals do what is required of them. He was not insensitive to problems stemming from anxiety, however. Witmer's indomitable will is depicted in a number of the orthogenic cases described by him or his students.

One case, Albert, was described by his mother as "lazy; he answers in any fashion to avoid thinking. He is gluttonous, sluggish, indifferent and thoughtless" (Parker, 1917–18). At age fourteen and a half, Albert's academic achievement level was judged to be no higher than that of fourth grade. He could add, subtract, multiply, and divide by rote, but he could not solve problems that required even simple addition or multiplication. He knew little or nothing of geography or history or current events. He read at a sixth-grade level and spelled at a fifth-grade level, but his comprehension was quite poor. He could not retain the plot of a movie and could not handle formal grammar, but he could compose stories rather well. Witmer's students described Albert as mentally lazy, not liking to work, and playing or bluffing his way through lessons. His attention wandered constantly; he had no self-control, and he could not work independently. He seemed to need someone at his side continuously to force him or coax him. He drove his teachers to distraction by making random, irrelevant comments and by restless fidgeting and inattentiveness. Albert behaved in this manner for five months under the care of two teachers who were unable to cope with him.

Then Albert encountered Mrs. G., the housemother. A large woman, she also possessed a "dynamic will, a will of immense energy to attempt the impossible, of extraordinary power to dominate the mind and behavior of those with whom she lived." Immediately on meeting Albert, Mrs. G. set herself to tame and transform him. Every aspect of his life was governed by a firm, demanding regime. He was forced to wash, to diet, to eat slowly and with good manners, to stop annoying children and adults in the house, and to dress tastefully and cleanly. He was bullied into exercising and performing manual labor. The following incident gives something of the flavor of Mrs. G.'s treatment of Albert:

> One hot afternoon in August, Albert had been sent to rake leaves from the shrubbery over the edge of the hill above the creek. Presently, he came into the cool of the living room, mopping his face.
>
> "I can't rake down there any more, Mrs. G. There's a yellow jacket's nest there."
>
> Mrs. G., unfamiliar with English, thought that "yellow jackets" were "little yellow birds." She pointed to the door with a long arm and compelling finger, and her answer was a single guttural monosyllable:

> "R-r-r-r-rake!"
>
> Albert raked. After a little, he returned to receive precisely the same answer:
>
> "R-r-r-r-rake!"
>
> A third time he came back and still Mrs. G. had for him but one word:
>
> "R-r-r-r-rake!"
>
> And Albert raked!
>
> There was a magnificent thoroughness in Albert's subjugation that, in spite of his resentment, bred in him a respect for the forces that had so mastered him. (Parker, 1917, pp. 132–33)

That Witmer concurred in such treatment that consistently demanded discipline is indicated by the following excerpt.

> Inside the classroom as well as outside, Albert was subjected to a new regime of vigorous discipline. During the first few days in June, Dr. Witmer himself worked with him. For once in his life the nervous wriggling boy did not move a finger. For the first time, he sat at his desk absolutely quiet.
>
> Dr. Witmer, from his examination, concluded that Albert had been pushed too far ahead in geography, history and arithmetic. Because of the picture of mental confusion which he presented, and the conspicuous defect in persistent concentration of attention, Dr. Witmer decided that some weeks of mental discipline must precede any attempt to give Albert additional information. Thoroughness and precision were the qualities which must be developed in him by the summer's work. Albert had, as Dr. Witmer remarked, a mind that "skims." He must be given tasks requiring exactness and completeness—tasks through which he could not skim. Above everything else, there was to be no "speeding up," no attempt to cover ground. Every single point must be known exactly and thoroughly. The teacher's motto, like General Grant's must be, "I'll fight it out on this line if it takes all summer."
>
> Dr. Witmer's outline for Albert's work in the next few months, therefore included:
>
> 1. The memorizing with absolute accuracy of the definitions in Webster's Abridged Dictionary, of words taken from Rice's Rational Spelling Book—Fourth Year—a very few words each day.
> 2. The memorizing, word for word, of the illustrative sentences in Rice's Speller containing the words studied.
> 3. The composition of original sentences containing the words studied.
> 4. Drill in penmanship in which precision must be rigidly insisted upon.
> 5. Simple stories like those in Aesop's Fables and Baldwin's Fifty Famous Stories; read and reread every word until every sentence was read with absolute accuracy.
> 6. Oral and written reproduction of each story until he could reproduce the content of what he had read in a logical and comprehensive manner.
> 7. The writing of solutions of simple arithmetic problems, to exercise his reasoning faculty within the limits of his very elementary comprehension.
>
> With this outline for work, Dr. Witmer handed Albert over to a vigorous young woman whose discipline was more adequate to the situation than that of the man who had first taught him. Dr. Witmer, too, continued his close supervision of the work throughout the summer and from time to time took over the actual teaching

> for one or two hours a day. Miss B. carried out the prescribed program energetically and faithfully. (Parker, 1917, pp. 133–34)

The remainder of the case study describes Albert's further encounters with Mrs. G., his continued treatment in the hospital school, and his later adjustment at a preparatory school. Albert was considerably changed, although he continued to have a variety of problems and by no means did well at school. But he was now functioning at something approaching his age level, and he was no longer so obnoxious that he was in danger of being thrown out of his regular school.

The excerpt from a rather long and fascinating case study reflects quite well Witmer's method of diagnostic teaching. On the basis of tests and observations, Witmer came to certain conclusions about the characteristics of Albert's mind. He did not focus on his infantile regressive behavior, nor did he concern himself with his provocative hostility. His attack on the problem was quite direct. The boy's infantile character was treated by persistent demands that he meet new standards. The adults who were successful with him were forceful in their determination that he comply, but there is no indication they ever punished him. Similarly, Witmer's approach to Albert's educational deficiencies was direct.

Witmer seemed to have little respect for academic degrees. In this case and others, teachers, nurses, and housemothers played important roles in the therapy of children in the clinic and the hospital school. Indeed, Witmer asserted that some of his best trainers had no formal background in special education, although they obviously had special personal characteristics.

Although one may question the faculty psychology (the emphasis on specific mental traits) that Witmer used, it is clear that he was trying to develop work habits and a level of frustration tolerance in keeping with his theoretical views that some instances of retardation were due to a lack of discipline in work and a lack of strenuous efforts to overcome frustration. Witmer did not concern himself with any possible psychodynamics in the learning inhibition, but rather, he attacked it directly and used the indications that the boy could learn as diagnostic signs that he could learn still more. There is no evidence in this case that any other symptoms appeared with the direct attack on his learning problem. The reader may compare Witmer's approach to the education of the emotionally disturbed child with the similar method found most effective by Haring and Phillips (1962) in their study of various approaches to the education of the emotionally disturbed child.

A case study in Witmer's article "Diagnostic Education" (Witmer, 1917–18), demonstrates his method. (This case is probably "Donnie," Witmer, 1919–20, reprinted in Sarason and Doris, 1969.) The case is remarkable not only for its achievement but also for its illustration of Witmer's indomitable spirit, his willingness to try, and his desire not to accept a diagnosis implying hopelessness. The case also shows methods that both teach and diagnose at the same time, and it demonstrates Witmer's ingenuity and sensitivity in interpreting qualitative aspects of performance in any situation.

> The case is of a boy, two years and seven months old at the time treatment began, who was initially diagnosed by Witmer and by an independent authority as fee-

bleminded. The child was born of normal, healthy parents. . . . His birth and development were uneventful until six months of age, when he had an attack of whooping cough. From then on, if placed on the floor he stayed put. If he fell on his face, he lay there until picked up. He made no effort to reach normally attractive objects, but spent most of his time in bed unresponsive and indifferent. He crept at 26 months only after his knees were moved by someone else. His walk at 31 months was still wobbly. He could not negotiate stairs. Given an object, he would hold it, staring at it by the hour. The hair on the back of his head was rubbed off and he had self-inflicted sores about his mouth and ears. He went into a violent temper tantrum if anyone attempted to take anything away from him or move him. He said only the words "kitty" or "daddy," but could respond to the question, "What does the crow say?" by responding, "caw, caw." Language comprehension seemed limited to pointing out a few of his facial parts when asked. He made no effort to imitate actions and did not seem to look at things. A lighted match passed through his field of vision seemed to startle him. He would hold a book and turn its pages and could hold a watch to his ear and say "tick tock." He did not feed himself and was still in diapers. Witmer undertook his training only because of the earnest pleading of the boy's parents. His first response was that the boy was feebleminded. (1917–18, pp. 71–72.)

The following quotations are from Witmer's account of his educational approach with this boy.

I put before him the formboard consisting of eleven blocks of different shapes, each of which had its corresponding receptacle. He would not make the slightest effort even to pick a block to put it back in place. I then tried him with the peg board, a board of 36 holes into which a corresponding number of pegs of the same size and shape can be placed. He could not imitate my action of putting a peg in its hole. He could not put a peg in the hole even when I placed the peg in his hand. I had to hold his hand, guide it to the hole and place it in position, but after having done this once, he put five or six pegs in successively. In all he put in fifteen before I stopped, although after the first six he put in each successive peg only when I said emphatically, "Put in another peg." I never knew him to fill the board with 36 pegs as the result of a general command. His attention appeared to wander and he always desisted. In this his behavior was exactly like that of a chimpanzee whom I taught, though not with the same ease, to put pegs in a board. Subsequent events proved that the reason he objected to putting 36 pegs into the board was because this action bored him, and not because he lacked persistent powers of attention. (1917–18, 74–75.)

Witmer then discussed how he taught the child to work at the formboard in the same fashion, at first by manipulating his hand and later by ordering him to place the forms. Witmer gradually gave him more and more forms to discriminate until by the seventh day of training, the boy could place six forms in their proper places without error. Witmer noted that the boy's coordination was good. Within two weeks, the boy placed eleven insets without error in eighty-five seconds. From this experience, Witmer concluded the child had a retentive memory, good imagery, and good analytic attentiveness and that he could be interested in a relatively difficult problem. When the boy's interest lagged in the formboard, Witmer gave him practice for several weeks on a more difficult performance task. Within three weeks the child mastered the task and again lost all interest in it.

Witmer then decided to teach the child to discriminate the letters of the alphabet.

> On April 17th I taught him the letter B, using for the purpose the large wooden block letter, saying, "This is B. Put B on the chair." In the afternoon he had forgotten it. After three separate periods of instruction he could pick out A, B, and C. He was asked to name them at the same time but would only name B. Some things you can force a child to do, but some things you cannot. I could compel him to pick out these letters but I could not compel him to name them, so I had to tempt him with new letters and new words. I got him to say "V" on one trial by dragging out the "V" sound. He loved the sound of the letter and the feel of making it, but it took nearly two weeks to get him to say "F." (1917–18, p. 76.)

Within three weeks, the child had learned to pick out all twenty-six letters, although he still confused M, N, and W. There was an interruption of one month, but in July Witmer began working to test his ability to learn combinations of letters, in effect to learn to spell. He began by giving him the three letters of the word *cat* and told him to arrange the letters in the proper order. Within one month, the boy learned to arrange the letters of the words *cat, boy,* and *pig* when these were presented to him. By August, he learned to arrange the letters for *cat* and *boy* when given the six letters, and he could spell *cat* or *bat* on command when given the letters BCAT.

> When a child is being taught, I always insist that he shall be taught at attention, on his toes as it were. Work is work and play is play. I even find that the same person cannot both teach and play with a child. Regularity of work is also an essential. The interruption of a few days usually means a great waste of time before the child's attention can be regained and held. In the early part of September, he showed a great gain when after an interruption of four or five days, he buckled down to work again without waste of effort. Beginning September 8th, words were printed on pieces of paper and passed to him to read. He began to read words by first spelling them. On September the 11th the words, "I see a cat," were put before him and he was asked to read the sentence. Dog, pig, etc., were on September 14th substituted for the word cat. By September 19th he could read the sentences, "A man can see me." "A boy can see a dog." On September 21st I tested his newly acquired ability by putting Monroe's primer into his hand for the first time. He read, "I can see a man. A man ran. A cat can see a rat." It was done haltingly, but it showed that from this time on the acquisition of reading was to be only a matter of drill. He can now both spell and sound words and will probably be graduated into the first reader by the first of next June.
>
> I do not care whether this boy can read or not. I have had him taught reading because it was the best way to engage his interest and train his attention, imagination and memory. He liked it, so far as anyone can be said to like work. Intellectual work I call this, and intellectual work, I say without hesitation, is an advantageous mental and hygienic stimulus to any boy of three years of age. (Witmer, 1917–18, pp. 77–78)

His achievement in teaching a child of less than three to read is itself remarkable, but his undertaking this task with a child who was probably psychotic and who appeared to be grossly defective is even more remarkable. The case reveals Witmer's sensitivity to qualitative cues, his ability to see the potential in his behavior rather than its pathology, and his patience and persistence in working with the child. His procedures were planned at every step and progressed from teaching elementary discriminations to developing complex sequential discriminations. Witmer started with the simplest step and moved gradually into more complicated material, using the boy's intrinsic interest in the tasks and his own social approval as the sources of reinforcement.

The Perfectibility of People

Witmer shared with other great educators a basic belief in the perfectibility of people (Mensh, 1966), accepting the principle that each individual has a potential for growth and that educational methods should be directed toward eliciting that potential. In his clinical work, Witmer assumed that a child could learn, despite apparent pathology and the apparent hopelessness of the situation. An integral part of Witmer's clinical methodology was his faith that if he expected that a child could learn, the child would learn. On a broader level, Witmer saw that if diagnostic education could produce such dramatic results with children who were clearly retarded, it held great promise for the education of normal children. Witmer's outlook can be compared with that of Maria Montessori, who started her work with retarded children. When she taught them enough to pass normal school examinations, she concluded that something was lacking in the teaching methods in the public schools that resulted in normal children's performing no better than did her retarded pupils (Montessori, 1964).

Witmer began with an attempt to apply the scientific psychology of that day to the practical problem of education, because he felt that both psychology and education would benefit. In keeping with the prevailing reform spirit, he viewed his approach as a means of changing public education to make it more responsive to individual needs. He did not start with the assumption that people were inferior or sick because they did not respond to the current teaching methods. Rather, he assumed that it would be possible to find ways of teaching that would take into account the characteristics of the people being taught. Witmer's educational approach held that it was indeed possible for any individual to respond to the proper conditions. In contrast with a psychotherapeutic approach that strives to cure the underlying deficit, Witmer seemed to follow the principle that each person contains the "spark of the divine," that if one expects a person to be good, then that good will be elicited. We believe it is no accident that Witmer emphasized such a conception when he did, nor was it an accident that he started with school problems. Both his basic philosophy and the social problems he undertook to correct were products of ongoing social change. We shall see similar viewpoints and similar therapeutic philosophies when we examine some of the other helping agencies that evolved during the same period.

In more immediately practical terms, Witmer's case studies and his educational approaches have much to say to those seeking new ways of helping. His experience should encourage those who may work through an educational modality and who use other than trained mental health professionals as the primary helping people. Moreover, the focus of Witmer's clinic gives the field of clinical psychology a much different view of the traditional. Those currently interested in educational and community-oriented approaches are continuing the earliest tradition of the field, a tradition that Lightner Witmer helped establish and one to which he made substantive contributions.

4

The Settlement House Movement: A Bridge Between Worlds

If history provides any model for the community mental health movement, it is the pre–World War I settlement house. The settlement house developed as a means of coping with the social disorganization induced by the rapid industrialization, urbanization, and immigration in the latter part of the nineteenth century. Its purpose was to promote the organic unity of society by reducing the distance between social classes, and it did this by providing an institutionalized form of help at the apparent point of the greatest disintegration of the social order. At its best, the settlement house initiated action to achieve sweeping changes in what were then viewed as the basic social causes of personal misery.

The early settlement house workers were highly educated young people with a bent toward social action. At their best, they created an exciting intellectual environment in the settlement houses, in which all manner of people could meet and freely exchange ideas. In addition to their helping purpose, the early settlement houses, often affiliated with a university, were designed as living laboratories for the "sociologists" of that day. In fact, they were the means whereby an enormous amount of social research was accomplished, the research providing the basis for programmatic efforts at sweeping social change. The best of the settlement house workers were social scientists working on the problems of the real world, in an attempt to understand their society and to make it into a more effective democracy, one in which their belief in the perfectibility of people could be realized.

Early Antecedents

The settlement house in the United States was a direct outgrowth of a movement established earlier in England. The Industrial Revolution in England preceded that in the United States by half a century, and the social problems it created called forth a new humanitarianism early exemplified by Robert Owen. He experimented with providing good pay, enlightened working conditions, a healthy family life, recreation, education, and a sense of community among his workers. Owen's experiments were not widely adopted, and as the industrialization progressed, the lot of working people worsened. Against the backdrop of social unrest that erupted in revolutions on the Continent in the mid-nineteenth century, philosophers, economists, historians, and social critics developed new ideas concerning the social order. Thomas Carlyle, John Stuart Mill, Karl Marx, Friedrich Engels, and later Leo Tolstoy propounded new views of society based on concepts concerning the distribution of wealth and the responsibilities of different groups in society for one another. Charles Dickens's novels and stories were instrumental in bringing the plight of the poor to the attention of the higher social classes and in paving the way for a variety of reforms.

In England, theological schools and churches developed the concept of the kingdom of God on earth, a concept embodied in the philosophy of the Christian Socialists. The Christian Socialists early and actively supported the labor movement through their writings, fund raising, and advice to labor leaders. Experience with the labor movement led the Christian Socialists to believe that higher education for the working person was an absolute necessity. A volunteer teaching staff of young, intellectually distinguished university men came to the fore and established a highly successful teaching enterprise in working-class neighborhoods, a forerunner of the university extension program.

After 1860, men who had been involved in the Christian Socialist movement or who had supported its tenets came into power at the major universities. These men—Matthew Arnold, a poet and critic; John Ruskin, a professor of fine arts at Oxford; Charles Kingsley, a professor of history; and Frederick Maurice, a professor of moral philosophy at Cambridge—critical of the cultural poverty of their time, called for the wide dissemination of the best of knowledge and culture in order to prepare the foundations of a true democracy.

As early as the 1860s, educated upper-class individuals started to live among workers in England, just as in Russia under the influence of Tolstoy, the Russian nobility attempted to share the lives of the peasants. Men such as Edward Denison and Arnold Toynbee, with the support and approbation of their professors, began to comprehend and to voice some of the problems of working-class neighborhoods. Samuel Barnett, a minister, requested to serve at the vicarage of St. Jude's, Whitechapel, described as the "wretchedest" parish in the London diocese. St. Jude's thus became a center for young Oxford men who wanted to serve.

Through Barnett's influence a university extension center was established in his parish. The center and its activities received wide publicity. Barnett planned to introduce a group of educated men as leaders in areas without constructive local leadership, in which the hard facts of industrial life had led to disorganization. He proposed that a house be rented where university men could "settle" for an extended period of time, studying the life and the problems of a working-class neighborhood. Barnett insisted that close personal acquaintance with the problems be a prerequisite to intelligent public action. He warned that no helping scheme would have a chance of success unless it brought the helper and the helped into a friendly, mutually understanding relationship.[1]

Toynbee Hall, named for Arnold Toynbee, who had died young, opened on Christmas Eve, 1884 and became an important model for the American settlement house movement. Jane Addams, Stanton Coit, Robert Woods, Charles Zeublin, and many other leaders of the movement either visited or were in residence at Toynbee Hall in the years preceding their work in the United States (Barnett, 1950; Woods, 1891; Woods & Kennedy, 1922).

Origins in America

The rise of an urban, industrial society in the United States after the Civil War was accompanied by what seemed to be a lowered standard of living for the laborer, which included crowded houses and tenements, unemployment, vice, delinquency, misery, and illness. Each successive wave of immigrants displaced the last in occupying the poorest areas of cities. The rise of the labor movement, the appearance of radicals of various persuasions, and widespread strikes and riots in the 1870s and 1880s called attention to the social problems of the working class. Henry George, Jacob Riis, Lincoln Steffens, and others wrote about the abuses suffered by working people under corrupt municipal governments and unrestrained private enterprise.

The churches also came under attack and were under pressure to change. As "foreigners" and Catholics arrived, the American Protestant churches moved out to the suburbs. But the churches found that the gospel of salvation and abstract theological issues were irrelevant to the problems of the day. One survey showed that less than 10 percent of the workers in any given area attended church regularly. They discovered what Charles Loring Brace had found out earlier when he tried to save street children with revival meetings.

Even though most churchmen supported the status quo, as early as the 1870s some ministers severely criticized the prevailing business ethics and talked of making Christian principles work here and now. American ministers also spoke about the kingdom of God on earth. Out of the ferment grew the Social Gospel movement, in which ministers—seeing themselves as the lineal successors to the Old Testament prophets—felt it their social and moral responsibility to denounce wrong, to explain right, and to lead people to the realization of their duty. These arguments would sound familiar indeed to socially conscious ministers today (see Callahan, 1967; Cox, 1967). Social Gospel novels, though maudlin, were extraordinarily popular and effective in directing the middle class toward their Christian duty to those less fortunate than themselves. (The extensive volunteer movement can be considered the forerunner of President George Bush's "thousand points of light" campaign rhetoric in 1988.) Many ministers took up the cause of the developing labor movement, as well as various other reforms.

The period during which the Social Gospel movement developed was also a period in which the physical sciences developed great influence. Religious thinkers found themselves under pressure to integrate their traditional teachings about society's morality with materialistic science, evolution, and the new biblical criticism. Christian sociology was born. The early academic sociologists were interested in social reform, and Christian sociologists saw it as their mission to determine scientifically the principles of social organization and social conduct that would best promote social welfare and the ethical life. In the divinity schools of the great universities, through courses in social ethics and sociology and an emphasis on field studies, religious thinkers influenced a whole generation of young people who then went out into the world to implement the concepts they had learned.

The Protestant churches' need for new methods to attract the working class led to the growth of the institutional church, the religious settlement, and the religious mission. The model for the institutional church was that of the YMCA and the Salvation Army. They were free, always open, and located in working-class neighborhoods. The institutional church had a resident minister and kindergartens, swimming pools, gyms, baths, clubs, libraries, dispensaries, clinics, open forums, classes, lectures, sewing and cooking schools, loan funds, penny savings banks, game rooms, soup kitchens, boardinghouse registries, employment services, and mutual benefit societies. Some of the institutional churches eventually contained hospitals and colleges. Temple University in Philadelphia, for example, is an outgrowth of a Baptist church. Some of the institutional churches evolved into nonsectarian settlements, but more important, the ministry of the Social Gospel made it possible and even necessary for the educated youth of the day to assume helping and reform roles (Abell, 1943; Hopkins, 1940; Miller & Miller, 1962).

By the 1880s books such as Henry George's *Progress and Poverty* and the works of Marx and Engels had been published. New thought in economics and diverse views in philosophy, political science, sociology, and anthropology were advocated in the universities (Curti, 1951; Goldman, 1956). Spurred by their professors to be critical of the social order, intrigued by the philosophy of self-sacrifice propounded by Tolstoy and Ruskin, told that the source of wealth of the nation was "tainted," and desirous of achieving a natural democracy, thoughtful students who saw only conventional careers for themselves were placed in conflict. The settlements, however, were a means of resolving personal conflict and putting into action the views of society to which they had been exposed in their college classrooms.

The Settlement House Workers

Although many young ministers and other socially conscious men eventually entered the settlement houses, of much greater importance to the movement was the influx of young, highly educated women. Seventy percent of the residents in the 250 most important houses were women (Woods & Kennedy, 1922). Jane Addams, Lillian Wald, Florence Kelley, Julia Lathrop, Alice Hamilton, Grace and Edith Abbott, Mary Simkhovitch, Vida Scudder, and Jane Robbins were brilliant women with master's degrees, doctorates, and law and medical degrees (Davis, 1959). They were in the first generation of women for whom higher education became a reality. The women's colleges had recently begun to grant degrees, and the women who received them, often ardent feminists themselves and often taught by socially conscious feminists, were determined not only to demonstrate that their intellectual productions were worthy but also to justify their own existence by making themselves useful in the world (Linn, 1935).

Davis (1959) hypothesized that the traditional professions of the law, teaching, and the ministry were losing prestige to the business executive. Those who entered settlement work frequently had parents in the professions, parents who were either sympathetic to or were themselves abolitionists. The new social service seemed a perfect way to carry on the reform tradition of their families and at the same time to enter a new profession, which, for women particularly, was open and promising. Woods and Kennedy (1922) assert that the outstanding achievement of the settlement movement substantiated women's claim to full acceptance in governmental affairs, an acceptance that might not have been achieved as readily through another field.

These new social workers had a variety of motives for entering the profession:

> All settlements are familiar with the sentimentalist caught by a shallow sympathy for and desire to help the poor, but who fails in fundamental democracy, humility and resource when brought face to face with normal people of an industrial neighborhood. Certain men and women are attracted by a supposedly ascetic flavor and undertake residence as a sort of moral scourging under which they hope to be unhappily happy. Closely related to this type is the missionary, the man or women enamored of duty for duty's sake, and the charity monger.
>
> A small number of men and women seek residence either to tide over an interim, or to find an agreeable place to stay, or to gain what they suppose will be a better social station. An occasional candidate labors under the delusion that he or she can in some way escape binding restrictions in another environment, but falls away after discovering that an industrial neighborhood is not in any sense a Bohemia, and that the very seriousness of experiments under way precludes the settlement from encouraging or tolerating a variety of irresponsible fancies.
>
> A certain number of men and women without consciousness of special vocation are attracted in the hope that actual contact with human life and need will discover them to themselves. An occasional person is received on this basis and allowed to test out his interests and powers in the widest and freest way. In their own self-education, settlement workers often apply the principle which governs so much of their class work, namely, that of allowing the individual to touch life at a sufficient number of different points to discover his mind. (Woods & Kennedy, 1922, pp. 428–29)

To use Erik Erikson's helpful term, for many the settlement house seemed to have served as the institution in which they could indulge in a *psychosocial moratorium*. To view the settlement workers as neurotic, guilt-ridden do-gooders, as some do (Lasch, 1965), misses the point. As Lynd (1961) shows, the best of the settlement workers were radical reformers with a definite social philosophy (Woods, 1906).

In her autobiography, *Twenty Years at Hull House,* Jane Addams tells how she traveled aimlessly for several years after graduating from college, trying to decide on a direction for her life. Through her travels, she became aware of the terrible poverty that existed, but she felt helpless to do anything about it.

> For two years in the midst of my distress over the poverty, which, thus suddenly driven into my consciousness, had become to me the "Weltschmerz," there was

> mingled a sense of futility, of misdirected energy, the belief that the pursuit of cultivation would not in the end bring either solace or relief. I gradually reached a conviction that the first generation of college women had taken their learning too quickly, had departed too suddenly from the active emotional life led by the grandmothers and great-grandmothers, that the contemporary education of young women had developed too exclusively the power of acquiring knowledge and of merely receiving impressions; that somewhere in the process of "being educated" they had lost that simple and almost automatic response to the human appeal, that old healthful reaction resulting in activity from the mere presence of suffering or helplessness. (Addams, 1910, p. 71)

In 1888 Addams decided to establish a settlement house and visited Toynbee Hall for ideas. She felt she had found a mission for her life. In reminiscing about those years of indecision, years she had thought of as "preparation," she concluded:

> It was not until years afterward that I came upon Tolstoy's phrase "the snare of preparation" which he insists we spread before the feet of young people, hopelessly entangling them in a curious inactivity at the very period of life when they are longing to construct the world anew and to conform it to their own ideals. (Addams, 1910, p. 88)

In an 1892 speech she further articulated the motives of the young settlement house workers. A better statement of the feelings and the problem of alienated youth can hardly be found.

> This paper is an attempt to analyze the motives which underlie a movement based, not only upon conviction, but upon genuine emotion, wherever educated young people are seeking an outlet for that sentiment of universal brotherhood, which the best spirit of our times is forcing from an emotion into a motive. These young people accomplish little toward the solution of this social problem, and bear the brunt of being cultivated into unnourished, oversensitive lives. They have been shut off from the common labor by which they live which is a great source of moral and physical health. They feel a want of harmony between their theory and their lives, a lack of coordination between thought and action. I think it is hard for us to realize how seriously many of them are taking to the union of human brotherhood, how eagerly they long to give tangible expression to the democratic ideal. . . .
>
> We have in America a fast growing number of cultivated young people who have no recognized outlet for their active faculties. They hear constantly of the great social maladjustment, but no way is provided for them to change it, and their uselessness hangs about them heavily. Huxley declares that the sense of uselessness is the severest shock which the human system can sustain, and that if persistently sustained, it results in atrophy of function. These young people have had the advantages of college, of European travel, and of economic study, but they are sustaining this shock of inaction. . . . They tell their elders with all the bitterness of youth that if they expect success from them in business or politics or in whatever lines their ambition for them has run, they must let them find out what the people want and how they want it. It is only the stronger young people, however, who formulate this. Many of them dissipate their energies in so-called enjoyment. Others not content with that, go on studying and go back to college

> for their second degrees; not that they are especially fond of study, but because they want something definite to do, and their powers have been trained in the direction of mental accumulation. Many are buried beneath this mental accumulation which lowered vitality and discontent. . . . This young life, so sincere in its emotion and good phrase and yet so undirected, seems to me as pitiful as the other great mass of destitute lives. One is supplementary to the other, and some method of communication can surely be devised. . . . . Our young people feel nervously the need of putting theory into action, and respond quickly to the settlement form of activity. (Addams, 1910, pp. 115–22)

The settlement house grew out of moral fervor and intellectual ideas, out of the need of the young educated person to find a useful place in the world, out of the complementary needs of the men, women, and children struggling to live in the massive urban slums created by an industrial might that produced both progress and poverty. In the settlement house, the two classes could meet and gain in fellowship because they were acting in a common purpose to satisfy mutual needs.

The fact that a predominant source of inspiration for the early settlement house workers came from the university gave the movement an intellectual bent that it retained in its emphasis on forming a scientific base for social reform. The broad intellectual character and the research orientation of this early movement in social work contrasts with the narrower emphasis on clinical method and psychoanalytic theory that later came to characterize the professional schools. In a real sense, the reforms instituted by the settlement workers in the early 1900s, and even the marked changes of the 1930s, initiated by many who had been in the early settlements, were born in the seminar rooms and lecture halls of the universities.

The Establishment of the First Settlement Houses

The first settlements were established in the late 1880s and the early 1890s in similar ways, many independently of one another. When Stanton Coit returned from Toynbee Hall, he became an assistant to Felix Adler who had some ten years earlier founded the Ethical Culture Society, a group deeply involved in social and religious reform. Taking his inspiration from Toynbee Hall, within a year Coit settled into a small tenement house on the Lower East Side of New York City. His initial reception by the neighborhood gives us something of the feeling of the times.

Coit intended at first to obtain an apartment in the worst tenement in the neighborhood, where one-tenth of all arrests for crime and one-half of all arrests for gambling in the city of New York took place (Freeman, 1961). But at the insistence of concerned friends, he visited the local police station. And when police officials could not offer even fair assurance for his physical safety, he decided to take a place in a somewhat quieter building nearby.

> The expressman called to move Dr. Coit's goods downtown protested at first that his client was in error concerning the address, and later was inclined to question his sanity. Neighbors were hardly less puzzled. A myth sprang up that he was a

> cast-off son of wealthy parents who had sought the East Side in the last descending stages of want. Popular sympathy was altogether with the supposed victim, and his family was hotly criticized for driving into such an environment anyone tenderly brought up. Only a dime novel plot seemed adequate to explain so unusual a situation as his presence in the district. (Woods & Kennedy, 1922, p. 42)

During the summer and fall, Coit cultivated the people in the neighborhood. He organized picnics for young people and later offered his apartment as a meeting place for a group of young men. From these beginnings came the concept of the club and then the Neighborhood Guild, which eventually grew into the University Settlement.

The second settlement grew from the first. Dr. Jane E. Robbins and Jean Fine had worked with Coit for a while, but within a year the two women rented their own quarters nearby and began to organize clubs for girls. Several years earlier, under the impetus of Vida Scudder, a group of Smith College alumnae had gotten together to secure support for such a project. An association for the ''support and control'' of college settlements for women was organized with representatives from Bryn Mawr, Vassar, and Wellesley. The project ripened none too soon, for the ward leader, ''the notorious 'Silver Dollar' Smith,'' their heartless landlord, dispossessed them on the pretext that the traffic of the young women of the neighborhood was wearing out the stairs in the apartment building. Coit had earlier started a campaign to obtain cleaner streets, and it may be that the shrewd ''Silver Dollar,'' quickly recognizing the settlement workers as potential enemies, wanted to get rid of them (Woods & Kennedy, 1922).

At any rate, in the fall of 1889, a group of seven young women moved into a house on Rivington Street to start the College Settlement. Their first caller was the local cop on the beat, who assumed that the only reason a group of young women would rent a house on the Lower East Side was for ''business'' reasons. He offered not to interfere, provided that the young women made an appropriate payoff (Davis, 1959). Undaunted, the women hung out a sign offering baths for five cents (no one came at first); they organized clubs for boys and girls, established a library, assisted at a Sunday school, took in a sick baby and nursed it back to health, and put up for the night a mother and her children who needed a temporary sanctuary from a husband in a drunken rage (Freeman, 1961). They made themselves useful in a friendly way, calling on and receiving calls from their neighbors.

Lillian Wald, a graduate of Miss Cruttenden's English–French Boarding and Day School for Young Ladies and Little Girls, entered the school of nursing at New York Hospital. Wanting to make herself useful after graduating, she taught nursing to immigrant women in a ''Sabbath school'' whose purpose was to select candidates for nursing schools. Encountering the disorder and the misery of the neighborhood, she was determined to find a place where she could be of greater assistance. She spent two months at the College Settlement with a friend, and then in 1893 the two women took an apartment on Jefferson Street, despite the anxieties of their parents.

The immigrant neighbors were ambivalent at first, concerned that the two were religious missionaries, but they were also pleased that ''Americans'' had come to live among them. The depression of 1893–94 increased the misery of the poor. Ministering to the sick, playing with the children, getting confirmation clothing for a little girl, sending an orphan to her mother's home in Rochester, ordering her nursing uniforms from a woman trying to support herself by sewing, and interceding with public agencies all established Miss Wald and her friend as people who wanted to help. In a short time the neighbors came to tell their troubles, to seek advice, or just to talk.

Within two years, the project expanded so that the small apartment was no longer sufficient. The enterprise, now taking on the characteristics of a full-fledged settlement, was established on Henry Street as the Nurses' Settlement. (A few years after the establishment of the Nurses' Settlement, the official name was changed to the Henry Street Settlement. Legend has it that the athletic clubs were at a disadvantage in intersettlement competition because at the height of battle, the opposition would unnerve them with the taunt, "Noices! Noices!"—Duffus, 1938.) Here the visiting nurse service, with its characteristic uniforms and arm bands, was established. Their nursing skills gave the women entry and safe passage everywhere and also gave them access to the problems of the neighborhood. Public school nursing and many public health efforts developed from the Henry Street Settlement. When Lillian Wald later sought public health legislation, she had a safe base of power in the neighborhood and, indeed, in the city (Duffus, 1938; Wald, 1915).

The most famous of the settlements was undoubtedly Hull House, established in Chicago in 1889 by Jane Addams. After a diligent search in the company of city missionaries, newspaper reporters, and officers of the compulsory education department and with the advice of an ex-mayor of Chicago, she selected Hull House on Halstead Street as a suitable site for her settlement. Her description of the area is valuable, for it not only gives a feeling of the neighborhood, but it also reminds us that things have not changed very much.

> Halstead Street is thirty-two miles long, and one of the great thoroughfares of Chicago. . . . For six miles . . . the street is lined with shops of butchers and grocers, with dingy and gorgeous saloons, and pretentious establishments for the sale of ready-made clothing. Polk Street, running west from Halstead Street, grows rapidly more prosperous; running a mile east to State Street, it grows steadily worse, and crosses a network of vice on the corners of Clark Street and Fifth Avenue. Hull House once stood in the suburbs, but the city has steadily grown up around it and its site now has corners on three or four foreign colonies. (Addams, 1910, pp. 97–98)

After describing the varied ethnic composition of the area, she continues:

> The policy of the public authorities of never taking an initiative, and always waiting to be urged to do their duty, is obviously fatal in a neighborhood where there is little initiative among the citizens. The idea underlying our self government breaks down in such a ward. The streets are inexpressibly dirty, the number of schools inadequate, sanitary legislation unenforced, the street lighting bad, the paving miserable and altogether lacking in the alleys and smaller streets, and the stables foul beyond description. Hundreds of houses are unconnected with the street sewer. The older and richer inhabitants seem anxious to move away as

> rapidly as they can afford it. They make room for the newly arrived immigrants who are densely ignorant of civic duties. This substitution of the older inhabitants is accomplished industrially also. . . . One of the most discouraging features about the present system of tenement houses is that many are owned by sordid and ignorant immigrants. The theory that wealth brings responsibility, that possession entails at length education and refinement, in these cases fails utterly. . . . Another thing that prevents better houses in Chicago is the tentative attitude of the real estate men. Many unsavory conditions are allowed to continue which would be regarded with horror if they were considered permanent. Meanwhile, the wretched conditions persist until at least two generations of children have been born or reared in them. (Addams, 1910, pp. 98–100)

Like the other settlement workers, Jane Addams and her companion Ellen Starr were greeted with a certain amount of suspicion by the neighborhood. Jane Addams quotes a man who said their presence in the neighborhood was the strangest thing he had ever seen. Luckily, the two women found they had an early ambassador to the community in the person of a girl who came to live with them to do housework; she became an important worker in the first few years. The Hull House group came to their neighborhood with the mission of sharing their cultural possessions with their new neighbors in order to establish the conditions for a fuller life in the area (an attitude some later critics regarded as cultural imperialism). One of their first activities was to start a reading party, in which a group of young women from the area met to listen to and to discuss readings from George Eliot and Nathaniel Hawthorne.

Another early activity was a kindergarten. The following anecdote suggests the problem of cultural difference and the attitude of the young workers toward their neighbors:

> One day at luncheon she gaily recited her futile attempt to impress temperance principles upon the mind of an Italian mother, to whom she had returned a small daughter of five sent to the kindergarten "in quite a horrid state of intoxication" from the wine-soaked bread upon which she had breakfasted. The mother, with the gentle courtesy of a South Italian, listened politely to her graphic portrayal of the untimely end awaiting so immature a wine bibber; but long before the lecture was finished, quite unconscious of the incongruity, she hospitably set forth her best wines, and when her baffled guest refused one after the other, she disappeared, only to quickly return with a small dark glass of whiskey, saying reassuringly, "See, I have brought you the true American drink." The recital ended in serio-comic despair, with the rueful statement that "the impression I probably made upon her darkened mind was that it is the American custom to breakfast children on bread soaked in whiskey instead of light Italian wine." (Addams, 1910, pp. 102–3)

From the beginning, the desire to help was expressed in concrete personal service. Jane Addams and Ellen Starr washed newborn babies, prepared the dead for burial, nursed the sick, acted as midwives at the birth of an illegitimate baby, and took in a fifteen-year-old bride who was desperate to escape the nightly beatings administered by her husband. They became true neighbors, exchanging visits, sharing meals, attending weddings and christenings, becoming godparents, and participating in the life of the community by their sheer enjoyment of human contact. Just as the other settlement workers did, Jane Addams and Ellen Starr made a place for themselves in the neighborhood by giving of themselves

Powdermaker's (1966) discussion of the anthropological technique of participant observation suggests that the early settlement workers shared the anthropologists' intuitions about how to enter and become part of strange communities. The later professionals seem to have lost to their professionalism, as Jane Addams said so well, "that old healthful reaction resulting in activity from the mere presence of suffering or helplessness." The early settlement house workers were not simply making people dependent on them; they were establishing a firm base in the community for later action.

Life in the Settlement Houses

Although the quality of the settlement houses varied, life in the best ones must have been an exciting experience spiced with vigorous intellectual interchange. At several of the houses, opinions of all kinds were heard and tolerated and subjected to argument. In fact, the settlements were bastions of free speech (Woods & Kennedy, 1922) whose excitement was enhanced by the continuous turnover of young residents, who came with a variety of purposes and from a number of diverse fields. There were young scientists and medical students, ministers and graduate students, wealthy dilettantes, ardent feminists, and social activists of many different kinds. Journalists, novelists, and muckrakers came to the settlements seeking material. All were absorbed in common work and had a common cause. From this intellectual ferment came hundreds of books and articles, both scientific and popular, and hundreds of research studies concerning the social problems of the day. Woods and Kennedy (1911, 1922) list extensive bibliographies.

The residential quarters of the settlements were simple, but by no means austere, as they usually were large houses formerly owned by the well-to-do. At first, the residents in the group rented apartments nearby or in the same building, but these often were uncomfortable. In the relatively small settlement buildings such as Hull House and Henry Street, the living arrangements at first were much like those in a family home. Later rooms resembled college dormitories. The dining room frequently served as a common room, and a few of the houses had a residents' living room, with a fireplace, piano, and sociology library. The rooms at Hull House were well furnished and contained paintings, photographs, and books.

Both male and female residents lived in the same building. The coeducational character of the settlements was the cause of some concern to the head workers, although it was considered desirable to have both young and old, men and women, and single and married people in the same center. A few settlements had accommodations for married couples and families. But few families remained in residence, primarily because they felt the immediate neighborhoods did not provide suitable environments for their children.

The residents tried to share the life of the neighborhood, but for some, differences in life-style and material wealth between them and their neighbors caused considerable personal conflict. An example can be found in the following excerpt:

> His family lived at the settlement of the Commons and he had a five months old baby. On the day that it was born a child was born in the family living across the street from the settlement. The two mothers had exchanged messages and afterward had become well acquainted, finding their motherhood a strong interest in common. When he and his wife noticed, however, that the other baby was not kept nearly so clean as their baby, they had to remind themselves that the mother across the street did all her own work and had a hard struggle to get on. Then it was difficult to get good milk in their neighborhood; even with a refrigerator and other conveniences the milk spoiled, and when the mother across the way asked how they fed their baby and they replied that they used———, she said, with a sigh, "yes, but that costs too much." It did cost. Here at one blow was cut away all the common ground between them. How could they advise, how could they confer, how could they pretend to help the family across the way when the conditions under which they were living were so different, when they were not on the same economic basis at all? For his part, the matter had given him sleepless nights, his heart torn by the thought that their pretense of being helpful in the neighborhood was a mere sham, that until economic conditions were changed, they were utterly helpless. (Richmond, 1961, p. 262)

In keeping with the democratic ideals of the settlements, most of the early houses operated according to cooperative principles. In weekly meetings the residents decided the details of cooperative housekeeping, house rules, and the nature of the work to be undertaken in the neighborhood. The members did much of their own work. Although Jane Addams did take in a woman to help out, the woman became an integral part of the household, participating in the work of the settlement. Jane Addams frequently took her along when she addressed middle-class groups, in order to have her respond to questions about life in the Hull House neighborhood. Although Jane Addams had two rooms for her own use, she insisted that her bath be shared by others. Lillian Wald and her companion did all their own work except for scrubbing and laundry. Residents took turns serving on the various committees responsible for running the household.

All residents were encouraged to accept the challenge of developing and working up to their highest standard. The basic responsibility for decisions and their implementation was left to individual workers. At Henry Street, each nurse was responsible for managing her own patients and arranged her time according to her own best judgment. A family council was available to discuss problems and difficult situations, as well as issues of interest to the group (Duffus, 1938).

Most of the settlements had periodic meetings to discuss the meaning of settlement work, which centered on broad questions of reform, problems of neighborhood organization, means of bringing culture to the people of the neighborhood, and interpretations of science, literature, philosophy, and religion in relation to the residents' personal outlooks. Invited speakers often came to talk about their research or their observations of other communities and foreign countries.

The seminar was not the only medium of intellectual exchange. The dinner table was an important setting, as well. Because the meals were inexpensive, young intellectuals would come to the settlements to dine. Rube Borough, an author and journalist, recalls that he and his friend Carl Sandburg frequently dined at Hull House in the early 1900s. Arthur Holden (1922), a settlement resident for a year or so, sets the scene:

> A favorite method of introduction is to get the stranger to take dinner with the residents. It will be a varied company. The women will probably be in the majority. There will be young women and middle-aged; there will be the well dressed attractive type and beside her the so called "New England schoolmarm!" There will be long haired men in soft collars, whom the uninitiated will instantly suspect of socialism, as well as short haired men in business suits with conventional neck gear. There will be the inevitable buzz of conversation. Almost everyone will openly avow a genuine interest in what everyone else is doing and real importance will be attached to discussion of general topics of the day. There will be talking across the table, questionings, banterings, hasty opinions snapped and well considered opinions weighed. It is very possible that the visitor may feel himself talked to, cornered, even patronized by some one whom he does not know, asked to come again and almost forced to accept an invitation for the following week. (Holden, 1922, pp. 35–36)

The intellectual caliber of the best of the houses can be judged (with some allowance for the adulation of authorized biographies) from the descriptions of some of the early residents.

> It is difficult to conceive of any attribute of Julia Lathrop that might be wished changed. She sparkled as did Florence Kelley; their talk was a firework. . . . Mrs. Kelley was a fighter; Miss Lathrop was a diplomat . . . when both were at Hull House together, arguing some problem of correcting a social injustice, and disagreeing as they often did on the best method of procedure, it is doubtful if any better talk was to be heard anywhere. Prime Ministers of Europe, philosophers of all doctrines, labor leaders and great capitalists and unpopular poets and popular novelists and shabby exiles from half the kingdoms of the world visited Hull House and dined there, and listened willingly and were glad to be there; and if they had only known it, in the "house meetings" afterward, which only residents attended, they would have heard more vivid discussions still, sternly practical, yet still enlivened by the same patient or impatient humor, as the case might be. (Linn, 1935, pp. 140–41)

Historians and others interested in the development of ideas have given renewed recognition to the settlement workers for both their intellectual and their social accomplishments. Jane Addams's contribution to the development of the Chicago school of social science is increasingly acknowledged. Indirect but important intellectual influences on Harry Stack Sullivan and John Dewey, for example, have been attributed to Jane Addams and to Hull House (Bernstein, 1967; Lasch, 1965, 1967; Lynd, 1961; Sullivan, 1964). As we shall see, legislation regulating child labor and women's work, tenement laws, sanitation codes, regulations promoting health and welfare, improvements in schools, the juvenile court movement, and similar reforms were based directly or indirectly on the research and activities of settlement workers. Still later, in the 1930s, much social and welfare legislation was written, or influenced in its form and content, by people who had been active in the earlier settlement house movement (Chambers, 1967).

The settlement workers' intellectual and research activities were among their most important contributions. William James said of Jane Addams's first book, *Democracy and Social Ethics* (1902): "The religion of democracy needs nothing so much as sympathetic interpretation to one another of the different classes of which society consists; and you have made your contribution in a masterly manner." Similarly, the editor of the *San Francisco Bulletin* commented on how Jane Addams's book had helped him understand and interpret for his readers the social and political conditions of the period.

As Davis (1959) summarized it, the settlement workers' research and their experiences put them in a position to know the problems firsthand, whereas their backgrounds enabled them to interpret the issues for the middle class and political leaders and helped effect sweeping social change through legislative action.

Professional activities were part of a full life and were not separated from the other aspects of living. The settlement workers intended to promote an intellectually and culturally rich democratic life among those they served, and they wisely began by applying their principles to themselves. It is unfortunate that as the houses grew, a more authoritarian structure became necessary, but in the early days, by design, the way in which the workers treated one another generalized to the way in which they treated those whom they served in the neighborhoods. Although there is a danger of oversentimentalizing the past, one cannot help but feel more than a touch of envy of those who had the privilege of sharing the fellowship of the early settlement houses.

How the Settlements Were Financed

Any group bent on social reform needs financial independence. The settlement workers soon learned that "he who pays the piper calls the tune." Most of the earliest houses were founded and financed by a group of unsalaried residents who rented their own quarters. Later, when groups moved into houses, the residents paid for the room and board. Although many worked full time in the settlement, others had jobs that gave them time for settlement work. The financial situation changed as the backgrounds of the residents changed. Not all who came later had the independent income of a Jane Addams or a Lillian Wald. In the absence of independent incomes, fellowships for young students and salaries for full-time residents and other employees became necessary.

The level of salaries presented problems. The residents' principles demanded that secretaries, domestics, janitors, and handymen receive living wages, but the settlements' finances often led to underpaid employees. Similarly, the salaries of the residents themselves were a problem. Too little money led to anxiety about illness and dependence in old age, but higher wages created guilt feelings in the workers. There was a tendency to set salary scales in accordance with what educators received.

Once the settlements started raising funds, they found that the flow of money sometimes depended on action, or inaction, that might compromise the settlements' independence and ideals. There are three major sources of funds for enterprises such as settlements: the government; private philanthropy; and subscriptions, memberships, and fees for settlement activities. Settlement workers rejected government (municipal) support, not just because they frequently were at odds with city governments but also because they experimented with a variety of services that they hoped the governments would adopt later (e.g., kindergartens, visiting teachers, probation officers). Settlement leaders felt that their freedom to innovate and experiment would be seriously compromised if they worked within the framework of governmental agencies.

This was before the days of massive federal support of social welfare projects, but current experience suggests that the settlement workers had hit on an important truth. No agency of government can afford to finance its critics, nor can it legitimately finance revolutionaries. For example, when the antipoverty agencies of the 1960s seemed to go too far in challenging the authority of city halls, new legislation undercut their efforts. Although the government might be benevolent and, in its benevolence, promote important changes, there probably are but a few changes that a government can tolerate at any one time.

Some of the settlement houses raised a substantial portion of their funds from the neighborhoods in which they served, but in most of the houses this neighborhood support amounted to about 10 percent of their budget. In a few instances all of the necessary funds were raised within the neighborhoods, and no other outside source of support was necessary or desired. It was the opinion of some that all support should be raised from the neighborhoods, but that proved difficult. The settlements served poor people who did not have money to contribute. In some instances, even when some money was raised from the immediate area, the bulk of subscriptions came from "uptown" members. Some of the settlements with larger budgets tended to disregard the money that could be raised from neighborhoods because they felt it was too difficult, or too unimportant, to be worth much effort (Holden, 1922; Woods & Kennedy, 1922).

Probably the two most important sources of funds were wealthy individuals and organizations such as the Junior League. Funds from large donors were important to building and developing endowments, although these funds were not without their problems. For one, the funds were often given for specific purposes and were not available for general expenses (Holden, 1922). For another, a settlement might become dependent on the goodwill of one or a few wealthy patrons. If the patron became displeased, the settlement would be faced with a sudden withdrawal of funds, and Woods and Kennedy (1922) point out that at one time or another, almost all of the active settlements took stands on controversial issues that cost them patronage.

High-spirited individuals such as Jane Addams and Lillian Wald did not back away from controversy, even though their activities cost them support. In 1899, a dozen black women, delegates to a congress held in Chicago, were invited to lunch at Hull House. As a consequence, southern editors "uninvited" Jane Addams to speak in the South, and speaking was an important source of income for her. In another instance, she defended a young man falsely arrested as an anarchist after the assassination of President William McKinley. Her defense of the young man cost her the support of a Chicago socialite and won her the castigation of several socially prominent ministers. Her later pacifist and antiwar stands also lost her considerable support (Linn, 1935). Lillian Wald, very active in antiwar organizations, also lost financial support because of her activities (Duffus, 1938). Later, after World War I when the political climate was anticommunist, the settlements, presumably hotbeds of radicalism, had great difficulty raising money for anything more than fresh-air camps or Americanization classes to teach English to immigrants (Holden, 1922).

Sometimes the funds offered were actually bribes. Hull House once turned down a gift of $50,000 that was dependent on the residents ceasing their agitation for a factory law in Illinois. It is a measure of Jane Addams that she did not congratulate herself for having turned down the bribe but, rather, felt mortified that someone had even thought to offer it to her.

The means by which wealthy patrons earned their money caused other problems for the settlements. There was a great deal of discussion about "tainted money," a salient issue when Jane Adams turned down a contribution of $20,000 for a building for the Jane Club, a cooperative apartment dwelling for working women, because the donor was notorious for underpaying his female employees (Addams, 1910; Linn, 1935). As Addams pointed out, the issues were often confused, but the fact that the issues arose at all suggests that the settlements, at least in their earlier days, were determined to maintain their independence and their freedom to question any aspect of society, including "that dubious area wherein wealth is accumulated."

Wealthy patrons who were morally committed and who had no fear of consequences were important to the work of the major settlements. Jane Addams's biographer provides a brief sketch of Mrs. Joseph T. Bowen, who was closely involved with Hull House for many years as well as with the juvenile court in Chicago (see Chapter 7). Mrs. Bowen was never a resident, but she joined the Hull House Women's Club, soon becoming its president, an office she held for seventeen years. As its president, organizer of the Juvenile Protective Association, and president of the Chicago Juvenile Court Committee, she was actively engaged in social reform. She could bring knowledge to her work as trustee and treasurer of the Hull House Association, as could many of the settlement board members who frequently had several responsibilities in a settlement.

The attitudes of patrons such as Mrs. Bowen are explained by Jane Addams's biographer:

> Rooted in conservatism, a patriot of the old school, proud of her long record of "Black Repbulicanism," Mrs. Bowen surveyed some of the activities, a few of the residents, and many of the visitors to Hull House with a severe though humorous eye. She liked things the way she liked them, and eccentric manifestations of radicalism she never liked; nor did she ever join the Women's International League for Peace. But she no more dreamed of forcible interference with the opinions of other people than she dreamed of the possibility that any one might successfully interfere with her own opinions, and took the same pride in the hospitality of Hull House to every shade of political expression that she took

> in her own hospitality to every sort of guest. To oddity of all sorts she opposed merely a kind of queenly acceptance. Jane Addams herself, in Mrs. Bowen's view, might think strangely, but she could never do wrong, any more than Mrs. Bowen herself could be mean; both conceptions were impossible. For four decades, Mrs. Bowen has retained a sense of feudality in connection with Hull House, a consciousness that noblesse oblige. She has laughed at herself, laughed at what she has regarded as centrifugal, laughed at everything except service and fineness; those two are the articles of her creed. (Linn, 1935, pp. 143–44)

In order to keep policies responsive to the needs of the neighborhood, it was necessary for board members to stay in close touch with the settlements' work. Women such as Mrs. Bowen were invaluable, and an attempt was made to have board members take on volunteer work and club leadership. Board members were encouraged to engage actively with the resident staff and the people of the neighborhood, to keep the board alert to local needs.

There seem to have been a few instances when there were neighborhood representatives on the boards, but generally it was difficult to interest working people in participating, and in view of the necessity that the board accept financial responsibility, it was simply not useful to have poor people represented.

The poor were, however, represented on the house councils, which were often given some power by the board (one group had the authority to impeach the head worker, for example). They were also represented on the boards of financial associations, groups consisting of those who contributed to the settlement or who were members by virtue of the fees they paid (Woods & Kennedy, 1922). Nonetheless, it was difficult to guarantee that the policymaking boards would be responsive to the needs of the neighborhood. Not all board members were as tolerant and open as Mrs. Bowen was. Woods and Kennedy (1922) and Holden (1922) allude to the various problems regarding the boards. Holden felt that the chief qualification for many board members was the money they were able to contribute, a fact that did not dampen their enthusiasm but that probably meant they did not have the knowledge and experience that the residents had. The problem of control by boards in later years is presented in a light that makes some of the settlements seem rather pathetic. Holden (1922) reprinted a "settlement catechism" written by Mary Kingsbury Simkhovitch in about 1912. It offers a view of the settlement in somewhat simplistic terms. With respect to the board, the catechism states:

> If the settlement is a family, what is the Settlement's Committee or Board of Managers?
>
> Such boards are friends of the family who help it to carry on its neighborhood enterprises and give it counsel. Such a board may properly refuse to give support to any enterprise in which it does not believe or may encourage by financial help that in which it does believe.
>
> Is such a board necessary?
>
> It or something similar is necessary unless the Settlement family is entirely self-sustaining, and this rarely is the case. (Holden, 1922, p. 192)

As the settlements' financial needs grew, the original democratic organization gave way to an authoritarian structure. Was the loss of self-determination entirely necessary? Woods and Kennedy (1922) argue that the best-financed settlements tended to be the least active and the most complacent and that the most politically active settlements had the most financial struggles. However, even though those settlements bold in their approach tended to lose some short-run support, over the long haul the more outspoken and active of the settlements found that they could depend on admiring followers to see them through financial difficulties.

The leadership of the settlements was also an important consideration. A strong leader with a moral commitment to settlement ideals could not conscientiously do anything other than attempt to live up to personal ideals. The later generations of settlement leaders, coming as they did from professional schools and having a careerist rather than a crusader viewpoint, were understandably less concerned with independence. They had programs, payrolls, and buildings to concern them, inherited successes of the previous generation of settlement workers. Professionalism in social work also influenced the new group coming to the settlement houses in the 1920s and the 1930s. As graduates of professional schools with training in scientific casework and group work, they were no longer concerned with social reform, particularly in the conservative 1920s. Sons and daughters of immigrants who became settlement workers and who had been brought up in the slums were not interested in living in the slums or in remaining there after hours. Settlement work became a job, and there was a conscious attempt to give the job a professional status. The enthusiasm and the commitment of the earlier era no longer existed in the same degree among the later workers (Davis, 1959). In part, the settlement house movement fell victim to its own success.

The Concept of the Community Organization

Stanton Coit, the man who established the first settlement in the United States, came to the settlement concept with a plan. He intended to help a neighborhood become organized for its own development:

> The first step in social reform, if my psychology be correct, must be the conscious organization of the intellectual and moral life of the people for the total improvement of the human lot. . . . Because of the lack of it, our ideals and schemes are cold, abstract, bloodless things, or at best, are impotent. . . . If this first step, the organization of the masses mentally and morally, were taken, the second step, the enlightenment of the people in social principles, could be easily made, and then the realization of the just state would not be remote, nor would it be brought in with violence. Now such a general organization of the life of the people, and such a general civic instruction, are the special field of action which the Neighborhood Guild would presume to assign to itself. (Coit, 1891, pp. 4–5)

Coit's Neighborhood Guild plan was guided by concepts concerning the nature of urban industrial life. Recognizing the tendency for the family's authority to decline and the tendency for ties between people to become role specific rather than diffused, Coit saw the guild as an extension of the family concept in which members banded together to pursue many goals. He specifically decried the tendency of modern-day groups to be formed for single purposes, noting that such single-purpose associations tended to overvalue the specific function and did not encourage people to know one another in more than a single dimension. The Neighborhood Guild, therefore, by drawing people together in an alliance to promote and to carry out "all the reforms—domestic, industrial, educational, provident, or recreative—which the social ideal demands," would bring greater advantage. "The supreme aim which it constantly keeps in view is the completest efficiency of each individual, as a worker for the community, in morals, manners, workmanship, civic virtues and intellectual power, and the fullest possible attainment of social and industrial advantages" (Coit, 1891, p. 11).

The guild was not to be limited "to the rescue of those who have already fallen into vice, crime or pauperism." It would also include the stable working class, for Coit saw himself engaged not only in preventive work but also in efforts to realize the highest ideals of the culture. "The way to save and prevent is often by educating the intellect, and cultivating the taste of the person in danger or already fallen; and, again, the superior development of one member of a family or of a circle of friends may prove the social salvation of all the rest" (Coit, 1891, pp. 11–12).

Coit recognized the numbers problem. The concept of mutual assistance and self-help was built in because he recognized that although it was possible for a small group to care for most of its own needs, it was impossible for any single association to serve everyone. He envisioned guilds that would propagate themselves through members organizing new groups in adjacent areas. Coit saw that only the force of self-help could begin to approach the magnitude of the problem. "If we consider the vast amount of personal attention and time needed to understand and deal effectively with the case of any one man or family that has fallen into vice, crime or pauperism, we shall see the impossibility of coping with even these evils alone, unless the helpers be both many and constantly at hand" (Coit, 1891, p. 19).

Coit also pointed out that organized self-help did not exist, that it arose spontaneously at a point of need. Neighbors know almost immediately when a family is in difficulty, and they understand the nature of that difficulty. Whereas an outsider comes necessarily as an inquisitor, a neighbor can know without prying. Neighbors do help one another in time of troubles; not only does the help take a concrete form, but it also is easier to accept from someone in similar circumstances. One does not become "a case." The giver of assistance and the receiver can identify with each other as having common needs. Coit comments: "It is terrible when men draw together only in suffering; whereas those who have laughed and thought together, and joined in ideal aims, can so enter into one another's sorrow as to steal much of its bitterness away" (Coit, 1891, p. 27).

The Club as a Unit of Organization

Within the settlements, the unit of organization was the club, which usually was a group of eight to twenty members, although some grew as large as seventy. The members of the club organized themselves formally and remained together for years, with the club serving as the reference group for its members. Each club had a leader who was not a member but who served as a guide and a counselor for the group. Following the principles of democratic organization, members elected their own officers, organized according to some set of rules, and conducted their business by parliamentary procedure. The club provided a laboratory for learning democratic process. The groups developed their own projects, decided on their own entertainment, organized for their own education, and in some instances helped in political and social action.

> When in 1902 New York's Reform Mayor Seth Low was running against Tammany, the American Hero Club (Henry Street's first club), made up of boys who had at first been disposed to throw decaying vegetables at the nurses under the impression they were missionaries, then had been won over to enthusiastic and lasting approval when they discovered they were friends, threw all their energies into the Low campaign and scattered campaign literature left and right. (Duffus, 1938, p. 87)

Although there were many clubs for children and young adults, the club concept extended to include other people as well. Clubs were established that brought together people with common concerns, but within the settlement, clubs of differing age, sex, and marital composition met together and worked and planned some activities cooperatively.

Holden (1922), who was a settlement house resident, describes how one club was started:

> A certain group of boys averaging about fifteen or sixteen years got the reputation of being "the toughest bunch on the block!" They used to stand around drug stores and side doors of saloons smoking very cheap cigarettes and cat calling at the girls who passed by. They had a scorn for the conventional type of hats and affected big caps pulled over their ears at curious angles. They had a peculiar way of spitting out of the corners of their mouths. They punctuated their sentences with words like Jesus and damn and hell, and others not so nice in their original meaning. They spent their evenings provoking trouble and hunting for excitement. One night they visited the neighborhood dance in progress at a settlement (Admission 5 cents). Two of them were kicked out for refusing to take off their caps, another was evicted for a roughhouse that ended in breaking a chair, a fourth was put out for using profane language. Three remained. They were engaged in conversation by a very large man they later learned had been a famous football player at Princeton. They were interested in the gymnasium equipment. The idea came to them that basketball could be played by boys who didn't go to high school. They asked if they could play. They were told that if they formed a club and had a director that they could play. They asked the big man to be their director and said they would get the rest of their "bunch." But the rest of the bunch resentful over having been put out, refused to come in. They asked the big man if he would come and talk to the others. He did. He spent an evening with them. Where they went he went also, but they noticed that he didn't cat call after girls and that he didn't wear a cap.
>
> They came to the settlement house again and asked him to spend another evening with them. The big man said he didn't much enjoy dancing on cellar doors and proposed that they should go to a show. Two of them hadn't any

> money and asked him to wait while they "swiped a nickel off the soda and candy man at the corner." He said he'd lend them the money. They said that would be all right that "they'd swipe it later." They noticed that he took off his hat when he went into the movie house. They asked if they couldn't form a club and play basketball. He helped them start their club. (Holden, 1922, pp. 67–69)

The section goes on to describe other ways in which the club leader helped the boys, getting them gym shoes and helping them to earn money for club dues. Holden enumerates the ways in which the group began to identify with him. Once the club was organized, it became a part of the settlement itself, engaging in a variety of activities.

Women's and girls' clubs followed the same general pattern. If anything, some of the settlement workers felt that these were even more successful because they had more leaders available. (Details of how the clubs were organized and the varied purposes they served may be found in Addams, 1910, Wald, 1915, and Woods and Kennedy, 1922). Unfortunately for our purposes, Woods and Kennedy (1922), the single most complete source describing the settlement houses before World War I, makes no mention of any formal studies of the various clubs' effectiveness in promoting the development of youth or preventing delinquency. Lillian Wald, however, noted:

> Although in the twenty-one years of the organized life of the settlement no girl or young woman identified with us has "gone wrong" in the usual understanding of that term, we have been so little conscious of working definitely for this end that my attention was drawn to the fact only when a woman distinguished for her work among girls made the statement that never in the Night Court or institutions for delinquents had she found a girl who had "belonged" to our settlement. (Wald, 1915, p. 174)

Wald went on to state that club members believed in the effectiveness of the club's control as a means of preventing delinquencies among boys.

It is certainly true that many of the children, and even the older people of the neighborhoods, simply would not have been exposed to a range of educational, recreational, and cultural activities had the settlement not existed. By its consideration of contemporary problems, by athletics, classes, dramatics, and trips to the country, the settlements brought something to the neighborhood that otherwise would have been unavailable. Perhaps it is true that they reached only those who were able to take advantage of what they offered, but who can measure the significance of the settlement bridge to the thousands who used it to pass from the ghettos and the slums into the larger society?

The Settlements and Education

The settlement movement attempted to educate both the larger society and the poor in the neighborhoods it served. Settlement workers believed that change would come when people knew the facts, and they eventually did achieve changes by educating people to the facts, facts they revealed through their own research. But their hypothesis, that social evils existed and persisted because of ignorance, needed considerable modification and qualification.

Contact with schools and homes showed the settlement workers they would have an uphill battle if they were to promote public education as a means of improving the lives of their neighbors. Many of their neighbors had neither interest nor faith in education. In fact, some of the older immigrant families saw the schools, with their emphasis on Americanization, as a means of stealing away their children's loyalties (Woods, 1898). Many of the families needed their children's help to earn a living and so kept them out of school to work even when it was illegal to do so. Many of the parents took most, if not all, of their children's earnings and expected to continue to take them. It was a tradition in many immigrant families to bring up children to believe it was their duty to care for their parents when they became too old to work. The schools' efforts to make the children independent therefore were viewed as extremely threatening, and it took great effort and great patience to change some of these attitudes (Lasch, 1965).

The schools themselves left much to be desired. Educators tended to regard the foreigners who could not speak the language as ignorant and unteachable. And the school system generally was dominted by ward politics (Cremin, 1964). "Schools in tenement neighborhoods showed the most serious fire risks, the most antiquated and insanitary quarters, the largest number of pupils to a class, and the least efficient teaching" (Woods & Kennedy, 1922, p. 275).

Local control of the schools did not necessarily mean that education was carried out in the interests of the neighborhood children. Woods and Kennedy observed:

> A committee of these [local school] trustees had power to appoint and remove teachers and janitors, to contract for supplies, and to engage buildings. Such power was, however, so much abused that settlements took an active part in securing a law which relieved local school boards of the largest part of their administrative work. Many residents continued to serve on the reconstituted advisory groups. (1922, p. 276)

Settlement residents almost immediately became involved with school administration through campaigns to "keep the schools out of politics," as part of municipal reform, and through participation on central school boards and local committees of school trustees. Jane Addams reported on her experiences with the Chicago school board in her autobiography (Addams, 1910).

Penny lunch programs were begun in the settlements after settlement workers gathered figures as a form of policy-oriented research, showing the prevalence of underfed children. These programs, later picked up by the board of education, resulted in school lunch programs and provisions for midmorning and midafternoon snacks for all children. Open-air schools and classes as a means of treating tuberculosis were established on an experimental basis in a number of settlements and were later adopted by the public schools.

The settlements were instrumental in starting the visiting teacher service, the forerunner of school social work. School nursing was established as an outgrowth of the visiting nurse service which was centered in the Henry Street Settlement.

Other more direct educational innovations also were instituted by the settlements. Many had kindergartens, and settlement workers persuaded the people in the neighborhood to demand that the public schools provide this service. College extension classes, held in the settlements, played a part in the request for public evening high schools. Classes for the physically and mentally disabled were supported by settlement residents, whose efforts were important to helping the public school system accept such special needs. Whereas other groups originated the concept of teaching handwork and homemaking in the schools, the settlements pioneered in gaining acceptance for curricular change. They introduced such programs into the summer vacation schools they operated and then pressed for their adoption in the regular curriculum.

Several settlements were established by schoolteachers for schoolteachers, and one on the Lower East Side was supported by Julia Richman, one of the first female school executives in New York. This settlement became a center for conferences on school and community relations and was highly influential among public school teachers (Woods & Kennedy, 1922).

Thousands of adults and children made full use of the settlements' extensive educational and recreational programs (Addams, 1910; Lasch, 1965; Woods & Kennedy, 1922). The settlement workers also encouraged the children, developed their talents, and supported their efforts to persuade their parents to make the sacrifices necessary for further education. Families unable or unwilling to provide for their own children were helped by the resources of the settlements and the workers.

Not only did the settlements provide libraries, but the libraries and living rooms also were places where residents helped with homework and organized small groups for remedial study. The settlements were available, and the residents were accessible as models for the children who wanted to become educated Americans. The profound effect of the settlements cannot be measured, but it certainly cannot be questioned and must be respected.

The Settlements, Social Action, and Political Involvement

Before we started this study, we believed that the settlement houses were glorified community centers that also provided some social services, that term being used in its contemporary professional sense. In fact, of some 321 settlement houses established before 1910, about 80 percent were engaged solely in educational and recreational activities, and only 3 percent of the settlements were heavily involved in social and political action. Many of the settlements, about 16 percent, did engage in research and investigations of local conditions that were relevant to social action. Those settlements that were politically active and doing socially relevant research were almost exclusively located in the big cities, in areas where recent eastern European immigrants predominated. They were largely independent or university affiliated; rarely did they have institutional affiliations with churches, and most of their funds came from private sources.

These data, compiled from the Woods and Kennedy "Handbook of Settlements'' (1911) by Levine and Levine (1967),[2] are a tribute to the power and the intellectual character of the early settlement movement. A handful of settlements, with strong leadership and mutual support were determined to forge tools of social research and political action. In so doing, they improved the life conditions of people throughout the country. These early settlements provide a model for contemporary community-oriented practice, about what forms of social and political action are effective.

Clean Streets and Garbage

Stanton Coit lived on Forsyth Street for less than a year before he tried to organize a campaign for clean streets. Dirty, dimly lit streets served as storage places for trucks; dark, dirty halls and stairways and unlocked hallway doors and cellars were invitations to crime. Slimy and crowded sidewalks, the only playgrounds for children, aroused anger in the settlement workers who lived in these conditions. More articulate than the typical slum dweller, they complained loud and long to municipal officials, to newspaper reporters, and to influential friends. They got some action, but the amount was small in relation to their effort and their need. What they discovered was that municipal officials felt they could safely leave slum districts until last when improvements were considered because there was no aroused public complaining of the poor conditions.

The settlements thus worked to arouse neighborhood interest in improvements in the area. Club members formed associations that assumed responsibility for cleaning the streets. Some settlements established brigades of children to help in this work. The job was not easy. One of the reasons that the immigrants did not complain was because their sanitary standards had always been low. Coming from villages where personal uncleanliness was neither offensive nor a danger to public health, the newcomers did not know how to use sanitary facilities and would deposit waste in places where it did not belong or just throw it out the windows. The residents of the settlement made some progress toward their goal through pleading, persuasion, protests, and complaints to the police, encouraging the latter to enforce the appropriate laws.

Keeping a neighborhood clean is partly a function of the garbage collection and disposal services. At that time, contracts for garbage collection were farmed out as political plums, and so service was careless and irregular. At Hull House, after the Women's Club had reported over a thousand instances of inadequate service and after Jane Addams lost out on a bid to become the local garbage contractor, she was appointed garbage inspector. She saw to it that the contractors did their duty. Although vigorous activity (keeping charts and records, following garbage wagons to the dumps, arguing with the contractor, having landlords arrested, and complaining to city hall) helped clean up the neighborhood, Addams became convinced that for permanent change, it would be necessary to oppose actively the local political boss.

Male residents were of critical importance in political action, and some, like Raymond Robins, a resident of Chicago Commons, were acknowledged past masters of ward politics (Davis, 1959). Women had not yet won the vote and were not well accepted in politics, as the following letter addressed to Jane Addams demonstrates:

> No man can love a woman who takes her place among men as you do. . . . And now for a little advice to help you defeat the good man you have so often tried to do without success of course, I can speak very plain to you, as your highest ambition is to be recognized as capable of doing a man's work. . . . Did it ever occur to you while on a tour of inspection, through alleyways old barns and such places where low depraved men with criminal records may be found (such a place a virtuous woman would be afraid to go.) You might for a small sum induce one of such men to sell you his pecker and balls. It would not be much loss to him, and will be your only chance to prove yourself a man. You could then go before a board of examiners chosen from the 19th ward prove yourself capable of filling a man's place, You could then have the privelage *[sic]* of casting a vote, and should that bold bad Johnny Powers challenge you you could produce your pecker cast your vote, and probably defeat him. (Lasch, 1965, p. xxi)

We shall return to the lessons learned from the several political campaigns in local politics, but for the moment it is important to sketch in some of the other accomplishments of the settlements in regard to sanitation, housing, and education. Residents were at the forefront of efforts to improve the disposal of stockyard refuse in order to get rid of the local stench; to replace a municipal garbage dump with an incinerator; to obtain public baths in neighborhoods where tenements did not have running water or adequate facilities; and to upgrade the schools.

Although other groups were seeking housing reforms, the settlement residents lived in tenement environments, close to the problems and the people. They were able to do the necessary research to show the effects of existing conditions on the health and welfare of the residents, on standards of homemaking, child care, and sexual morality, and on family structure. In many cases, it was the settlement residents who documented the problems and provided the basis for corrective legislation. The laws were not ends in themselves, for the residents learned that vigilance in enforcing the laws was necessary to obtain change. They found that enforcement really depended on the community because neighbors ''conspired together'' to secure and to maintain standards.

The details of these fights are not important to our purposes. What is important is that these efforts at social reform were made on the basis of the findings of participant observation and survey research, that each step of the research led to a broader view of the neighborhood's problems, and that the procedure represented a method for the empirical study of the community's social organization. In effect, each effort at intervention could be characterized as an experiment designed to test a hypothesis concerning the social causes of observable phenomena, and each failure at intervention led to reformulating the hypothesis of causation. Thus the settlement house workers developed a basis for arriving at an understanding of the community and of how people lived in it.

The efforts at intervention led the settlement workers into conflict with the local ward boss. Most of the time, they found they really did not have the strength or the know-how for effective opposition. In some instances, the ward leader's opposition was sufficient to cause the settlement to close or to move. In fact, he was a master of the game of becoming important by providing direct assistance to people. Even the most corrupt of bosses served important and socially beneficial functions in their neighborhoods (Lasch, 1965; Woods, 1898).

That political bosses and settlement workers saw themselves as rivals is clear. The following passages, quoting George W. Plunkitt of Tammany Hall, show that their methods of reaching people were highly similar.

> There's only one way to hold a district; you must study human nature and act accordin'. You can't study human nature in books. Books is a hindrance more than anything else. If you have been to college, so much the worse for you. You'll have to unlearn all you learned before you can get right down to human nature, and unlearnin' takes a lot of time. Some men can never forget what they learned at college. Such men may get to be district leaders by a fluke, but they never last.
>
> To learn real human nature you have to go among the people, see them and be seen. I know every man, woman and child in the Fifteenth District, except them that's been born this summer—and I know some of them too. I know what they like and what they don't like, what they are strong at and what they are weak in, and I reach them by approachin' at the right side.
>
> For instance, here's how I gather in the young men. I hear of a young feller that's proud of his voice, that he can sing fine. I ask him to come around to Washington Hall and join our Glee Club. He comes and sings, and he's a follower of Plunkitt for life. Another young feller gains a reputation as a baseball player in a vacant lot. I bring him into our baseball club. That fixes him. You'll find him workin' for my ticket at the polls next election day. Then there's the feller that likes rowin' on the river, the young feller that makes a name as a waltzer on his block, the young feller that's handy with his dukes—I rope them all in by given' them opportunities to show themselves off. I don't trouble them with political arguments. I just study human nature and act accordin'.
>
> But you may say this game won't work with the high-toned fellers, the fellers that go through college and then join the Citizens' Union. Of course it wouldn't work. I have a special treatment for them. I ain't like the patent medicine man that gives the same medicine for all diseases. The Citizens' Union kind of a young man! I love him! He's the daintiest morsel of the lot, and he don't often escape me. . . .
>
> I tell you frankly, though, how I have captured some of the Citizens' Union's young men. I have a plan that never fails. I watch the City Record to see when there's a civil service examination for good things. Then I take my young cit in hand, tell him all about the good things and get him worked up till he goes and takes an examination. I don't bother about him any more. It's a cinch that he comes back to me in a few days and asks to join Tammany Hall. Come over to Washington Hall some night and I'll show you a list of names on our rolls marked "C.S." which means, "bucked up against civil service."
>
> What tells in holdin' your grip on your district is to go right down among the poor families and help them in the different ways they need help. I've got a regular system for this. If there's a fire in Ninth, Tenth, or Eleventh Avenue, for example, any hour of the day or night, I'm usually there with some of my election district captains as soon as the fire engines. If a family is burned out I don't ask whether they are Repbulicans or Democrats, and I don't refer them to the

> Charity Organization Society, which would investigate their case in a month or two and decide they were worthy of help about the time they are dead from starvation. I just get quarters for them, buy clothes for them if their clothes were burned up, and fix them up till they get things runnin' again. It's philanthropy, but it's politics, too mighty good politics. Who can tell how many votes one of these fires bring me? The poor are the most grateful people in the world, and, let me tell you, they have more friends in the neighborhoods than the rich have in theirs.
>
> If there's a family in my district in want I know it before the charitable societies do, and me and my men are first on the ground. I have a special corps to look up such cases. The consequences is that the poor look up to George W. Plunkitt as a father, come to him in trouble and don't forget him on election day.
>
> Another thing, I can always get a job for a deservin' man. I make it a point to keep on the track of jobs, and it seldom happens that I don't have a few up my sleeve ready for use. I know every big employer in the district and in the whole city, for that matter, and they ain't in the habit of sayin' no to me when I ask them for a job.
>
> And the children—the little roses of the district! Do I forget them? Oh no! They know me, every one of them, and they know that a sight of Uncle George and candy means the same thing. Some of them are the best kind of vote getters. I'll tell you a case. Last year a little Eleventh Avenue rosebud whose father is a Republican, caught hold of his whiskers on election day and said she wouldn't let go till he'd promise to vote for me. And she didn't. (Riordan, 1905, pp. 46–53)

Settlement workers engaged in local politics to different degrees. The political leader in Robert Woods's district was apparently honest and benevolent, and Woods found that he could cooperate with him in many projects. In other areas, settlement workers ran for local office, often unsuccessfully. In some instances, the settlement workers discovered that it was not wise to run for local office because they thus closed off important avenues of advancement for young neighborhood men. Some settlement workers felt that it was more effective not to run campaigns themselves but to try to educate the local populace to demand more of the political boss. When the bosses could be shown that neighborhood improvement was in their interest, they would fight for and obtain improvements that the settlement workers wanted. Often the settlement workers actively supported reform candidates, or good men, regardless of party (Davis, 1959; Lasch, 1965; Woods, 1898; Woods & Kennedy, 1922).

Since the inception of the contemporary community mental health movement, some have argued that mental health workers and behavioral scientists must enter politics in order to achieve social change. Jane Addams's experience at Hull House with the local political leader suggests that such a view should be examined with great care.

Over a period of years, Hull House residents clashed with the local ward leader on a number of issues in addition to the problem of garbage. Although Hull House maintained what was tantamount to a supplementary school system and although research by a resident showed that there were three thousand fewer seats in the schools than children in the ward, the local ward boss did not favor a new school. A proposal for a new neighborhood school was blocked in the city council, even though the settlement workers had received approval for a new school from the Chicago school board.

Given these obstacles, Jane Addams encouraged the Hull House Men's Club, a group of residents and young neighborhood men interested in politics, to run an independent candidate for alderman, much to the amusement of the local boss. Even though their candidate actually won, the newly elected alderman was unable to resist the opportunities of the situation; and it was not many months before the reform alderman fell into line with the machine.

In the next election, Hull House decided to take on the boss himself. Their candidate was a middle-aged Irish immigrant who was the former president of a bricklayers' union. Hull House sponsored a vigorous campaign, focusing on the evils brought to the district by the rule of the ''prince of boodlers.'' Although their candidate was easily defeated, he did manage to cut into the overwhelming majority that the boss usually won. In a third election, in 1898, Hull House's candidate was again overwhelmingly defeated despite its every effort. Similar campaigns were held when other settlements fought at the local level, although in some areas the settlements did become powerful factors in local ward politics by holding the balance of power.

Every attempt at intervention is an experiment that reveals the social order, and Hull House discovered a great deal about the realities of life in its ward. By crusading for better streets or against vice and prostitution in the area, it discovered a powerful combination of property owners, bankers, churchmen, and journalists in league with the local political boss.

In one district, defeats at the polls led some of the workers to study the local political sociology. Here the adolescent street gang was shown to be both a training ground and a source of support for later political leaders. Potential political leaders made their reputations among their peers who later, as adults, joined the local political clubs (Woods, 1898).

Settlement workers also learned that running a worker in overalls as the people's representative was not the wisest tactic. Many people in the district felt that their representative should stand with the best, and they felt more favorably inclined toward the man who drank champagne and wore a diamond stickpin. When their man received campaign contributions from successful businessmen, it was not a sign of corruption to the voters but of his good connections.

Moreover, they learned that people who supported them during the campaign later expected them to do what the ward leader did: help families get husbands and sons out of jail, provide jobs, and hand out other favors. They discovered that the private system of political favors served functions that impartially administered laws could not offer to the people.

Direct political action on the ward level was successful in limited ways and in special circumstances. What the settlement house workers learned about the problems of introducing social change was far more relevant. Efforts toward political action revealed the organic nature of a community. As the politically active settlement workers came to realize that the neighborhood, the ward, the city, the state, and the nation were inextricably one, the arena of action expanded accordingly.

The Settlements, the Labor Movement, and Welfare Legislation

The settlement workers not only participated in local municipal reform but also were an important influence in the often-violent labor struggles of that era. Living in working-class neighborhoods, the settlement residents saw that poor working conditions, low pay, and long hours violated human dignity and human welfare. Although they did not intend to become involved in the labor movement, it took little time for many to become convinced that the human waste and misery they observed were intimately connected with the conditions of industrial capitalism.

The large settlements in Chicago and New York were particularly helpful in supporting workers in their attempts to organize. Initially, the workers were not always enthusiastic about receiving help from educated, upper-class youth. Resentment, distrust, and differences in viewpoint concerning what was helpful characterized the relationship between labor leaders and settlement workers. The settlement workers frequently wanted to play the role of mediators in labor disputes, and the laborers wanted allies who were unquestionably on their side (Addams, 1910; Davis, 1959; Woods & Kennedy, 1922).

The settlement workers made distinct efforts to allay labor's suspicions by helping new locals organize, allowing labor unions to meet in the settlements, and organizing discussions, lectures, and conferences; they demonstrated their support of the rights of organized labor in speeches, articles, and books. In some instances they actively supported the strikers by collecting money, feeding strikers, and obtaining economic help for blacklisted labor leaders. For example, Hull House sponsored a cooperative residence for women—the Jane Club—so that women who wished to participate in strikes need not fear they would be put out on the streets (Addams, 1910).

Activity that went beyond simple mediation cost the settlements the financial support of some of their ''uptown patrons,'' who were shocked at what the radicals were doing and also earned them the enmity of some newspapers who attacked them for harboring anarchists. Some of the settlement leaders were outspoken socialists, but many others, though wanting the workers to receive a bigger share of the wealth, held no brief for redistributing the ownership of the means of production (Davis, 1959; Woods & Kennedy, 1922).

The settlement workers were partly responsible for nationwide publicity favorable to the labor cause. In the Chicago stockyard strike of 1904, for example, articles written by settlement workers caused President Theodore Roosevelt to order an investigation of conditions there. Further investigations resulted when Upton Sinclair's muckraking novel *The Jungle* was published in 1906. Sinclair lived close to the University of Chicago Settlement while he was working on the book, and he frequently ate at the settlement house and relied on residents for advice and material for the novel. President Roosevelt also consulted settlement workers before he established a commission to investigate conditions, and the commission's work resulted in the federal inspection of meat-packing plants (Davis, 1959).

Settlement workers played a leading role in the fight for laws regulating child labor and laws improving working conditions generally. They were able to arouse public opinion by books, articles, and speeches to the powerful women's clubs and church groups, and they also were active in gathering data concerning industrial abuses.

Although some felt that settlement workers relied too heavily on numbers, their numbers produced results because they did not stop with the collection of data. Florence Kelley, an expert on child labor, lived at Hull House beginning in 1891 and at Henry Street in 1899. Her research produced the statistics that influenced legislators to correct child labor abuses. When she was appointed special investigator for the Illinois State Bureau of Labor, she had an official base for her work. Under the provisions of one 1893 bill regulating child labor, Florence Kelley, Alzina Stevens, and Mary Kenney, all settlement workers, were appointed factory inspectors by reform Governor John Peter Altgeld (Davis, 1959; Woods & Kennedy, 1922).

This formidable group set out against the odds to enforce the bill, which limited the employment of children under age fourteen to eight hours of work, to be done in daylight. This and other state laws passed later proved to be weak, however, and the settlement workers wanted to obtain stronger federal legislation. In 1904 a National Child Labor Committee was formed, with the settlement movement well represented. They drew up a model child labor law and encouraged state campaigns, but they encountered political and philosophical opposition and constitutional obstacles to bills regulating child labor. Learning the variety of techniques necessary to achieve meaningful change, they became experts in the law and in the use of pressure and publicity. It was not until much later (really not until the 1930s) that their efforts were successful (Davis, 1959).

Eventually, Jane Addams and her group were instrumental in gaining enforcement of the law through complementary compulsory school attendance laws. The Hull House reformers influenced settlement workers throughout the nation, and the fight for laws regulating child labor was carried on in many other states, from Massachusetts to California (Woods & Kennedy, 1922).

Lillian Wald suggested the possibility of a federal children's bureau as a coordinating agency to gather data and make recommendations. The Children's Bureau was born after nine years of labor and a White House Conference on the Care of Dependent Children. It was finally signed into law by President Howard Taft in 1912. Julia Lathrop, a Hull House resident, was appointed its first chief, and Grace Abbott, another Hull House resident, its second.

The settlement workers were instrumental in improving working conditions for women and in promoting the women's trade union movement. The details of the story are not important to our purposes, but suffice it to say that their efforts were fundamental. As one example, settlement workers encouraged President Roosevelt and Congress to pass a bill calling for a federal investigation of the problems of women and children in industry. The Department of Commerce and Labor conducted the investigation, and published the results in nineteen volumes. These volumes were the chief source of information on women and children in industry for many years and a major resource for reformers in their continuing fight to better living and working conditions for everybody (Davis, 1959).

Settlement workers played a major role in developing several social welfare measures (Woods & Kennedy, 1922). Reform of county poorhouses, relief for the aged, help for the unemployable through sheltered workshops, widows' and mothers' pensions and day nurseries for their children, studies of living standards, workers' compensation, and unemployment insurance all were programs either initiated by or supported vigorously by settlement workers. The laws were passed in part because of the settlement residents' lobbying, and the laws' contents were suggested by settlement residents who functioned as experts and consultants to legislators and other governmental officials (Cohen, 1958; Davis, 1959; Pumphrey & Pumphrey, 1961).

The Settlement and the Progressive Party

For fifty years after the Civil War, but particularly from the 1890s on, there was a realignment of social forces toward the establishment of a more far-reaching democracy. On the political scene there were demands for reform ranging from the fight against corrupt local political bosses to the popular election of senators, a right that was finally achieved in 1913. The campaign for women's rights met increasing success after 1890 (Flexner, 1966). During those years the labor movement held strikes and won concessions from the industrialists. In 1890 the Sherman Anti-Trust Act, although used initially against labor unions, established means for social control of the huge corporations' activities. An income tax leveling inequalities of fortune was proposed, supported, and instituted, and the definition of what could be supported on a national level under the concept of general welfare was extended.

By 1912, for more than twenty years, settlement workers had been in the forefront of the struggle. Many were convinced that change could be effected only at a national level, and in former president Teddy Roosevelt they felt they had found a candidate who spoke their language. Roosevelt, who had lost the Republican nomination to Taft, became the presidential candidate of the newly formed Progressive party. He drew a collection of reformers and idealists, including leaders of the settlement house movement and other social workers who by now were cooperating with the settlement workers in efforts to achieve reforms. At the Progressive convention in August 1912, Jane Addams delivered one of the keynote speeches and seconded Roosevelt's nomination. Raymond Robins, another settlement leader, seconded the nomination of Hiram Johnson as Roosevelt's vice-president.

After the convention, many of the settlement workers joined the Progressive party's political organization. They served as politial researchers, ward leaders, committee chairpersons, speakers, and candidates for a variety of offices. Jane

Addams was a member of the party's national committee, state committee, and county committee, and stumped far and wide in support of Progressive candidates.

It was no surprise when Roosevelt lost in 1912; the social workers were disappointed but ready to continue the fight. Many felt the campaign's primary purpose was to educate the public and prepare the way for a new and more perfect democracy. After the election of 1912, the party did continue, and many residents continued the work of political organization in preparation for the next steps. In fact, the Progressive party gained new adherents shortly after the election.

Within months, however, the Progressive party was wracked by factional disputes and personal squabbles. Jane Addams removed herself from the strife, and others followed her example. Some continued with the Progressive party through the election of 1914, when it lost heavily everywhere. The settlement house workers' venture into national politics came to an end. For many, the party's collapse was a sign that they could not achieve social reform through direct political action. Later, when the involvement of many of their leaders in the antiwar movement led to their popular vilification, the settlements lapsed into a policy of providing educational, recreational, and group work services without the social and political action components (Davis, 1959).

Settlement Leaders, the Pacifist Movement, and World War I

When the war in Europe broke out in 1914, many settlement leaders, particularly Jane Addams, Lillian Wald, and Graham Taylor, became concerned with peace and antiwar movements. Although not all settlement workers agreed that international problems were their concern, enough participated to provoke popular reaction against them.

Although Jane Addams was awarded a Nobel Peace Prize in 1931, during and after the war, she was widely regarded as akin to a traitor and was subject to near-violent abuse. The American Legion and the DAR (Daughters of the American Revolution) called her un-American. Her protest against conscription and her defense of conscientious objectors, her defense of the "aliens" who were her neighbors in Chicago, and her insistence that war was no way to settle human problems cost her the support of the Women's Clubs. Her actions on behalf of the peace movement and other causes far removed from those that the developing social work profession had adopted resulted in her isolation even within the field (Addams, 1910; Linn, 1935). Lillian Wald had similar experiences (Duffus, 1938; Wald, 1915).

After the war, Lillian Wald and Jane Addams were listed as subversives in a report prepared for a Senate committee (Duffus, 1938; Linn, 1935). With popular fears stirred up by the Bolshevik revolution, the settlements—which supported and provided forums for radicals of all kinds—were seen as nests of potentially dangerous subversives. The conservative atmosphere that would not support criticism of the United States during the war years persisted and intensified in the immediate postwar period. Efforts to change social conditions were therefore no longer acceptable (Allen, 1959; Curti, 1951; Davis, 1959).

World War I presented a number of other problems for the settlements. Money was no longer easy to raise. War-related causes, the Red Cross, and similar overseas relief agencies had captured the attention of philanthropists, and accordingly the settlements received less. Settlement workers spent a great deal of time raising money; eventually many joined the Community Chest and lost their identities as special agencies (Woods & Kennedy, 1922).

Funds for fellowships were cut drastically. No longer could an eager young reformer find support and a place to work through the settlements. Some observers noted that the settlement movement even earlier had begun to lose its power to attract the young rebels. Greenwich Village, overseas service, and newer, independent radical movements attracted those who might have gone to live in the settlements. The newer workers coming into the settlements were no longer sparked by deep convictions regarding social justice. They tended more and more to be professional social workers trained in casework, social welfare, and community organization. They wanted to help, but most were no longer concerned about social and political action (Davis, 1959).

After the war, some of the leaders turned their attention to international affairs. Jane Addams was involved with the Women's International League for Peace and Freedom and in the next decade spent a great deal of time in overseas travel and at international conferences. Lillian Wald traveled to Europe, Russia, and Mexico during the 1920s. With the older leadership losing its interest and with the changes in the population of the settlement neighborhoods, in the political and social climate, and in the new workers, the settlements—the former "spearheads of reform"—became social work agencies in the narrower, contemporary sense of that term. Reform in social work went into its "seedtime" (Chambers, 1967).

Summary

The settlement movement began with social need and reached the height of its influence while society was undergoing important changes. The movement lost influence when the social order was no longer open to accepting reforms designed to assist the poor and the working classes. As a helping form, the settlement movement was firmly based among those it sought to help; settlement workers entered the community by providing immediate, concrete, freely available services that the people who were served both wanted and needed. It maintained its vitality for more than twenty years by drawing on the energies and ideals of young intellectuals eager to grapple with the meaningful problems of their time. The settlements had important effects through their local programs, but they had more far-reaching effects on the larger society through both their writings and their efforts to obtain legislative changes at all levels of government. Finally, the settlements had a delayed effect as a training ground in reform for those who later took important positions in government. The settlement workers were less successful in their efforts to achieve change through direct political action.

Some may argue with the role of the settlements in achieving change, about whether or not they might have supported more revolutionary social action. Some may contend that the settlement workers were paternalistic and were uplifters and agents of deviance control who supported a value system and a morality that stifled human development. It can be argued that they were too much involved in Americanization and too interested in seeing the poor immigrant enter the mainstream of American life to be concerned with the ultimate fate of the mass of people in the hands of big business and big government. Perhaps they could not really develop local leadership that would have its own class interests at heart. Rather, they served to remove the leadership and to direct the interest of the talented toward individual upward mobility. The settlements, it is sometimes said, had a one-way view of help, with their own social position and their own moral standards unquestioned and unchanging.

These are important considerations, particularly for mental health professionals and social scientists turning today to the problem of social change and how they can best participate in change (e.g., Gouldner, 1968). The Black Power movement of the 1960s for example, argued that there is no place for whites in the black struggle, that whites will protect their own interests, and that their very presence emphasizes the blacks' inferiority and inability to help themselves (Carmichael & Hamilton, 1967). The issue here was not just one of black self-development versus white repression. Rather, it seemed to reflect the more complex issue of the degree to which power can be expected to be used benevolently for the interests of dependent people, when the necessary changes involve a loss or a sharing of power and position.

The early settlement workers were consciously representing the best of their own social, cultural, and political traditions. They were uplifters, and they saw their own cultural forms as superior. But it is clear that they respected the traditions of their immigrant neighbors and urged the immigrant children to value the Old World culture. It is true that they urged upward mobility, but this was, after all, what the immigrants came here to seek. Large numbers of the immigrants were eager to learn from the settlement workers, did learn from them, and advanced educationally, socially and economically. There were radicals of all persuasions in the neighborhoods that the settlements served, and the radical movements gained their own adherents. Choice was available to meet the people's differing needs. It is a moral absolutism of its own and a gross oversimplification that would assert that only a thoroughgoing revolutionary program then would have led to some modern Garden of Eden today or that such a course was undercut by the settlement workers' commitment to the American system as they understood it.

Moreover, those who believe that revolutionary change is the only feasible means of achieving true change need to examine the time perspective of social change. Over time, urban population density, health standards, educational standards, living conditions, and other material benefits have improved considerably. That we continue to have serious concerns about social inequities and the quality of life means that some old problems continue and new ones are added. Such questions will always be with us as long as we do not live in a static world (Sarason, 1978).

Helping forms are concerned with the quality of life. The settlement workers understood this and acted to improve the quality of life as they understood the concept in terms meaningful to their own day. The settlement movement faltered and lost momentum when it tried to continue its concern with issues that no longer had the significance of earlier years. Perhaps it is asking too much of any social group to recognize that its own central values and goals are not absolute and immutable but need to change as conditions change.

The lesson is that things do change, that it is vital to recognize change, and that it is necessary to accept the diminution, if not the demise, of the power of older helping forms as newer forms with newer values evolve to deal with contemporary problems. When helping forms gain power and contribute to cultural lag—become a cultural drag, so to speak—then the helping forms open themselves to assault. In historical perspective, we cannot see that we can argue with how a helping group defined its problem. We can only learn from it, in order to help future forms serve social need more effectively.

5

The Visiting Teacher: Forerunner of the School Social Worker

The visiting teacher service arose in the same set of social conditions that resulted in Witmer's clinic and special classes: a multitude of social problems, a reform climate, and community action. The visiting teacher originated as a means of providing the child with an agent to link school and community.

In the distant past, children were socialized and educated in the family. For a time, in colonial America and well into the nineteenth century, close contact was maintained between home and school (Atkinson & Maleska, 1964; Bailyn, 1960). Teachers were selected by the community and lived with families, often on a rotating basis, receiving board and room as part of their salary. They were part of the community but often regulated by it in ways that other members of the community were not; for example, they were barred from smoking, drinking, marrying, and other forms of behavior considered natural rights by most adults (Hollingshead, 1949; Waller, 1932). The goals of formal schooling were far different from those of today. Until the second half of the nineteenth century, with the exception of men training for the ministry, law, and a small handful of occupations, most people who attended school were not educated beyond the three Rs; women, minorities, and the poor rarely had even that much formal education.

With the United States' social and economic development, profound changes occurred as formal schooling became more important to society. Schools became more structured; universal public education laws resulted in heterogeneous school populations forced to attend school; and teachers became professionals who related to their pupils almost solely in their formal roles. As a consequence, the teacher became psychologically and sociologically isolated from the community.

By the late 1800s, the isolation of schools and teachers from the surrounding community was exacerbated by the rapid growth of the urban population and the influx of immigrant families with different attitudes toward the schools and schoolteachers. Whereas the school population was once essentially self-selected, it was now a matter of public policy to provide educational services for all children. The compulsory school laws, and the increasing demand that these laws be enforced, brought and kept children in the school system. Class sizes of sixty or more and the rapid turnover of teachers meant that the opportunities for teachers to know and to understand the children's families and world outside the school building were greatly limited. After the advent of universal public education, those who taught immigrant children were generally from a previous generation of immigrants and differed in ethnicity or religion from the children they taught. In New York for example, 68 percent of the teachers lived more than fifty blocks from their schools (Henderson, 1910–11), and home visits by teachers were infrequent (Oppenheimer, 1925).

The Need for Visiting Teachers

There is no better way of describing the need for the visiting teacher than to quote from a report by a visiting teacher describing conditions in the schools before 1915:

> The extension of the influence of the school through the visiting teacher is particularly needed in a city like New York, where the population is heterogeneous and shifting, where the schools are large and congested, and where so many schools are included in one system. The parents of our school children are, many of them, like Angelina's mother, ignorant of the value of education and of the ideals which the school is trying to inculcate. The crowded schools prevent the teachers from giving sufficient individual attention to the children in the classrooms or from influencing to any appreciable degree their homes. The size of the system has made it seem necessary in the past to organize all schools alike, with uniform methods and standards, regardless of the racial and national characteristics which represent widely differing school needs. (Johnson, 1916, pp. xiv–xv)

> The work of the visiting teacher is not radically new, but rather a very natural extension of the function of the public schools as a child welfare agency adapted to meet the needs of children in a large municipal organization.
>
> There was a time when the school regarded itself as concerned only with the academic instruction of children under its care, and when social interests and needs were foreign elements not relevant to the big task of education. An inevitable change, however, has come with the changing times.
>
> The doctors and nurses have found their way into the school at the call of the child who is physically unfit for school work; the curriculum has been made to yield to the needs of children who are mentally disqualified; and still there is a group of pupils who seem unable to take their training in wholesome fashion, but need more individualized treatment. They are below standard in scholarship without belonging to the mentally sub-normal class, they are difficult in conduct without being disciplinary cases, and though they avoid truancy they are not in constant attendance.
>
> There are, moreover, the adolescent girls, irritable and neurotic, who are getting poor marks in scholarship and conduct because they need country care or medical advice[1] or perhaps only the understanding sympathy of a friend. There are the slow girls who have had an increase of home duties placed on their shoulders at the time when school demands are also increasing. There are the restless children who have begun to strain at the tether, whom school does not interest because it is not as real to them as life outside its walls and whom the great world of industry will seize if activities and interests are not provided. There are the retarded children who are reaching the limit of their mental development and

> paying the deferred bills of early malnutrition or heredity, and who are needing special guidance. Adolescence, individual departure from the accepted average, mental retardation and the urge that sends the boy and girl out into the world of accomplishment are not phenomenal problems. They are ever recurring factors and must be reckoned with by the schools if opportunity and a chance for development are to be offered. These, then, are the children for whom the help of the visiting teacher is enlisted. (Johnson, 1916, pp. 1–2)

The visiting teacher concept grew out of settlement house work with children. Sometimes the workers felt the need to be in close touch with a child's teacher. Located in the neighborhood, the settlement house workers could use their relationship with the children and their families to assist the school in obtaining the families' cooperation. The teachers found they could better understand the children when a settlement house worker was able to discuss a child in another context. As a result of this informal experience, a resident in each of several settlement houses called on the families of children who had special educational, social, or medical problems. This worker came to be known as the school visitor or the visiting teacher.

The First Visiting Teachers

The first visiting teacher was a former schoolteacher, and her work was planned in cooperation with Julia Richman, the first woman district superintendent of schools in New York. Some schoolteachers visited on their own and gave personal charity to a few children. Volunteers from churches and clubs also visited on an informal basis. When the philanthropic societies offered help, principals and teachers were more than happy to refer children to them (Report of the City Superintendent of Schools, 1913–14). However, the main organizations concerned with formalizing visiting teachers were the settlement houses and the Public Education Association of New York.

One of the guiding spirits of visiting teaching was Mary Marot, a former teacher who believed that the schools were a prime medium for far-reaching social reform. In the fall of 1906, the settlement houses formed a committee to supervise the work of two visitors in three school districts. A few months later, the Public Education Association of New York became interested in the movement and absorbed the Visiting Teacher Committee started by the settlement house workers. In the fall of 1907, the Public Education Association hired Jane Day as a full-time visiting teacher. She worked closely with Julia Richman. By the academic year 1911–12, seven workers were employed. The Public Education Association publicized the visiting teacher work in order to persuade the board of education to adopt the service in the schools. In 1913, the New York Board of Education allocated funds to establish a visiting teacher service in the city system.

The visiting teacher service began in a somewhat similar fashion in about 1906 in several other cities. In Boston, the Women's Education Association first employed a home-and-school visitor for one school. Over the next few years, neighborhood associations, the Boston Home and School Association, and other private groups placed visitors in additional districts. Settlement houses continued the work by providing staff members to carry out home-and-school visiting in their immediate districts. In Hartford visiting teachers assisted the school psychologist in obtaining histories, and they carried out the recommendations of the psychological clinic at home and in school. Witmer had long used social workers in a similar capacity in his clinic. In Philadelphia the White Williams Foundation (earlier the Magdalen Society), organized to help delinquent and wayward girls, was instrumental in bringing school counselors to the Philadelphia public schools (Oppenheimer, 1925). Other private associations in that city provided workers to do home investigating, sometimes as part of the attendance service, in an effort to enforce the compulsory education laws. In other communities, women's clubs and private citizens were instrumental in beginning visiting teacher work and influencing boards to adopt the service permanently. Vocational and educational guidance services were established as part of the same movement in Philadelphia and Chicago. From 1913 on, many public school systems appointed their own visiting teachers.

An impetus to the spread of the visiting teacher was the Commonwealth Fund's Program for the Prevention of Delinquency, adopted in 1921, which we shall discuss in greater detail in connection with the establishment of child guidance clinics. This program was organized on the premise that early contact with children in the schools could be used to prevent delinquency and maladjustment. The Commonwealth Fund, working through the Public Education Association, established the National Committee on Visiting Teachers, which placed thirty visiting teachers in thirty communities for a three-year demonstration. Additional funds were provided for visiting teachers by the Bureau of Children's Guidance, a psychiatric clinic in the school system connected with the New York School of Social Work. The Bureau of Children's Guidance and the New York School undertook the professional training of visiting teachers during the bureau's existence.

The Public Education Association continued to maintain a demonstration staff that conducted research on children's problems in the schools and on the work of the visiting teachers. Concern for evaluating the work was present from the beginning (Johnson, 1916), and evaluation research was part of the effort to interest schools in adopting the service. In addition, as we have pointed out, the settlement workers were committed to research as a tool for stimulating social change. Job descriptions for the visiting teacher thus frequently included research to determine the needs of children in the schools (Oppenheimer, 1925).

Visiting Teachers in the Public School System

Within a relatively few years after the beginning of the service, the public school systems took over the demonstration programs. From that point onward, the history of the visiting teacher movement illustrates the problem of innovating services in the school system. Let us consider the work of the visiting teacher. Clinical work brought the visiting teacher into direct contact, and sometimes direct conflict, with the school system's organizational structure. The conflict stemmed in part from the visiting teachers' theories of behavior disorder and therapy, which were largely situational in nature.

> The function of the visiting teacher is the adjustment of conditions in the lives of individual children, to the end that they may make more normal or more profitable school progress. These adjustments may be made in the school, in the home, or in the environment, wherever there proves to be an adverse condition responsible for school conduct, scholarship or attendance, or influencing it to a greater or less extent. (Johnson, 1916, p. 3)

When she began her work with a school (and most were women), the visiting teacher would become familiar with the neighborhood's ethnic character, the standard of living, the attitudes of the people toward education, the housing conditions, the predominant industries in which the parents worked, the recreational opportunities, the special educational programs in school and out, and the availability of the services of public and private agencies. These were the neighborhood's resources and liabilities, resources that the visiting teacher would use in making adjustments in the child's interest, and liabilities that she would explain to the school in order to elicit a more sympathetic view of the child's problem. ("Selling a child back to the teacher" appears to be a common tactic among school consultants [Sarason et al., 1966]).

The effectiveness of the visiting teachers seemed to be based on their understanding of a particular school situation. The literature contains many allusions to problems that the visiting teachers encountered in the schools, some of which may have been related to their goals. The early visiting teachers made suggestions concerning curriculum, school organization, and methods of teaching, and some complained that the system was too unyielding to individual needs (National Association of Visiting Teachers, 1921). Oppenheimer (1925) states that principals in difficult areas were chosen for their socialmindedness and that such social-minded principals were quick to recognize the value of the visiting teachers' work. By implication, some were not socialminded and did not quickly grasp the significance of the work. There were suggestions that visiting teachers should not be placed in schools where the principals were unreceptive. Despite conflicts, word of the service's value spread through the New York system until—within ten years after its beginning—there were requests for these teachers from over 175 school principals (Johnson, 1916).

The visiting teachers had "port-of-entry" difficulties (see Sarason et al., 1966), but the pressure of school problems facilitated the development of the service. It was no accident that the visiting teacher service was initiated in what we would today call an inner-city school. (Sarason and his colleagues [1966] also found that a clinical service was more readily received in an embattled urban system than in a suburban system.) Once the visiting teacher entered the school, she had to establish working relationships with the faculty. It soon became apparent it was better for all concerned if the visiting teacher concentrated her efforts in one school and was viewed as a member of the staff. Her ability to influence the situation, however, depended on herself:

> She could never work out her plans for the children under her care by the force of any authority she might be given over them or over the teachers. She can be effective in a school only as she can make the teachers realize that she understands their problems, and that if her work is primarily to help the children and give them a fuller opportunity, that help must be given in such a way that it will be felt to be an assistance, not a hindrance in the classroom and in the principal's office. (Johnson, 1916, p. 8)

This admonition, statements that above all a visiting teacher needed tact, the veiled and not-so-veiled criticisms of aspects of the school program, and instances in which the visitor's recommendations differed from the teacher's all imply that there were difficulties in the relationships with the visiting teachers. By and large they were able to establish good working relationships within the schools, but they were not explicit enough when writing about their problems to give us helpful examples (Allen, 1929; Ridenour, 1948).

Social Treatment

How did the visiting teachers operate? First, they developed programs to deal with general problems of the school and the community, and second, they dealt with the needs of individual children. The first method of operation appears to have been prominent only in the early years, and by the 1921 report (National Association of Visiting Teachers, 1921), little further mention is made of organizational work in the community. By 1921, concepts of casework were becoming current, and the visiting teachers, identifying themselves as social workers, began to seek training in casework, psychoanalysis, psychiatry, mental hygiene, and even mental testing. What was thought of as social service by the settlement house workers seems to have disappeared in favor of casework with individuals.

The visiting teachers engaged in diverse activities, for the work was an experiment and could be flexible. They created a variety of clubs and classes to meet the needs of both the children with whom they had been working and other children. Classes included special assistance in schoolwork, dramatics, athletics, and handicrafts, and there were reading clubs, dancing classes, and excursions as well. These clubs fostered social relationships between younger and older children. Housekeeping classes were established for some younger girls. The visiting teachers also helped arrange systematic programs of clinical examination, medical treatment in the school, and even special diets for poorly nourished children. In some schools, programs for the psychometric evaluation of retarded children were initiated through the visiting teachers' efforts. Educational efforts in the community were carried out through lectures at parents' meetings and appearances at teachers' conferences. Some visiting teachers formed school and neighborhood associations. The extension of special classes, the wider use of the school building, and the school lunch service were wholly or in part the consequence of the visiting teachers' activities.

The early visiting teachers, reflecting as they did the orientation of the settlement house, envisioned in each school a social service department to supervise the visiting teachers and to be responsible, in an organizational sense, to the school principal. All community-oriented activities—including the formation of student groups, parent groups, neighborhood associations—and all relationships with outside social agencies and community organizations would be the responsibility of this department. Experimentation with the school as a social center, the development of after-school clubs and classes, and even vocational programs would fall under the aegis of the service. Research, including social surveys of the neighborhood and the school, would be another function. In an important sense the Gary school (see Chapter 6) was the fulfillment of this dream, but in later writings about the visiting teachers little further mention is made of the broad social service department.

In work with the individual child, the visiting teacher used a situational theory. She would obtain a referral from the teacher through the principal, discuss the problem with the teacher, observe the child in class, interview him or her in school, and then visit at home. The visiting teachers, at least in their early days, had no working hours but would call on families at night or on Saturdays, if need be, to see working parents. The visitor tried to determine "which of the conditions found have been most instrumental in causing the difficulty, or in hindering the necessary adjustment" (Johnson, 1916). The fundamental difficulties were divided roughly into seven areas:

1. *School maladjustment* ("when school conditions seem to be definitely responsible for his failure to come up to a standard or when a modification of requirements adjusts his difficulty") (Johnson, 1916, p. 55).
2. *Lack of family cooperation* (neglect, suspicion of school, poor parental educational standards).
3. *Economic stress.*
4. *Ill health.*
5. *Immoral family conditions* (drunkenness, physical violence, sexual irregularities).
6. *Adverse neighborhood conditions* (gangs, disorderly houses, cheap shows).
7. *Individual peculiarity* (social maladjustment, mental unsuitability for school life, exceptional instability). (The cases of "individual peculiarity" constituted about 4 percent of referrals in the school year 1913–14; see Johnson, 1916.)

The visiting teacher explained the child's behavior to the school in order to modify the school's behavior toward the child (Johnson, 1916, pp. 9–13).

Example 1: A teacher found an eighth-grade child cheating at her lessons. The visiting teacher investigated the situation and discovered the child was waiting tables and washing dishes in the evenings to earn money for a graduation dress. She told the teacher, who realized the child's behavior was a means of doing her schoolwork, not avoiding it. The visitor then made arrangements to obtain money for the child and helped her to get a tutor and a scholarship, which enabled her to go on to high school.

Example 2: Angelina, a third-grade child, was referred for constant lateness and indifference in school. The visitor discovered the child was working at home sewing, not because the family was in need, but because the mother believed this was proper for an Italian girl. The visitor helped to persuade the mother that her child's education was important, and she encouraged the teacher to be lenient about the lateness while mother was being educated to education. Instead of being punished for lateness, Angelina was commended for improvement, by the teacher. She started her days in school in a better frame of mind. With continued attention and stimulation from the teacher and the visitor, the child's work improved until she was placed in a rapid-advancement class where she did two years' work in one.

In these and other examples, the visiting teacher was able to obtain information that put the child's behavior in a new light for the classroom teacher. Although it is apparent that the visiting teacher had the time to seek explanations for the child's behavior and that the classroom teacher did not, it is also true that the classroom teachers tended not to think in terms of causes of behavior.

The visiting teachers sometimes recommended a change of class as a way of handling a problem.

> Example 3: Anna had fits of temper in her fifth-grade classroom, refusing to work and throwing her books on the floor. The visiting teacher enlisted the child's interest and that of her teacher in keeping a "self-control" book, which the child brought to the visiting teacher each time they met. The procedure worked well for two months, when again friction arose in the classroom. The child was transferred to another class where she worked well and gave no trouble.

There are several cases of precocious children whose problem was solved by promotion.

> Example 4: For three years, Sam gave all his teachers except the dramatic teacher the impression he was dull. He was a disciplinary case frequently reported to the principal. His teachers thought him feebleminded, but when the visiting teacher's investigation showed that out of school he made friends with engineers and mechanics who taught him about machines and allowed him to run engines and motors, she gave him an intelligence test. He rated so high that she suggested he be advanced two grades. This was done and he immediately began to improve. There were no more shamings in school or trips to the principal's office except to show a good report card. Sam had found his proper level. (Johnson, 1916, p. 32)

Special-class situations were sometimes used as a temporary means of helping a child with a problem. When no special class existed, visiting teachers often encouraged the institution of such classes to provide for children with a variety of problems, not just the retarded.

The method most frequently used by visiting teachers was called *supervision,* which required more than seeing the child for a few interviews. The visiting teacher took a personal interest in the children, met them at school and at home, shared their interests, and helped them realize that she wished to share their successes and prevent any failures. The relationship often continued for several years. The help provided was concrete, personal, and exhaustive.

> Example 5: Nine-year-old Sadie had a bad reputation, was neglected and disheveled, and often fell asleep in school. Home visits soon established that her mother lived away from her home at her place of employment. Her father drank heavily, and he and the child slept together in the only bed. The visitor helped Sadie's mother be home more often and arranged for the child to have lunch money. The visitor taught the child to fix her hair, gave her hair ribbons and dresses, and made it a point to inspect her frequently. She arranged for a tutor at the nearby settlement and helped program the hours after school. The neighborhood librarian took an interest in Sadie; she joined a club at the settlement house and was admitted to a gym class after the visitor obtained the proper shoes and costume. A kindly neighbor took the child in when the parents were away. All of this attention resulted in a marked transformation. Sadie's appearance changed, and she began to bring perfect papers for the visiting teacher to see.

The teachers intervened with the courts, social agencies, and welfare services on behalf of the children. They helped obtain employment, grants for special purposes, clothes, and scholarships—including funds that enabled the children to feel they were contributing to their families while they stayed in high school. The visitors arranged for participation in organizations such as the Boy Scouts. The teachers often acted as school counselors, helping the students select their courses. The visiting teachers apparently used tutors, including high school students, to help younger children individually or in groups. Sometimes the tutor was instructed to minimize the formal academic content of the sessions. Volunteers from Big Brothers and Big Sisters were used extensively as companions for the children.

Psychotherapeutic methods had not yet been developed, but some of these cases indicate that the visiting teachers permitted some ventilation of personal conflicts, such as guilt over masturbation. If the visiting teacher felt the problem called for a specialist's help, she made a referral to the appropriate source. The visiting teachers cooperated closely with specialists in the child guidance clinics, providing observations about the child at home, in school, and in the neighborhood and carrying out the specialists' directions.

Many different symptoms and complaints were treated by the situational approach. Not only were neglected and dependent children treated in this manner; children exhibiting behavioral disorders also seemed to be helped. Temper tantrums, incorrigibility, shyness and inarticulateness, apparent feeblemindedness, underachievement, indifference, impertinence, neglect of personal appearance, parental overprotection, truancy, lying, stealing, fearfulness, excessive daydreaming, and adolescent oversensitivity were also treated successfully.

Such outcome studies as there are in the early literature indicate that some 92 percent of the cases were improved or partially improved, according to the judgment of the visiting teacher. There was a significant improvement in grades among children helped by a visiting teacher, compared with those from the same classrooms matched to the treated cases and those neither referred nor treated. Because the treated cases were referred as problems and the controls were not, this finding is remarkable (Oppenheimer, 1925). This early experience, if supported, should have challenged those in the mental health professions who later came to regard office psychotherapy as the preferred treatment, if not the only effective way of working with children's problems.

We conclude this section with a typical case illustrating the methods of treatment used over a period of years:

> The case of Miriam is one in point. She was in a 6B grade. The principal reported her as incorrigible, with a tendency toward immorality, unruly in the classroom, untruthful, and untidy in appearance, and asked the visitor to take her out of school and send her to work. When the visiting teacher called at the home she found that Miriam's mother had died a short time before, leaving Miriam in charge of the household which consisted of her father who was out of work and two brothers. She cooked the meals, washed the clothes, and took the place of the mother.
>
> Of a highly sensitive nature, very retiring and backward, she made few girlfriends. She was untruthful, but she told tales to win sympathy. She was on the street at night, and while she did not seek companions of the lower type, they came to her, using her as a shield to cover some of their wrongdoings.
>
> The visiting teacher became very friendly with Miriam and found new friends for her, and the old ones were given up. Through the assistance of a relief organization, the family was moved to better quarters. Work was secured for the father, and the younger brother was placed in a Hebrew class in a neighborhood organization.
>
> When Miriam was promoted to the seventh grade the visiting teacher watched her closely. She asked that the child should be given to an especially sympathetic teacher to whom she told the story of her home life.
>
> Throughout two years, the visiting teacher followed her progress. The child came to her with all sorts of problems, now a discouraging mark in school work, now household cares that needed school help for their adjustment and again financial difficulties caused by the unemployment of her father or brother. Tutoring was provided, arrangements were made to excuse her a little early so that she could prepare the evening meal for the Jewish Sabbath; and plans for tiding the family over a period of stress were worked out with the agency for relief.
>
> Gradually, Miriam showed the result of this friendly supervision. The dime novels which had been her choice and rough friends ceased to satisfy her, and when she graduated she had won the affection of the finest girls in her class, and the genuine respect of her teachers. All trace of immoral tendency disappeared. (Johnson, 1916, p. xi)

Later History

The remainder of the history of the visiting teacher movement is the story of changes after its introduction into the public school system. Clinical services are shaped as much by their settings as they are by the guiding concepts the professionals employ in their work. The period of the visiting teacher movement's greatest growth was in the 1920s, when the predominant social forces and the mood of the country were vastly different from those of the twenty years preceding World War I. The fact that the visiting teacher movement was encapsulated by the system that it meant to change should be considered by those in community mental health who hope to influence other systems and agencies to change.

From the beginning, the sponsoring agencies (settlement houses, Public Education Association, Commonwealth Fund) conducted demonstrations of the service to be adopted by the public school system once it had proved its worth. What was not considered was the effect the new context would have on the service. In this case, as public support increased, much of the literature and interest shifted from a consideration of the work itself to a consideration of the place of the work in the school system. Johnson's 1916 report contains little discussion of administrative problems, whereas the later reports of the National Association of Visiting Teachers in 1921 and of Oppenheimer in 1925 devote a great deal of attention to administrative, organizational, and professional issues and relatively little attention to the theory and the details of professional practice.

In the case of the visiting teacher, the settlement houses and the Public Education Association supported the service for about seven years after it started. Although the superintendent of schools in New York was in favor of the work, he had to include requests for funds in his budget for several years before the board of education was willing to hire its own visiting teachers. The first seven visiting teachers, supported by the board of education, were placed under the direction of an associate superintendent of schools. Almost immediately, discussion of supervision, cooperation with other departments, hours, caseloads, record keeping, and the qualifications of education and experience began.

After the National Association of Visiting Teachers organized in 1916, it considered standards of training and practice. (Eventually the group was absorbed into the National Association of Social Workers as the School Social Workers Section.) By 1921, the visiting teachers themselves felt that they had created "a real technique of social case work as applied to the schools and that this technique is only acquired by training plus experience" (National Association of Visiting Teachers and Home and School Visitors, 1921). The training plus experience considered desirable by the National Association leaders included graduation from a four-year college with a background in education, one year's training in social work, and two years' experience in teaching grade school. The salaries in most systems were based on the scale for grade school teachers, and so someone with such training might find herself earning less than did a high school teacher in the same system.

The National Committee on Visiting Teachers, which attempted to conduct demonstration programs in thirty cities, apparently had difficulty recruiting people with the requisite background. It recognized that from a practical viewpoint it would be almost impossible to obtain people with the desirable qualifications and so proposed that minimum qualifications should include graduation from a two-year teacher training course, one year's teaching experience, and one year's work in an approved school of social work. Many systems, in their desire to have a home-and-school visitor, selected classroom teachers from within the system or employed people with the qualifications to be appointed as teachers.

The realities of the situation resulted in a shift toward people who would follow an accepted course of training and away from the "mavericks" who were attracted into the work in the earliest days. In discussing the settlement house movement, we noted the motivations that seemed to be characteristic of those who entered the settlement house. A sizable percentage of the earliest visiting teachers must have been people similar to the settlement workers. One survey of their educational backgrounds—when there were no more than a hundred people in the field—showed that of those with college degrees, many had received them at schools such as Barnard, Bryn Mawr, Simmons, Smith, Vassar, Wells, Teachers College, Columbia, Clark, University of Chicago, and New York University; just a sprinkling of the state universities were represented (National Association of Visiting Teachers, 1921).

It seems inconceivable that women entering the field after attending normal schools in pursuit of conventional careers as teachers would be of the same intellectual caliber, or have the same social outlook, as did the educated upper-class women of the early 1900s who were the backbone of the feminist and other reform movements of the day.

By the 1920s, efforts toward social reform had all but died. In the field of social work, professional training had become influenced by psychiatric and psychoanalytic thinking because of the development of schools of psychiatric social work. Casework was formalized, and the "social service" of an earlier day became less important to the professionally trained worker (Lubove, 1965). No longer was the visiting teacher an agent of social reform or of reform within the school system itself. With changes in the times and professional practice, social workers had narrowed to a focus on the inner psychological problems of individuals.

6

The Gary Schools: The Prevention of Alienation

The problem of alienation was foremost among the concerns of social reformers and intellectuals at the turn of the century. Their efforts to combat the social problems of the day by changing the institutions or creating new ones were, at least in their early days, closely intertwined with community life. The visiting teacher represented one effort to change an institution by linking home, school, and community. The progressive education movement, a product of the nineteenth and early twentieth centuries, was designed to make education more relevant to an industrial society. It is not our purpose to recount the history of the movement or the reasons for its success. That has been done very well by Cremin (1964). But in the sense that the progressive education movement was designed to help the individual operate more competently in the social environment and achieve personal fulfillment, the movement can be said to represent a mental health effort in the community. Though it may be going a little far afield to discuss the Gary schools in detail, as they cannot claim even a distant ancestry to later clinical services, the Gary schools are relevant to us today. That is, in the Gary story we find a systemwide organizational cure for the malaise that grips urban schools, and we find ideas far in advance of those contained in many contemporary proposals for community schools.

Progressive education and the school reform movement produced no more complete model than did the public schools of Gary, Indiana. Begun in 1906, under the leadership of the superintendent of schools, William Wirt, the Gary schools reflected John Dewey's influence. They were to be both an intimate society and an extended community. Within the school, teachers and pupils would be closely related; the school would also function as a social and cultural center for the surrounding population.

William Wirt sold the idea of what came to be known as the Gary school as a practical expedient for solving some of the education problems of Gary, Indiana. At the time Gary was a steel town, growing rapidly from the barren sand dunes and struggling to assimilate the immigrant laborers required for industrial growth. Because Wirt had the advantage of developing a new system rather than rebuilding an old one, he was able to propose extensive curricular and organizational innovation. The school board liked his plans for a better education and a more efficient use of the educational dollar.

Wirt built schools that contained, in addition to regular classrooms, a number of specialized facilities, including auditoriums, gyms, swimming pools, shops, laboratories, art studios, and luncheon facilities. The surrounding acreage was suitable for gardening and farming, parks and playgrounds. Wirt pointed out that by doing away with the concept of a classroom for each teacher and a seat for each child, by using a platoon system, and by using all of the classrooms and specialized facilities all of the time, a school could accommodate twice as many children on a full-time basis. To reduce discontinuity in education, the school building contained cultural facilities such as libraries and museums. The school was open to adults in the community for recreation and education and was tied to other community agencies such as the YMCA and the churches. Wirt used the released-time concept for religious education as part of his program of involving the school with the community. The school day, the school week, and the school year all were extended, making the school a year-round community facility.

Within the school, all work, study, and play converged to create not a preparation for life but a life in itself, much as the household was a life in itself. The concept was implemented by providing all facilities at the school and by making the actual academic work of the school real and vital. In the Gary school, repair, decoration, and improvement of the school were part of the educational program. Skilled workers were the shop teachers and were chosen for their abilities in carpentry, plumbing, cabinetry, sheet metal work, printing, and electricity and for their personal traits. These men did all the maintenance, repair, and building, assisted by older students who functioned almost as apprentices. Each of the shops was used as a teaching facility for younger children who came to help the older ones and to be taught by them. This system extended to academic classes as well, with the distance between faculty and students reduced by the use of student assistants.

Shop projects were the school's real work and its method of education. In the lunchroom the children learned to cook by preparing meals that were sold on a profit-sharing basis to faculty and children. The children learned arithmetic, accounting, business, and secretarial skills by keeping records of the school's business activities. In some instances, grades were replaced by token wages that the children earned. In sewing classes the children made clothes for themselves and their families. They also learned to make and repair shoes because it was discovered that many poor parents could not provide them. In botany and zoology classes, they took charge of the gardens, the lawns, and the school zoo. Children of all ages participated in the work. All the shops and laboratories had generous expanses of window so that everyone who was curious could see what was happening inside. Nothing was hidden; every activity was open to all in some form.

The auditorium was a central aspect of the Gary school system. In one building the auditorium contained a stage large enough to hold an athletic contest. Each child spent one hour a day in the auditorium, younger and older children at the same time. The teachers were responsible for arranging each day's program in cooperation with their pupils, and some part of the auditorium program was devoted to presenting an aspect of school life. Specialty teachers discussed their work, and the children debated about school life, exhibited schoolwork, or put on plays, some of them written by themselves. The auditorium was meant to be a community theater in which the children shared all the school's activities.

Within the limits of a large school system, the administrators and teachers tried to individualize the program. Dewey and Dewey (1915) described how children could emphasize their strengths, compensate for weaknesses, and proceed at an individual pace. The attempt to tailor the programs was in part to provide for "even the most difficult pupil," so that instead of dropping out or failing, each could find something of value in the school.

Discipline was to come primarily from the sense of involvement with the school itself and only secondarily from the teacher's authority. Interest in the work was to help sustain order, and the emphasis on the teacher as a helper was to eliminate the teacher's role as an arbitrary disciplinarian. Some observers felt the children wanted to be in school so much that the threat of being sent home was sufficient to enforce order. The school attempted to create an atmosphere in which the children could enjoy learning, but they did not come to school solely to play, as some critics charged. Many came to school voluntarily on Saturdays, Sundays, and during the summers to do their schoolwork under the guidance of their regular teachers.

For a few, the more vigorous attention of the school principal was necessary. Some boys who could not adapt to the school were sent to the school farm, which had a model dairy, orchards, and numerous farm buildings on a hundred acres. The farm was run by a young agriculturist who employed scientific methods. The boys who went there built their own living quarters and clubroom, worked for wages, and paid for their room and board out of their earnings. A teacher integrated formal schooling with the farm work. Eventually the boys either returned to the school or got jobs.

There were no extracurricular activities in the schools and no fraternities. All of the school teams, clubs, the school paper, the orchestra, and glee clubs were part of the formal educational program. Students participated in active political campaigns and elections, student councils, and civic improvement societies.

Because the schools were completely equipped, they served as community educational and recreational facilities on evenings and weekends. The Gary evening schools' attendance was two-thirds that of the day schools, a figure confirmed even by the schools' critics. All courses authorized by state law were offered, and all shops and laboratories, studios, and classrooms were open to the public in the evening. Special courses were even designed to correlate the evening-school work with the needs of local industry.

Both adults and children had full and free use of the recreational facilities. The gyms, pools, and playgrounds were almost always open, and the playgrounds were lit so that they could be used at night. Social, political, and neighborhood organizations used the schools' facilities, presenting films and lectures and holding community forums, group meetings, and conferences. Parents brought their children with them when they attended evening classes or meetings in the schools, and the children were allowed to play freely. In the truest sense, the public schools became schools for the public.

Teachers were relatively self-directed because they were less compelled to follow the rigid curricula common in other school systems. The overall head of the school system was the superintendent, with an executive principal in charge of each school. This executive principal was responsible for scheduling, supervising pupils' schedules, and maintaining discipline and for other ordinary administrative work but not for the educational program.

The supervisors of instruction, with the teachers, drew up the curriculum and promoted or demoted children. All the industrial and manual training shops were under a director of industrial work who also supervised repair and maintenance. The teacher-workmen in the shops were responsible to the director. Each building had a head manual training teacher who both supervised the industrial classes and acted as a vocational adviser to the children. The teachers in the various academic divisions were designated as assistant supervisors of instruction and were responsible for coordinating the courses in their subjects. Those teachers under them were specialists in the subject matter, and many taught both elementary and high school classes. Each taught a subject to a different group each hour.

In addition to the subject-matter specialists, there were application teachers who met their classes in the shops and studios. They were expected to consult with the subject-matter specialists so that they could teach a subject through application. An application teacher was the head of a group of eight teachers and was responsible for correlating the work of various classes.

Teachers were also classed as head teachers and assistants according to their level of experience. The head teacher in each classroom supervised the assistant teacher. New teachers were initiated in this fashion, so that a constant spirit of in-service training was maintained.

Instead of employing visiting teachers, the Gary system gave responsibility for the work to a register teacher. This register teacher was the faculty adviser, the disciplinary and sociological overseer for a geographic area with some fifty families. All the children in the area met with the register teacher for a weekly general conference. Monthly reports, problems in discipline, attendance, and other issues were channeled through the register teacher, who was expected to meet with the parents in school or to make home visits if necessary.

The burden on the teacher was large, but it was relieved in a number of ways. First, the teachers were given time to do their preparation and other administrative work during the school day. Pupil assistants helped with many classroom chores such as grading papers.

Second, the teachers were paid additional money whenever they taught in the evening, on the weekend, or in summer school. And their salaries were substantially above those of New York City teachers.

Third, the importance of teachers was emphasized when they were asked to train people from other schools who wished to learn the Gary system. A visitor was assigned to a teacher as an assistant and learned by participating in the classroom. The visitor paid a fee that was given to the instructing teacher.

Fourth, the teachers were given a great deal of leeway in the conduct of their classes. Educational innovation was encouraged, and rigid adherence to a curriculum or to specific materials was not demanded. Supervisory control was minimized; teachers were permitted to work out their own creative methods. Considerable overlap in function was allowed, and auditorium programs informed the teachers about what was happening in the school. Thus the system encouraged communication, involvement, and maximum self-direction.

The Gary system matched the spirit of the times and received a great deal of popular and professional attention. Educators from all over the United States and some two hundred foreign countries visited the Gary schools. Numerous school systems throughout the country adopted some of their features. Our description of the system is taken largely from Randolph Bourne's (1916) book, and Bourne also published a series of articles about the Gary schools in the *New Republic*. John Dewey, who wrote about the Gary schools in *Schools of Tomorrow* (Dewey & Dewey, 1915), gave the system his personal blessing. Wirt himself wrote and spoke extensively about them (Wirt, 1912). In 1914, he became a consultant to the New York City schools and devised a program that was actually implemented in a number of schools in the Bronx. A preliminary report of the experience in one New York school is provided by Taylor (1916).

The attempt to introduce the Gary system into New York City provides one of the important case histories of the problems of large-scale social and institutional change. The "Garyizing" of the New York schools became an issue in the mayoralty election of 1917. In October 1917, just before the election, there was a week-long riot in the Bronx, Brooklyn, and upper Manhattan. The children broke windows, fought the police, and, supported by their parents, tried to spread the strike to the uninvolved schools. At the height of the disturbance, ten thousand people demonstrated on the streets. John Purroy Mitchel, the incumbent mayor, was badly defeated, partly because he advocated the Gary plan (A. Levine & M. Levine, 1970; Levine & Levine, 1977).

Although the Gary experiment was greeted enthusiastically by many, some of whom had observed it only superficially, the experiment also had its critics. Some of the criticism came from professional educators who were appalled at what they saw as a lack of discipline in the classrooms. They objected that many of the shop teachers had no formal training. Some critics, interested only in the system's economics, did not warm to the spirit of the venture. In making recommendations to their own board, a committee from Syracuse, for example, suggested that various parts of the Gary system, such as platooning, could be adopted in Syracuse, but they did not recommend using the whole system (Committee of Syracuse Board of Education, 1915).

In 1918, a more thorough examination of the Gary schools was published by the General Education Board of New York City, a Rockefeller-funded agency interested in the study of education. The General Education Board was commissioned by the Gary Board of Education and by Superintendent Wirt to make a thorough evaluation of the Gary schools. The report was published in eight volumes, and the volume written by Flexner and Bachman (1918) contains an overview of the survey and a summary of the General Education Board's findings and recommendations.

Although the report praised the schools' concepts, it was extremely critical of many phases of the plan's execution. In particular, Flexner and Bachman felt that the loose supervisory and administrative structures made for confusion rather than progress. They felt that the execution of such a bold plan required exacting attention to detail, attention not present in the Gary schools. As a result, the academic work was not productive; too many of the teachers, left to their own resources, simply reverted to teaching by the methods they had been taught in their normal schools. Even though Flexner and Bachman cited many examples of inspired teaching, they felt that the typical classroom was really no different from that in other schools. Though they found many teachers who were trying to accommodate themselves to the Gary concept and had faith in it, they also reported that many felt confused and overburdened. Moreover, when compared with national standards of academic achievement (Courtis, 1919), the Gary students fared poorly. Even those who attended the Gary schools throughout their school years did no better. No more pupils seemed to complete school in Gary than pupils did elsewhere, and they were held back at about the same rate as in comparable cities. Dewey and Dewey (1915), however, claimed that there was greater power of retention for Gary schools and that an unusually large number went on to college. Although the Flexner and Bachman study is admirable in its breadth and depth and its care in exposition, it does not speak of the community's feelings about the school, nor does it emphasize the school's students.

Flexner (1940) claimed that the Flexner–Bachman report effectively squelched interest in the Gary system. That claim, as Cremin (1964) points out, is unquestionably wrong. Interest in the Gary system declined after the disastrous effort to introduce it into New York and with the entrance of the United States into World War I, but the Gary school concept continued to be influential in American education well into the 1920s. Literally hundreds of communities adopted Gary's ideas and modified aspects of their school system accordingly.

We have presented the criticisms of the Gary system in order to provide a balanced picture. Besides its comments on the Gary schools, the Flexner and Bachman report is important because it discusses some of the problems in innovation and institutional change, and its background provides pertinent information about issues in evaluation research (Levine & Levine, 1977). Founded on a concept of the social and institutional causes of personal distress, the Gary schools were designed to prevent or relieve social and mental health problems through a programmatic approach. The distress was to be countered by an enhancement of individuality and by an attempt to make schoolwork real and relevant. The early visiting teachers were aware of the need for such a program, and some of them worked toward it. The Gary concept, developed by an educator, shows the school's potential as a model preventive mental health service, as well as demonstrating the problems of changing a system.

7

The Chicago Juvenile Court and the Juvenile Psychopathic Institute

Urban poverty in the 1890s was accompanied by a high degree of familial and social disorganization. The clients of the social agencies, the mental hospitals, and the jails were mostly the foreign-born urban poor. The American conscience supported some form of care for dependent children in orphan homes, and houses of refuge, but under the influence of a conservative social Darwinism, efforts to ameliorate the conditions of the poor were sporadic and inadequate. The dominant viewpoint of the 1870s is well expressed in the following paragraph from Herbert Spencer, the philosopher whose interpretation of Darwinism as a sign of continual progress made him a revered figure among American businessmen:

> Fostering the good for nothing at the expense of the good, is an extreme cruelty. It is a deliberate storing up of miseries for future generations. There is no greater curse to posterity than that of bequeathing them an increasing population of imbeciles, idlers and criminals. To aid the bad in multiplying, is, in effect, the same as maliciously providing for our descendants a multitude of enemies. It may be doubted whether the maudlin philanthropy which, looking only at direct mitigations, persistently ignores indirect mischiefs, does not inflict a greater total of misery than the extremest selfishness inflicts. (Spencer, 1873, p. 312)

The Origin of the Juvenile Court

Conservative social Darwinism applied to children as well as adults; few special efforts were made to assist children in need, especially those who came into conflict with the law. Before 1899 only a few places treated differently adults and children who committed crimes. A few states authorized separate hearings for juveniles, but in most instances there were no special facilities for children who had been arrested (Ryerson, 1978).

As early as 1824, a reformatory was established in Elmira, New York, so that after being convicted of a crime, children would not have to be confined with adults. In most states, however, they were jailed with adult offenders, particularly while they awaited trial, and if found guilty, children were punished or placed on probation exactly as though they were adults. Reformatories operated on the principles that productive labor, education, and a religious experience would reform the prisoner (Bordin, 1964).

The juvenile court represented a marked change in conception, for not only was it established to deal with juveniles separately, but it also was established with a sweeping mandate for child welfare. Furthermore, because it was a service that resulted from concerted community action, the juvenile court reflected the humanitarianism that flowered in the last decades of the nineteenth century.

The first true juvenile court was established in Chicago on July 1, 1899, although Judge Ben Lindsey of Denver has some claim to precedence (Lindsey, 1925a). The regular courts in Chicago had treated juveniles separately from adults since 1861. The juvenile court was the result of the efforts of two groups, the Chicago Women's Club, under the leadership of Mrs. Lucy Flower, and Jane Addams's Hull House (Platt, 1969).

The women's clubs of that era had been established by upper-class women for their own cultural enrichment. Then, when they became caught up in the moral and humanitarian mood of the 1890s, these clubs became a potent force for reform. They did not offer direct charity but, instead, publicized the plight of the urban poor and lobbied for reform legislation. The General Federation of Women's Clubs was organized nationally in 1890 and numbered a million members by 1912. The federation championed a vigorous program of reform to assist women and children, to improve the schools and the conditions under which they worked, to reinforce consumer protection, and to beautify their communities (Hays, 1957). The Chicago Women's Club had a jail committee as early as 1883, which called the public's attention to the fact that juveniles were treated as criminals, and it proposed a separate juvenile court as early as 1891 (Platt, 1969).

In 1898, the Illinois State Conference of Charities devoted its annual meeting to "The Children of the State." The conference, at which members of Chicago's Women's Club were well represented, agreed that there ought to be a children's court in Chicago. The committee that drafted the act included, among others, Mrs. Flower, president of the Chicago Women's Club; a judge who coped with the constitutional problems that the act entailed; and the state representative who introduced the bill. The support of the Chicago Bar Association was enlisted. Mrs. Flower and her fellow clubwomen arranged for the judges of the various courts to be entertained at a luncheon at which the bill was announced. Platt (1969) points out that the new juvenile court was less a radical reform than a political compromise. It served the interests of sectarian organizations which objected to placing children in religiously operated institutions, the Board of Public Charities which favored the broad powers of the court to reach others besides those accused of crimes, and the administrators of industrial schools and reformatories. The Chicago Women's Club and other interested people further supported the Juvenile Court Bill by raising private funds to pay for the services of the probation workers (Lathrop, 1925).

The women's clubs were actively involved with the court's operation later on as well. A detention home was attached to the court under the supervision of a member of the Chicago Juvenile Court Committee, the latter composed of representatives from the various women's clubs. The role and the power of the Chicago

Women's Club are depicted by Mrs. Joseph T. Bowen, the second president of the Chicago Juvenile Court Committee:

> We had at that time, a fine body of men and women who were most anxious for the success of the Court and for the good of the children, and we finally secured passage of a law which provided that probation officers be placed on the payroll of the county. I well remember how that law was passed, because it gave me a great feeling of uneasiness that it was so easy to accomplish. I happened to know at that time a noted Illinois politician. I asked him to my house and told him I wanted to get this law passed at once. The legislature was in session; he went to the telephone in my library, called up one of the bosses in the Senate and one in the House and said to each, "There is a bill, number so and so, which I want passed; see that it is done at once." One of the men whom he called evidently said, "What is there in it?" and the reply was, "There is nothing in it, but a woman I know wants it passed" and it was passed. I thought with horror at the time, supposing it had been a bad bill, it would have been passed in exactly the same way. (Bowen, 1925, p. 301)

Mrs. Bowen, a formidable Chicago grande dame long associated with Hull House, was a driving force in developing the court, training probation workers, selecting judges, and organizing the Juvenile Protective Association (Linn, 1935; Platt, 1969). The Chicago Women's Club members made frequent, even daily visits to the detention home, to inspect its facilities and guarantee standards of cleanliness and care. A new courtroom was obtained because the women encouraged businessmen-husbands and friends to lobby the appropriate municipal officials for them. The Juvenile Court Committee participated in selecting probation workers and interviewed and made recommendations for the appointment of the juvenile court judges. In the court's earliest days, two members of the Juvenile Court Committee regularly sat with the judge to advise or to assist in the disposition of cases (Bowen, 1925; Linn, 1935). Although this voluntarism involved the community with the court, in many other places, voluntarism proved problematic and was a factor in the demand for professional probation workers (Ryerson, 1978).

The Settlement House and the Nonprofessional Probation Worker

Hull House played an important role as a helping agency in the development of the juvenile court. Jane Addams tells the story:

> From our earliest days, we saw many boys constantly arrested, and I had a number of enlightening experiences in the police station with an Irish lad whose mother upon her deathbed had begged me to "look after him." We were distressed by the gangs of very little boys who would sally forth with an enterprising leader in search of old brass and iron, sometimes breaking into empty houses for the sake of the faucets or lead pipe which they would sell for a good price to a junk dealer. With the money thus obtained they would buy cigarettes and beer or even candy, which could be conspicuously consumed in the alleys where they might enjoy the

excitement of being seen and suspected by the "coppers." From the third year of Hull House, one of the residents held a semiofficial position in the nearest police station, at least the sergeant agreed to give her provisional charge of every boy and girl under arrest for a trivial offense.

Mrs. Stevens, who performed this work for several years, became the first probation officer of the Juvenile Court when it was established in Cook County in 1899. She was the sole probation officer at first, but at the time of her death, which occurred at Hull House in 1900, she was the senior officer of a corps of six. Her entire experience had fitted her to deal wisely with wayward children. She had gone into a New England cotton mill at the age of thirteen, where she had promptly lost the index finger of her right hand through "carelessness" she was told, and no one then seemed to understand that freedom from care was the prerogative of childhood. Later she became a member of the typographical union, retaining her "card" through all the later years of editorial work. As the Juvenile Court developed, the committee of public spirited citizens, who first supplied only Mrs. Stevens' salary, later maintained a corps of twenty-two such officers; several of these were Hull House residents who brought to the house for many years, a sad little procession of children struggling against all sorts of handicaps. When legislation was secured which placed the probation officers upon the payroll of the county, it was a challenge to the efficiency of the civil service method of appointment to obtain by examination men and women fitted for this delicate human task. As one of five people asked by the civil service commission to conduct this first examination for probation officers, I became convinced that we were but at the beginning of the non-political method of selecting public servants, but even still and unbending as the examination may be, it is still our hope of political salvation. (Addams, 1910, pp. 323–25)

Placing the selection of probation workers under the civil service proved to be the kiss of death for the juvenile court in Chicago. The public treasury was niggardly in its support of probation workers, and the overloaded probation staff of the Chicago court functioned relatively inadequately. Indeed, the record of children on probation in the Chicago court was quite poor compared with that of other cities (see Chapter 8).

The quotation from Jane Addams indicates that she was uncertain about the relative costs and benefits of civil service for selecting probation officers. Judge Lindsey, William Healy, and many others responsible for the development and operation of the early court recognized that the personal qualifications of the probation officers were vital to determining whether or not the individual would function effectively with children as a probation officer. Mrs. Bowen states the case:

I think the first probation officer was Mrs. Alzina Stevens, perhaps the best example of what a probation officer should be. Her great desire was to be of use to her fellow men. Her love of children was great; her singleness of purpose and strength of character so remarkable that she exerted a great influence over the children committed to her charge. I find among some old papers the following concerning the duties of probation officers: "They must be men and women of many sides, endowed with the strength of a Samson and the delicacy of an Ariel. They must be tactful, skillful, firm and patient. They must know how to proceed with wisdom and intelligence and must be endowed with that rare virtue—com-

> mon sense.'' These qualities would seem to be needed just as much today as they were twenty five years ago. (Bowen, 1925, pp. 299–300)

The ''unbending'' civil service examination did not attempt to select people who had the necessary personal qualifications to do the job. Why did the juvenile court not seek more people like Mrs. Stevens?

Lincoln Steffens's muckraking classic, *The Shame of the Cities* (1904), described the political corruption rampant in cities across the United States. Political bosses maintained their control in part through patronage appointments to municipal agencies. Municipal reformers favored a civil service system as a means of curbing the ward bosses' power. That the tactic was effective can be seen in Tammany boss George W. Plunkitt's complaints that the civil service was a curse on mankind, undermining the loyalty of one man to another and destroying the patriotic feeling that brought good Americans out to Fourth of July picnics and political orations (Riordan, 1905).

To be consistent with their principles, municipal reformers had to support civil service appointments for probation workers, even though then, as now, the objective civil service test could not measure personality qualifications. Although those who operated the court might have been able intuitively to select workers with the necessary traits, this method of selection would have laid them open to charges of inconsistency and of operating their own patronage system. This is an excellent example of how larger issues in the society influence the development of a helping service in ways that may have nothing to do with the service's needs.

The Juvenile Court as a Helping Agency

The juvenile court is not generally thought of as a helping agency, but the legal principle under which it was established had enormous significance for the concept that child welfare was indeed the responsibility of the state. The legal principle not only took the child's case out of the criminal courts, but it also established that the court was to deal with the children under its power of *parens patriae* and in chancery or civil jurisdiction. The legislators and the judiciary may well have used an expansive definition of *parens patriae* and chancery powers in drawing up juvenile court legislation, but the aim of serving and protecting children was so strong that there was little opposition on these grounds (Ryerson, 1978).

The significance of this change was discussed by Judge Julian Mack, the first judge of the Chicago juvenile court, Chancery procedure, he pointed out, established

> the conception that a child that broke the law was to be dealt with by the state, as a wise parent would deal with a wayward child. . . . That is the conception that the State is the higher parent; that it has an obligation, not merely a right but an obligation, toward its children; and that is a specific obligation to step in when the natural parent, either through viciousness or inability, fails so to deal with the child that it no longer goes along the right path that leads to good sound, adult citizenship. . . . The State always stepped in when there was property involved,

> took charge of the property and appointed a guardian, and when it appointed a guardian for the property, it appointed a guardian of the person of the child. But until the Juvenile Court law was enacted, nobody seemed to think of the possibility of extending this principle of the whole parenthood of the state, the ultimate parenthood of the State towards its little ones. No one before that seemed to think that the State ought to be and ought to act as the wise parent or the natural parent, or to undertake successfully to guide the child along the right way. (Mack, 1925, pp. 311–12)

The statutes under which the juvenile courts were established included exceedingly broad definitions of delinquency, in keeping with the new concept. A complaint against a child in the juvenile court is not an indictment; it is usually called a "petition in behalf of the child." According to Sutherland and Cressey (1960), juvenile court law rarely defined delinquency precisely. Jurisdiction was established over those whose "occupation, behavior, environment or associations are injurious to his welfare" and over those who violate any state law or municipal ordinance. The Colorado statute, for example, included children who were "vicious, incorrigible, or immoral in conduct," who were "habitually truant" (an offense created by the passage of compulsory school attendance laws), or who were deemed to be "disorderly persons."

The conception underlying the juvenile court law, in effect, gave the juvenile court full responsibility for child guidance and child welfare. The juvenile court statutes defined delinquency for the mental health professions, and they established the court as the institution with responsibility for the care and amelioration of various interpersonal and emotional difficulties of childhood. It is this concept that explains why Healy's clinic was later established as part of the juvenile court.

The trade-off for assuming a broad role in child welfare was the loss of procedural protections for youths coming before the court. There was some recognition of the problem early on. Some complained of the sweeping powers given to juvenile court judges and the lack of definition or even concern with the proof of an offense which characterized the informal proceedings (Ryerson, 1978). But, it was not until 1966 that the U.S. Supreme Court recognized the trade-off and admitted that it had been a failure because the youths received neither the protections of due process nor adequate treatment promised in the establishment of the juvenile court (*In re Gault,* 1967).

Probation and Prevention

In the early days of the Chicago court, the probation workers were supported by private, not public, funds. The Juvenile Protective Association worked with the court. Their membership consisted of twenty-two privately supported probation workers, many of whom were also settlement house workers, and the executive committee of the citizens' group that financed them. The Juvenile Protective Association thought in preventive terms, identifying conditions in the city that adversely affected the lives of children and young people and acting to correct these conditions. The association worked with the Druggists' Association to induce its members to stop selling indecent postcards; with the Saloon Keepers' Association to stop selling liquor to minors; with the Grocers' Association to stop tobacco sales to minors; with department store managers to supply matrons to help reduce shoplifting; with watchmen in the railroad yards to persuade them to report boys to the association instead of arresting them for trespassing; with theater owners to provide wholesome entertainment; and with film makers to develop educational materials that presented information about health and morals in an entertaining and instructive manner. The association worked to have social centers opened, to turn unused buildings into recreation centers, to turn vacant lots into gardens, to organize hiking parties, to develop bathing beaches, and to open public schools for social purposes. In fact, it was partly through its efforts that Healy's clinic was initiated.

Whatever one might think of the desirability of these measures, it is noteworthy that the members worked both to correct social conditions they believed contributed to the development of delinquency and with the individual delinquent. Civil service probation workers did not see preventive, programmatic work as part of their job; in fact, independent community action by probation workers probably would have been prohibited. The difference in the contemporary courts' approach reflects the problem of the institutionalization and the professionalization of innovation.

The Origin of the Juvenile Psychopathic Institute

The probation work of the Juvenile Protective Association was largely carried out according to the philosophy that the children who came before the court were not bad but rather, in need of satisfying, purposeful activities. It was felt that education and appropriate activities would prevent further difficulty. The court workers, however, quickly encountered many cases that defied such simple understanding and treatment. For example, a seven-year-old cherub with long curls down his back, who was in a court-associated detention home, poured kerosene over all the beds and set fire to them. Other children repeatedly stole, lied, or committed sexual offenses for reasons that remained obscure.

Encountering these cases, the Juvenile Protective Association, stimulated by an active member, Mrs. Ethel S. Dummer, called for scientific research into the causes of such behavior. Mrs. Dummer promised financial support for five years for such a project. John Dewey's influence, however indirect, is evident. Because of Dewey, the Chicago public schools had established a child study department to help with children's problems. His interest, and the interest of men such as James R. Angell and George H. Mead established a precedent in Chicago for the scientific study of the real problems of the day (Addams, 1929). A committee chaired by Julia Lathrop, a Hull House settlement worker, met at Hull House to select someone for the job. The man who was selected—with recommendations by Angell, William James, and Adolph Meyer—was William Healy.

Healy, who was born in England, had studied psychology with William James at Harvard, and he was at Harvard when Meyer was at the nearby Worcester State

Hospital. Meyer had written about child psychiatry from 1895 onward, and given Healy's interest in pediatric neurology, it is likely they knew each other (Healy credits his approach to the case history to Meyer). Healy obtained his M.D. degree at Rush Medical School in Chicago in 1900. His psychiatric training consisted of experience as an assistant physician at the Wisconsin State Hospital from 1900 to 1901 and a postgraduate year in Vienna, Berlin, and London in 1906–7. No convenient record indicates whether Healy had any contact with the early psychoanalytic groups; but, his interest in psychoanalysis and its methods was apparent in his first large report on his work with the Chicago court, *The Individual Delinquent* (Healy, 1915).

Healy practiced in Chicago as a neurologist and developed a reputation for being able to treat children, some of whom had been referred to him by the juvenile court. His public statements against the punitive handling of children in the home and the court brought him to the attention of the Juvenile Protective Association. The association had discovered many puzzling cases that it could neither understand nor help. It was also concerned that the juvenile court judge had to make decisions about the disposition of children without adequate knowledge. Indeed, the judge admitted he was "often in a fog" about given cases and that he would welcome any help he might receive from the careful study of individual cases.

Healy decided to establish a research project into the causes of delinquency, coupled with a clinic to treat children's behavior problems. Before establishing his clinic, he toured the country and found that with the exception of Witmer's clinic and Goddard's laboratory at Vineland, New Jersey, no facilities employed both a psychological and a physical examination in the diagnostic study of children. Healy named his clinic the Juvenile Psychopathic Institute because of the prevailing psychiatric opinion that serious antisocial behavior implied serious psychopathology. This theoretical viewpoint was based on very little firsthand study, and it was a view that Healy soon discovered to be incorrect.

The Juvenile Psychopathic Institute was organized in March 1909, with a five-year endowment from Mrs. Dummer. Healy's institute had the benefit of a distinguished advisory council, with Angell, Mead, and Meyer among its members. Jane Addams and Julia Lathrop, later chief of the Children's Bureau in Washington, were members of the executive committee. John Wigmore, the dean of Northwestern University's law school and a member of the advisory council, was instrumental in bringing Healy's work to the attention of the law schools.

The influence of the source of funds may be surmised from Mrs. Dummer's comments. Healy's discussion of the unconscious and his attempts to help by means of psychotherapeutic methods apparently elicited criticism from the local intellectual community. "Some Juvenile Protective Association members wanted a committee to pass on Healy's findings before publication. Holding the purse strings, I insisted his results were his own and he was free to publish as he pleased. This caused a break with the Association" (Dummer, 1948, p. 10).

The Juvenile Psychopathic Institute did not develop as an isolated entity but was a part of the reform movement and was clearly influenced in its direction and operation by the predominant intellectual views. For example, although Healy does not cite George Herbert Mead or other early sociologists in the Chicago School of Social Sciences, his discussion of the psychology of the delinquent in relation to the problem and its treatment seems to have been influenced by the prevailing sociological views (Healy, 1925). However, his focus on the characteristics of the individual, coming near the end of this period of reform, seems to be a portent of the postwar era.

Healy's Clinic

In April 1909, Healy, psychologist Grace Fernald, and a secretary constituted the first staff of the clinic. The clinic was housed in the detention home in the same building where the daily sessions of the juvenile court were held. Augusta F. Bronner, who became Healy's longtime associate and eventually his wife, joined the clinic in 1913. Bronner had been E. L. Thorndike's assistant at Teachers' College, Columbia University, and she brought a strong interest in educational problems as they related to delinquency.

In the first years of the clinic the profession of psychiatric social work did not exist. Social histories and the treatment of families were the responsibility of social workers from other agencies and of well-trained probation workers. Professional psychiatric social work did not come to the Juvenile Psychopathic Institute until after 1917. By that time, Dr. Herman Adler had succeeded Healy, who had left to establish the Judge Baker Guidance Center in Boston.

Healy's task was to study the individual delinquent, and to do so he devised a highly detailed and thorough examination, which included a family history; a developmental history; a history of the social environment; a history of mental and moral development, including school history, friends, interests, occupational history, and bad habits; and a history of the individual's contacts with law enforcement agencies or institutions. Then there was a complete medical examination, from psychiatric and neurological standpoints, as well as anthropometric and psychological studies. For purposes of the psychological study Healy and Bronner wrote a number of performance tests of their own. (In this, they were apparently advised by Angell, then chairman of the psychology department at the University of Chicago.) The tests, the examinations, and their purposes are discussed in detail in *The Individual Delinquent* (1915). It is interesting to see how Healy anticipated later psychological studies of delinquency in his tests and in the traits he chose to study.

Healy kept a careful record of the delinquencies committed by each subject and careful follow-up records. His data included a diagnostic and prognostic summary and an attempt to judge, insofar as he could, the most important causes in the case. His record forms were drawn up by Mr. Dummer, a business executive. There are probably few child guidance centers today that offer as complete a diagnostic workup as did Healy or Witmer in those early days.

The Clinic's Relationship to the Court

Healy was interested in studying delinquency, but on a clinical level he was also performing a service. For some time he sat in on the daily sessions of the juvenile court and was asked his opinion as cases came up. He soon realized that he had little more basis than the judge did to offer an opinion on a child he had not studied, and so he abandoned the practice. It became his job to make recommendations to the court and the probation workers for the treatment or the disposition of various cases, but only after a thorough study of the child. But from his comments, it is clear that the courts did not always follow his recommendations, despite their requests for his services (Healy & Bronner, 1926, 1948).

There were a number of dispositional alternatives, although resources were indeed slim at that time in Chicago. First, a child might be placed on probation under the supervision of a probation officer. Healy commented on how the differences in the probation workers' personalities influenced their effectiveness with the children. Second, a youth might be placed in a foster home, often with a family on a farm, or, as Healy put it, "placed in country life." A third possible disposition was commitment to a correctional institution or to an institution for dependent children.

These institutions had widely varying characteristics, and the one chosen could make a considerable difference in the child's life. The Chicago Parental School was maintained by the Chicago Board of Education. Children of school age were committed there by the court for an average stay of five months. The school was organized on the cottage plan, offering regular school subjects, manual training, and excellent recreational facilities. A second short-term institution, abandoned soon after Healy's association with the court, was the John Worthy School, noted for its strict disciplinary control. A third institution was the state training school, serving the state of Illinois. Its chief drawback was the parole system, which employed only one officer for the entire state. In addition there were a number of church-related or privately owned institutions for dependent children which offered residential care, general education, and shop work. Finally, there were a few institutions for the care and training of girls.

Healy viewed the problem of placement as an empirical question requiring careful research. He made recommendations for disposition on the basis of intensive studies, and as a scientist, he was concerned with following up the children in their different settings. The follow-up reports of cases he studied later in Boston reflects the care and attention to detail that Healy displayed in his clinical and research work (Healy & Bronner, 1926).

Healy's Views of Therapy: Programmatic Approaches

Although Healy's reputation is based on his application of psychotherapeutic methods and psychoanalytic concepts to delinquency, he had a broad outlook toward treatment. He did not arbitrarily exclude any form of intervention that might produce a desirable behavioral change. Although uncertain about the efficacy of a punitive approach, he understood that a judicial system could not function without punishment. He even wrote about some cases that benefitted from exposure to harsh treatment in an institutional setting.

Healy was fully aware of the necessity for programmatic approaches, arguing that the availability and quality of resources for recreation, education, foster home placement, and other child welfare measures were vital to the causes and treatment of delinquency. Evidently there already were difficulties with cooperation among the services, for he also recommended trying to coordinate them. An organization to monitor the social conditions causing delinquent conduct was a part of the program he recommended.

In his view, a full program would also take into account the characteristics of the court and the police, for Healy felt that the way the police handled and the way the courts treated an offender would affect what happened to him or her later. In some cases Healy sensed that an element of sport entered the delinquent's relationship with the police, for he or she risked penalty against the possibility of escaping detection. Healy implied that some modification of the relationship between the law-enforcing agencies and the individual might be beneficial not only to the rehabilitation of a given individual but also to the prevention of delinquent acts.

A full program would attempt to change public attitudes toward crime and the criminal. Healy felt that in many groups the delinquent and the criminal were held up as culture heroes and that crime was viewed as a form of adventure. This comment, found in a book written in 1926 (Healy & Bronner, 1926), may reflect the ethos of the Roaring Twenties.

Psychotherapeutic Methods

Environmental manipulation, with the help of church, school, and social agencies, was one part of Healy's treatment of children. In addition, he worked psychotherapeutically with individuals and also in what he called the reeducation of families. Unfortunately, although Healy's writings describe the characteristics of individuals and delinquent populations, they fail to specify his approach to individuals and families. He produced much case material from the delinquent's "Own Story" but left us little to help us understand his conduct of interviews. (In the following material, we had to take bits and pieces from his writings in order to illustrate his methods of working.) Healy even criticized *The Individual Delinquent* himself, because it does not sufficiently discuss methods of treatment.

Although Healy debated with himself whether the physician or the psychologist is better equipped to handle problems, he believed the psychologist's approach to be more important:

> The only attitude to be assumed with much profit is that of shrewd but sympathetic inquiry into an unsolved problem. We have insisted that the examiner should have no special nose for the pathological and should be willing to survey all the

facts, and to be guided in his conclusions by no special bias. The question for him must be: What is the cause in this person or in his experience and how can it be altered?

Often I have stated the following fact, which has become increasingly apparent to me. Just as soon as the offender and his relative realize there is someone who takes the attitude of the friendly family physician, to whom they can go with their secret troubles, the case frequently undergoes the most remarkable transformation from the fighting aspect actually seen in the court room or while the interested ones are in contact with the police or other authorities of the law.

The opening of the interview with some such friendly and reasonable statement as the following has been found in itself to have a rationalizing effect. One may say: "Well, you people do seem to have a difficult affair on your hands with this boy. Let's sit down and talk it all over, and study it out together—how it all began and what's going to happen. I'm at your service. Did you ever think it all out carefully?" (Healy, 1915, pp. 36–37)

The timing of the interviews was important, and Healy defined the "golden moment" for intervention during a period of crisis and the value of being a part of the court:

The offender must approach you willingly before you can do anything for him. Now when will he exhibit this willingness? Certainly not when he is "on the outs" and feeling it quite unlikely that he will recommit offenses or at least be caught again. No, the golden moment is when he feels himself to be a problem, and his relatives feel it, and all want a promising solution of the difficulty. It is after he has been caught, and while he is either detained or on probation, and has not already been sentenced that is the best time of all for inquiry. Then parents will come many miles in search of a solution, not by any means always desiring the softest outcomes for the offender. Then the offender will himself strive hardest to achieve with the "doctor" some fundamental explanation of the causes of his delinquent tendencies. (Healy, 1915, pp. 40–41)

The problem of initial resistance was handled in the following manner:

Over and over from relatives and others we have heard of the difficulty in getting their problem individual to come and see us. It seems to be hard to get it understood that because there is delinquency there must be need of study. The answer is given, "There is nothing the matter with me. I know what I'm doing," so it comes about that a collateral explanation is offered. "We want you to go to see the doctor to find out if you are healthy," or, "We want to find out what you are best fitted for." This latter explanation indeed makes a truthful form of entrance that we have come to use most frequently as offering the chance of developing the greatest amount of interest. The question of vocational diagnosis is really part of almost every young person's thoughts, however crudely apperceived. . . . In all this tact is of greatest service and one learns to develop an elastic method which best of all subserves scientific as well as practical interests. (Healy, 1915, pp. 45–46)

Healy preferred to use a series of interviews rather than a single one because he recognized that the positive and negative conclusions based on a single interview were likely to be erroneous. He also preferred to deal with the patient and his relatives one at a time, for he believed that in private interviews he was better able to discover friction in the household and that in any event people wanted to be alone with their physicians when telling them their troubles.

Apparently Healy approached his cases with great openness and directness. When dealing with pathological lying, for example, he recommended direct confrontation:

> In discussing treatment great emphasis should be placed upon the primary necessity for directly meeting the pathological liar upon the level of the moral failures and making it plain that these are known and understood. It is very certain that frequently this type of prevaricator has very little conception of the social antagonism which his habit arouses. . . . When it comes to specific details of treatment, these must be educational, alterative and constructive. In cases 1 and 3 under treatment we know that when the lying was discovered or suspected the individual was at once checked up and made to go over the ground and state the real facts. The pathological liar ordinarily reacts to the accusation of lying by prevaricating again in self-defense, but when with the therapeutist there has been the understanding that the tendency to lying is a habit which it is necessary to break, the barricade of self-defense may not be thrown up. (Healy & Healy, 1926, pp. 273–74)

Healy practiced psychoanalysis himself and, with others, wrote an important textbook on it (Healy, Bronner, & Bowers, 1931). He did not disapprove of brief psychotherapies, as many practitioners did later during the 1950s and 1960s (Healy, 1915, p. 119). Rather, he felt that brief therapeutic interviews were effective:

> With these [few cases], however, there were some startlingly good results which proved much to us particularly when in the interview some youngster recalled a traumatic episode and verbalized the attendant and often continuing emotional upset. . . . There were failures, perhaps due to not using more skilled techniques or perhaps to not being able to alter reality situations, but there is no gainsaying the fact that comparatively few interviews involving no deeply penetrating interpretations did work wonders in some cases of serious misconduct, and still do. (Healy & Bronner, 1948, p. 29)

Although enthusiastic about the possibilities inherent in psychoanalysis, Healy was also sensitive to the limitations of a solely psychoanalytic approach:

> The therapeutic effects of the application of the psychoanalytic method to the study of offenders prove in some instances nothing short of brilliant. . . .
>
> By merely showing to the subject, through hauling up the contents of his own mental reservoirs, what his failure is based on may not prove sufficient, if environmental or physical conditions which serve as two of the three instigating causes, are still irritating as of old. Various reasons will readily suggest why this should be so. Habits and thoughts and tendencies of years' standing are not to be lightly overcome if nothing but added knowledge is to stand up against them. Reeducation and helpful new interests from the outside are also frequently necessary. For energies which previously found outlet in socially undesirable behavior, ''substitution'' must be made possible by discovery of a junction point where now by conscious volition, shunting on to another track can take place. (Healy, 1915, pp. 120–21)

Child Guidance: Working with the Parents

Healy worked in child guidance; that is, he worked with the parents of the children he saw. He preferred to work with both the children and the parents himself, rather than have one person see the parent and another the child, as was common in later child guidance practice. Although he was interested in the parent's Own Story, too, and although he recognized the role of parental friction and discord in producing mental conflict and delinquency, he did not seem to blame the parents. Sometimes he worked with parents to help them express a particular personal problem, but he treated them as intelligent persons who could, if given sufficient direction, help solve the problem.

For example, one case was that of a ten-year-old girl who had engaged in petty stealing and had been expelled from two schools. Healy discovered that the child was obsessed with sexual thoughts. He thus advised the mother to give her child sexual information and suggested measures to keep her busy at school and at home.

In another case, that of a sixteen-year-old boy who had run away from home and who had at various times forged his father's name to get money, Healy made specific recommendations for treatment:

> In this case a definite tendency to delinquency was under consideration, a tendency that had not been modified by admonitions or threats. The outlook now, with understanding of beginnings, was altogether different. The coffee and smoking should be stopped, the evening reading, which had been somewhat opposed, should be allowed at home, but above all there must be modification of parental behavior. The scolding, even though justified, and especially the speaking of the boy behind his back, should be abandoned. The mediation which our discoveries led us to offer, was well received, and the father's negotiations for placing the boy in a reformatory situation were broken off. For the six months elapsed since we saw this case, the report is that there has been an entire change of conduct. (Healy, 1915, p. 397)

The Necessity for Follow-up

For Healy, the essence of treatment (and of research) was the follow-up:

> As one makes more and more studies of the formative period of life, and watches cases go on to success or failure, one sees clearly that a great feature of treatment is the careful carrying over of offenders through the period of adolescent instability. A little touch here and a little touch there to the young individual is not sufficient; there must be that prolonged studying of the case that offers the best chance of forfending the growth of delinquent tendencies. . . .
>
> A very weak point in practically all social and moral therapy is the lack of follow-up work. Criticism may be extended to parents who have no patience to deal systematically with a problem child, to court admonitions which imply the ability of human nature to change itself in a trice, to public administration which sends back old offenders from institutions to an environment where they are almost sure to fail again. (Healy, 1915, p. 178)

Residential Treatment

Healy did not operate a residential treatment center or a correctional institution, but he studied and acted as consultant for many such institutions. In 1915, Healy and Bronner outlined a model correctional institution, based on firsthand observation and also their diagnostic and follow-up studies of individuals treated in various institutions. Portions of that outline are reproduced here, both because the program of milieu therapy is so modern in tone and because it reflects the educative approach so well. After discussing goals (to fit the individual to cope with all phases of an ordinary social environment), the physical equipment, the selection of staff ("The selection of working personnel is the most important single consideration. The influence of a man or woman of good understanding upon youthful delinquents, whenever in contact with them, is not to be overvalued"), and the urgent need for adequate follow-up after release, Healy and Bronner describe the treatment program for an institution to serve delinquent children:

> Treatment in general: Before discussing specific phases of treatment, its general moods and aims should be taken up.
>
> a. The entire institutional life should be adjusted with the ideal that it is treatment, that it is educational, and all to the end that the delinquent shall be better fitted to meet an outside environment.
> b. This requires high individualization. One of the arguments against the advisability of a set system is found in the successes which are actually obtained by a rational and understanding approach to the problem of the individual. Both education and work must be adapted to the individual needs.
> c. Three things to avoid are any kind of deceit, the show of pedantry, and any demonstration of irrationality. It is most desirous to make the individual rational and honest, and this can only be done by showing a good example in these respects.
> d. The method should be elastic in all ways, particularly in institutions for girls, where allowance must be made for outbreaks and explosions of pent-up emotions and energies, either occasional or periodic. Of course, physical fluctuations must be allowed for.
> e. Punishments: These must be highly individualized according to personalities involved. There is no doubt that stimulus to doing better is more apt to result from the promise of rewards than the administering of penalties. There must be goals toward which the delinquent is to work as the reward of good behavior. With constructive treatment the problems of discipline largely tend to disappear. It should be remembered that coercion and punishment by inflicting pain are the lowest levels of control.
> f. Above all things, mental vacuities, either on week days or Sundays, must be prevented. "The empty mind is the devil's workshop." There should be abundant opportunity for good conversational reactions. This may be as important as formal instruction, and always the mental life should be the first and foremost consideration.
> g. The whole institutional equipment should be used with the sole idea of its social and moral worth.

h. General social and educational life should include the planning of service and of rendering helpfulness to others in the institution. Cultivation of this is worth much, and from it can be built up larger ideas of social relationships. Perhaps the best way to avoid jealousies is to inculcate the idea of service, one to the other in the institution.
i. Intimate social life: One of the best helps toward a better life is an understanding friend and advisor with whom the cause and the help for trouble may be discussed.
j. In considering treatment in general it must not be thought that building up is always the point, or that positive habits are the only good; the inhibitions of bad impulses must be also considered. In some cases excessive physical vigor, or obstinancy of will, make special forms of modification necessary.

Dress: A moot question is over the dress of the institutional inmates. One point stands out clearly proven; namely, that any self-expression that is practicable in this matter should be cultivated.

Work: The arrangement of work to be done by the inmates has its economic and also its social and moral values. The immediate economies must not conflict with the aims of the institution. If the work has a deteriorating effect, or is interfering with treatment, it should be done by outsiders. But this does not mean that difficult or even so-called menial work should be neglected. The idea of duty on the part of the spelling [work shift] may be cultivated, although perhaps with difficulty in early adolescence, through the understanding that the institution ought to be largely self-sustaining. Work of all kinds is done, chiefly for common welfare. If it is merely assigned as a matter of routine or punishment without this feeling, work is apt to be detrimental and cause a grudge.

Very much of the housework and other work can be done in the spirit of scientific training. There may be attention to skill and success in many household occupations. It must be shrewdly recognized that there may be great benefits accruing to selected individuals through their engaging in hard labor, either physical or mental, or both.

Religion: Religious training would be out of place to discuss here. In general, we may say that religious training which takes the individual as one of a group and does not meet special problems is not apt to get results. Then it must be remembered, in all common sense, that natures differ greatly. The religious appeal is very strong in some, and others are oblivious to it. (Healy & Bronner, 1915, pp. 307–9)

Much of the remaining outline is a detailed curriculum for a school program. Dewey's influence is not acknowledged in the article, but the influence of progressive education is seen in the authors' emphasis that in each instance the subject matter was to be taught in a manner relevant to the adolescent's life situation and to therapeutic goals. Thus, the subject matter might center on the female's role in the home and the male's role in the working world. There also was to be an emphasis on vocational possibilities and on life in the community. The curriculum would include work in local political organizations, social welfare agencies, and educational and recreational resources in the community. All of these items in the curriculum reflected the goals of the educational reformers (Cremin, 1964).

Healy and Bronner pointed out that a standard curriculum would have little utility for adolescents with a great deal of living experience and little formal educational training. They argued that any standard reading series would be useless. Reading materials should be interesting and be applicable to the immediate life situation in order to stress the enjoyment and utility of reading. Music, art, and dramatics were to be part of the curriculum and to include participation and "spontaneous and joyful expression." Dramatic performances were seen as opportunities to develop cooperative effort and practical skills in making costumes and designing and building stage sets.

Arithmetic, writing, and spelling should underscore what is needed and used in everyday life. For example, instruction in arithmetic should demonstrate its application to keeping household and other accounts.

Language, according to Healy and Bronner, should be taught specifically to develop powers of expression and to exchange opinions. Healy and Bronner suggested that verbal expression be encouraged through "free and frank discussion of actual living problems that arise daily or that have arisen in the past. There should be no specified period for this, but it should be done as the occasion arises" (Healy & Bronner, 1915, p. 311).

Physical education was viewed as an opportunity not only for exercise but also for social participation, for the opportunity to develop cooperative effort, team spirit, and self-discipline.

Healy and Bronner recommended that special educational resources be available, but they stressed that the formation of special classes to meet individual needs was undesirable. They also spoke of the consulting role of the psychiatrist or psychologist in the educational process.

> No one of these [special educational or therapeutic programs] necessitates the withdrawal from the ordinary classroom during the whole day or the formation of special classes. The special teacher should give daily such time as the individual case requires in the special field, whereas other work can be carried on in the regular classroom. The cooperation between teacher and diagnostician should be particularly close in these cases. (Healy & Bronner, 1915, p. 315)

Healy and Bronner favored some form of self-government, but they noted in a brief comment that external control by the staff was still necessary. One wonders whether Healy and Bronner had some opportunity to observe a development such as that described in William Golding's novel, *Lord of the Flies* (1954), in which unsocialized children legitimize their authority by means of physical strength.

This outline, surprisingly complete in its presentation and modern except for its comments on the temperaments and future housewife role for girls, stresses the educative, therapeutic, and rehabilitative potential of every aspect of the institutional setting. The tone of the institution and the importance of all staff members as therapeutic agents are emphasized, as is the relevance of the curriculum to the adolescents' needs and interests. Healy and Bronner realized that there would always be problems in management. In too many institutional settings, the staff seem to respond with the expectation that no management problems should arise, and if they do, they are seen as staff failures rather than as expected incidents or therapeutic opportunities.

This educative approach is also instructive in that it does not center on psychopathological manifestations. There is no indication that Healy and Bronner found it necessary to obtain an admission of illness as a prerequisite to receiving help. The individual was placed in a situation in which the staff, the peer group, and the program itself were the therapeutic or change agents, and it was assumed that the appropriate responses such stimuli elicit would result in behavioral change. Healy and Bronner recognized that in some instances, "delinquencies which arise on the basis of obsessive mental imagery" would require "special counteractive aids," but even these were integrated into the special educational environment. They favored a "living through" in a structured environment rather than a "working through" on a verbal–conceptual basis in a psychotherapeutic relationship. Healy saw psychotherapy as having a special place in the program, but also recognized other means of helping.

Healy and Bronner anticipated criticism of their proposed program, for their outline closes with the following paragraph:

> We would firmly contend that the above scheme is not Utopian; it is thoroughly practicable where a group of intelligent and well trained workers can be gathered in an institution. The basis of our survey of the subject is simply, as we have said, experiences with the needs of delinquents, and successes and failures of treatment. If our ideas of constructive efforts appear complex and difficult, it must be remembered that they are not any more so than the details of education and home life in any well conducted school and family. As for ultimate values accruing from such efforts—well, we are told that in Heaven there is much rejoicing over even one delinquent saved. (Healy & Bronner, 1915, p. 316)

Summary

The juvenile court was an instrument and a product of the reform movement that flourished at the turn of the century. The problem of delinquency was prominent because of the issues created by urbanization, industrialization, and poverty with its attendant social and familial disorganization. In concept, the court's responsibility changed from determining guilt and setting punishment to rehabilitating and guiding "as a wise parent would guide a wayward child." The social action and legal principles underlying the establishment of the juvenile court made it explicit that the community was responsible for the welfare of all its inhabitants.

The early juvenile courts, those not only in Chicago but elsewhere as well, were firmly embedded in the community, with lay people taking much of the responsibility for the courts' work. In Chicago, volunteer organizations and a settlement house that was part of the community worked to establish the court. What we now call the power structure was instrumental in its growth, development, and early operation. Ben Lindsey's Denver court, as we shall see in the next chapter, Charles W. Hoffman's Cincinnati court, and Judge Stubbs's Indianapolis court all were closely tied to their communities. In Philadelphia the early probation workers were supported by the churches and synagogues, mothers' clubs, settlement houses, and private philanthropy (International Prison Commission, 1904). Healy's clinic was opened as a result of community initiative, and it too was integrated into the work of the court. Consciously or not, those who planned this early clinical service understood the value of a strategic location near the manifestation of the problem. The clinical service in the court did not wait passively for the child or the family in need to seek out the helping agency.

The ideals of the juvenile court movement were fulfilled in only a few places. In many other places, the resources were not available, or the court was not established with a full understanding of what it might be able to do. Many juvenile court judges worked only part time and thought of the position as having a low status in the legal community. In other places, the judges seemed to exercise arbitrary authority. The volunteers were not always suitable for the work of relating to youth. Many criticized the reliance on them to do the basic probation supervision. Moreover, rates of delinquency, or even of recidivism, were unaffected. Many observers believed the juvenile court had failed, but it is also possible that it was not fully implemented in the ways intended by its designers (Ryerson, 1978).

Healy's clinical methods and his thinking about delinquency are noteworthy. Although he was familiar with the concepts of psychoanalysis, he adopted a flexible approach in his thinking and methods, recognizing the importance of creating personality change by permitting the individual to live out his or her problems in a new environment. Healy's psychodynamic approach, his willingness to use psychological methods, his departure from seeking merely to establish a diagnosis, and even his willingness to work in the community were remarkable innovations, considering the hospital psychiatry of his day. Running throughout all of his therapeutic work, however, is an emphasis on what the person might become and less concentration on his or her psychopathology. According to Healy, "the examiner should have no special nose for the pathological. . . . The question for him must be what is the cause in this person or in his experience and how can it be altered?" (Healy, 1915). He seemed to share the mood and attitude of the social and political reformers, who also had boundless faith in human potential if only everyone were given the proper conditions in which to fulfill that potential.

8

Judge Ben Lindsey and the Denver Juvenile Court: An Institution of Human Relations

Am I a judge now? I AM NOT, HEAVEN DEFEND ME AGAINST THE HERESY! I may be a doctor or some new kind of specialist that will someday receive an appropriate name—a specialist in the human heart and human behavior. But not a judge. (Lindsey & Borough, 1931, p. 315)

So wrote Judge Ben B. Lindsey, who singlehandedly created a juvenile and family relations court in Denver. He thought of his court, which began functioning almost simultaneously with the Chicago court, as an "institution of human relations (misnamed a court)" designed not to punish people but to help them. In his court, "the human artist succeeded the executioner" (Lindsey & Borough, 1931, p. 315). Not only did Lindsey think of the court as a strategically located helping agency, but he also succeeded in making it a place where children voluntarily came for help. He sent hundreds of juveniles to reform school and prison without escort and without the loss of a single prisoner; his rate of commitment to institutions of those on probation was 3 percent; and there is some suggestion that at least in the early years of his court's operation, the number of delinquent acts in the community fell.

Lindsey was internationally renowned as a founder of the juvenile court system, and although his clinical contributions have been forgotten, his many contributions to law have been retained in statute and precedent. Lindsey was a powerful clinical therapist because he was also a powerful social therapist. His power to reach individuals came about because his court was part of his community. He understood his community and the culture of the children with whom he worked, and he used both as helping tools. A shrewd, able politician of unimpeachable integrity, he enhanced his role as a judge by working selflessly and fearlessly for political and social reforms to benefit the people of his city.

In this chapter we describe Lindsey's methods, for many aspects of his approach exemplify important social and clinical principles. Lindsey had an informal manner; he modified many court procedures; and he placed many children on probation. His desire to help was coupled with an understanding of the children with whom he worked and with an intensive and personal follow-up of the work he started in the court.[1]

Denver at the Turn of the Century

Lindsey's Denver court cannot be divorced from either the man or the city. The city came into being during the Pike's Peak gold rush in 1859, and the prospectors used Denver as a headquarters to purchase supplies, to conduct other business, and for recreation. After the Civil War the railroads came west, and as early as 1872 Denver was a center for four railroad companies; Lindsey's father came to Denver to work for one of them. Whereas the population of Denver was only 4,759 in 1870, by 1880, when Lindsey's family arrived, the population had jumped to 34,555. The gold rush had petered out earlier, but in 1880 silver was discovered, and from that point on Denver's growth in population and in industry was spectacular. By 1890, the population was 106,000; by 1900, 133,000; and by 1910, it was 213,000.

Denver was a mining territory (Colorado did not become a state until 1876), peopled by tough pioneers and adventurers, squatters and claim jumpers eager for riches. A loose but somewhat complex legal system operated until the 1890s when the state legislature began to revise it. Outlaws were common in the area, and legally constituted authority was often supplemented by vigilante law. As late as 1878, the "Indian agent," an army general, and a major were killed in a minor Native American uprising. The city catered to the prevailing desire for excitement; bars, dance halls, and gambling were the principal local industries. Even the *Rocky Mountain News,* in an anniversary edition meant to celebrate Denver romantically, characterized the theaters of the time as "somewhat rowdy"; that is, they were basically drinking establishments with theaters on the second floor.

The discovery of silver in 1880 made mining and metal refining major industries. Private mints for coinage had been constructed earlier to solve the problem of transporting gold dust. In addition to mining and related industries, a wholesale trade grew, and railroading became an important source of employment. By 1926, Denver was served by nine railroads. After 1880, as the only large city within a five-hundred-mile radius, Denver was established as an industrial and commercial center.

The population growth that accompanied industrialization and urbanization came largely from migrants going west, but there were also a number of immigrants. In 1900, when the population of Denver was 133,000, there were about 25,000 foreign-born individuals, most of them from England, Ireland, Scotland, Germany, and Sweden. There were only some 1,500 Russian-born individuals in Denver at that time and an insufficient number from other eastern or southern European countries to be listed separately, even though Lindsey did speak of sections of Denver called "Little Israel" and "Little Italy." There were only 3,300 blacks in Denver, for it had never been involved in the slave trade. Cultural differences between the majority and minority groups were probably less pronounced here than on the eastern seaboard. This factor may have helped Lindsey gain support for his work in the community, for there was no American-born majority to complain about ''foreigners'' or ''outsiders'' who created problems in their city. Many were recent arrivals, whether or not they were foreign born.

The newness of many of Denver's formal institutions may also have been important to Lindsey's success in gaining widespread community support. With the growth of industry and population, Denver began to settle down. The Denver Bar Association was formed in 1891, at about the time that a series of reforms were introduced into the judicial system. The state reformatory was established in 1890, and a state industrial school for boys in 1891. The public school system was also quite new, dating only to the early 1870s. Again, the absence of a long tradition may have made it relatively easy for the schools to adopt modern practices. Denver's public schools in the 1880s and 1890s already included health education, home economics, industrial arts, music, art, and modern languages in the curriculum, subjects that schools in the rest of the nation did not include until years later. A school law designed in part to strengthen the school's hand in dealing with truants and disciplinary problems was passed in April 1899, and it was this law that Lindsey used in establishing his court.

The city of Denver did not receive its municipal charter until 1902, shortly after Lindsey instituted his juvenile court. In Denver, the juvenile court was not only part of the general reform movement of the period; it seems also to have been part of the establishment of legitimate forms of institutionalized social controls in a city that was ready to stabilize.

Lindsey the Man

Throughout his forty-year career, Lindsey lived the ''dangerous life,'' by which he meant that he exposed himself to great professional and personal risk to support what he believed was right. At different times in his career and in a variety of combinations, he took on the trusts, corrupt politicians, both the Republican and Democratic parties, the bar association, the Colorado Supreme Court, the sheriffs and police, the penal system, social workers whom he called ''scientific robots,'' civic groups, patriotic societies, the Ku Klux Klan, and the church.

He braved personal insult and threats to his life and livelihood and fought legislative investigations, smears, and frame-ups until 1929 when his political enemies succeeded in disbarring him because he had acted as a mediator in the case of a disputed will in New York City while he was a judge in Colorado. The case was not in his jurisdiction in Colorado and had nothing to do with his court. After his disbarment in Colorado, he practiced law in California and later was elected to a six-year term as judge of the Los Angeles Superior Court. The Colorado Supreme Court offered to reinstate him if he apologized for statements he had made about the court in one of his books. But characteristically, Lindsey refused, saying the court's offer itself vindicated him. He was nonetheless restored to the Colorado bar several years later (Lindsey & Borough, 1931; *New York Times*, March 27, 1943).

In the 1920s he achieved notoriety for his advocacy of companionate marriage, a legally sanctioned form of monogamous, childless marriage that could be ended more readily than could marriages involving children and whose contractual relations could be altered with the decision to have children. These concepts concerning marital and sexual relationships derived from his experience in the juvenile and family relations court, in a day when contraception was all but legally prohibited. He provoked fervent opposition with his proposals for liberalized and realistic divorce laws, for sex education in the schools for children and adults, and for the repeal of anti–birth control laws (Lindsey & Evans, 1925). Lindsey was even accused by clergy, physicians, and others of promoting "trial marriage," a concept that he actually opposed.

Never one to back away from a battle, Lindsey used public relations weapons to express his views and to promote his causes. Because of his ingenuity in gaining publicity, however, he acquired a reputation as a grandstander. But, such an assertion overlooks the many reforms in law he actually enacted and, as we shall see, is false with respect to his clinical accomplishments while on the court.

Lindsey was born in Tennessee in 1869. His father, a Confederate captain and an intellectual, managed the Western Union office in his town. His mother was a true Southern belle, noted for her Irish beauty and wit. She was born and raised on a plantation, whose great manor house was the gathering place for young officers of the Confederacy and their ladies. Married after the Civil War, Lindsey's parents lived on the plantation with Lindsey's grandfather, a powerful, bluff man of impeccable standards. As foreman of the grand jury in his area, he denounced what was crooked or underhanded in public life. Lindsey's grandfather considered himself a freethinker and described his own ancestors as "free booters." Apparently the Civil War did not drastically affect the family's style of life. Lindsey and his brother were cared for by a black nursemaid, played and lived with black children, and enjoyed the farm, the animals, and the countryside. The family was apparently well-to-do, even in the Reconstruction period, for Lindsey states that his mother lived an easy life with many servants.

His father, a restless man, converted to Catholicism after Lindsey was old enough to recall his father's participation in the Episcopal church. His mother also converted, and Lindsey was sent to a parochial school. The family's Catholicism was sufficient cause for the family to be almost ostracized in a Southern Baptist community. Therefore, when the West opened up and the railroads needed telegraphers, his father accepted a position in Denver and moved the family there in 1879, when Lindsey was ten.

For a while the move went well for the family. Lindsey and his brother were sent to the Notre Dame prep school where he learned to debate and experienced several religious crises. He studied the lives of the saints and was considered a promising candidate for the priesthood. But, while the boys were at school, their father lost his job, became ill, and fell into debt. Lindsey returned to Tennessee to the plantation and continued in a Baptist school, where he persisted in both his Catholicism and his debating. A few years later, he rejoined his family in Denver, but his father's health was poor, and so Lindsey went to work to help support the family. He was not even able to attend the Denver high school.

Lindsey made up his mind to study the law, and at age seventeen he secured a position with a lawyer as an office boy. In addition to working and studying at the law office, Lindsey supplemented his income by selling newspapers and doing janitorial work at night. Shortly afterward, his father died penniless, one day after his insurance lapsed because he had neglected to pay the last premium. At this point, Lindsey was so despondent that he attempted suicide and was saved only because the cartridge in his revolver failed to explode.

The failure of his suicide attempt enabled Lindsey to face life with renewed vigor. He returned to his work and the study of law, and in 1894, he was admitted to the bar. As a young lawyer he was appointed as the attorney to defend two small boys who had been arrested as burglars and who were in a cell with adult criminals. His defense of the boys included a public indictment against the state's treatment of these children and won him some popular attention.

Later Lindsey entered into a law partnership with a man who became a state senator, and while in the partnership, he participated in Denver's politics. Lindsey proved to be an excellent party worker, although he soon became disillusioned with the deception he encountered. He advanced in Democratic party councils to the point that he was considered for the district attorney's office. Although he did not receive that nomination, he was appointed instead as county judge to fulfill the remaining months of the term of a man who, with Lindsey's help, had won election to the state supreme court. In January 1901, Lindsey took office as judge and terminated his law practice.

The Beginning of the Juvenile Court

While Lindsey was hearing a civil case, the assistant district attorney asked him whether he would interrupt the proceedings for a few minutes to dispose of a larceny case. A boy had been caught stealing coal from the railroad, and he had no defense. Under the law, Lindsey was obliged to sentence the boy to the state reform school. As he did, the boy's mother, who was in court, began to scream. Lindsey was so touched by the woman's anguish that he arranged for a suspended sentence. That night, Lindsey went to the boy's home, and there discovered the family's utter poverty. The father was sick from lead poisoning, an occupational disease he had acquired while working in the smelters, and the whole family was cold.

With the injustice of the case still in mind, Lindsey heard another case of three boys charged with stealing some pigeons from a man whom Lindsey knew from his own youth to be a grouch and a target of boyish mischief. The boys had been brought into court on criminal charges. Investigating, he learned that officers of the court had been paid for each conviction. When they wanted more money, they simply rounded up more children and brought them into court on criminal charges. Lindsey's indignation aroused, he began looking into the problem of the children's treatment by the police and the courts. The more he found, the more concerned he became. He visited the jails, compiled statistics, and searched the statute books for laws that might help children in trouble. He did find a section of the Colorado School Law of 1899 according to which it was possible to treat children in trouble as "juvenile disorderly persons" and not as criminals. The district attorney agreed to file all complaints against children under this law in Lindsey's court, an action that marked the beginning of the juvenile court in Denver.

In part, Lindsey decided on a course of probation because of the abominable conditions in the prisons and reform schools. He worked hard to get the school boards to appoint truant officers whom he could also use as court probation officers. His use of the school law to justify the court probably helped involve the schools in the court's work. In the next few months Lindsey spoke to countless groups of schoolteachers, women's clubs, church groups, charitable societies, and community organizations to bring to the public's consciousness the problem of the children. He attracted a great deal of attention to his court. His desire to help won him the cooperation of parents whose children got into trouble, and Lindsey wanted to use popular support to gain leverage for several reform laws he was drafting.

Lindsey as a Social Reformer

Earlier Lindsey made political enemies when he refused to participate in shady propositions involving his court and when he refused to cooperate in making patronage appointments to the court's staff. A few months before the 1901 election he heard the case of a saloon keeper who was selling liquor illegally and who was involved with prostitution. The man paid protection to the political powers and fully expected to be acquitted. But Lindsey convicted him. Even though he had come to the favorable attention of the reformers who were trying to help Denver become established, he became identified to politicians as a man who would not play the game. The machine politicians could not prevent Lindsey's renomination in 1901, but they attempted to elect his Republican opponent. Lindsey, however, had built a broad base of popular support, and the people of Denver rallied to save "the kids' court." He ran two thousand votes ahead of his ticket and thus found himself in a position of political power and independence.

From his vantage point in the county court, Lindsey began to press for reforms. He wanted a law to prevent politicians from exploiting the estates of widows and orphans and another to forbid the collection of fees for prosecuting children. He eventually obtained a contributory juvenile delinquency law to prosecute neglectful parents and other adults who got children into trouble; a juvenile court law with probation officers who had police powers; and a detention school for children. He urged improved child labor laws and the enforcement of the compulsory school attendance law. He worked for public baths, public playgrounds, and trade schools.

In his fight for these laws, Lindsey went into gambling dens, saloons, houses of prostitution, and factories to see for himself what was happening there and he publicized what he saw in speeches, newspaper interviews, and occasional columns. When he found the district attorney's office would not prosecute the gamblers, Lindsey invited the board of police commissioners to his court. He informed the newspapers and then publicly accused members of the board of knowingly permitting the ruination of children, of being personally responsible for the appalling immorality that came before his court. The newspapers spread the story; the ministers took up the cry; and as the worst of the dives shut their doors, Lindsey went to the legislature to obtain his contributory delinquency laws and police power for the probation officers of his court.

Lindsey wanted a detention home-school because he found the conditions in the jails almost as abominable as they were in the dives. Boys learned from adult criminals to become criminals; they were subject to homosexual assault, engaged in homosexual activity among themselves, and managed to contact women and girls who were also confined in the jail. Lindsey believed that when they were released, the children returned to their schools and introduced others to what they had learned. Lindsey also found that children were frequently abused physically by the police. As a result of their confinement, many came to regard the law as a hated enemy.

His attempts to obtain prison reform laws were sabotaged in the legislature not by open opposition but by legislative skulduggery, and so Lindsey again went to the public. An interview with a friendly reporter resulted in a front page story detailing the horrors that children encountered in the jails. When the police commissioner denied Lindsey's charges, he demanded an open investigation and set a meeting in his court to which he invited the governor, the mayor, fifteen prominent ministers, the police board, and members of the city council.

When Lindsey learned that the subpoenas for witnesses had not been officially served, he turned to Mickey, "the worst kid in town." Mickey, a leader among the children, was working with Lindsey to organize a newsboys' association under the court's supervision. (Lindsey frequently used children as unofficial agents of the court and used their gangs to prevent delinquency. Lincoln Steffens [1909] stated that Lindsey was "the leader of every kid's gang in Denver.") When told of Lindsey's dilemma, Mickey rounded up twenty boys who had been in jail, herded them into the court for the hearing, and, at Lindsey's invitation, acted as an assistant by selecting the order of witnesses to be heard by the investigating body. The boys detailed their experience in the jails and with the police. As a result of the investigation, Lindsey gained further support from the ministers and the governor of the state. With the subsequent publicity, he was able to get the state legislature to pass his juvenile court bills.

Lindsey's fight for public playgrounds took somewhat the same course. Two years of public agitation via newspaper articles, speeches, appeals to the city council, and the formation of a juvenile improvement association were necessary to win the playgrounds. The case for public baths was won when Lindsey encouraged his boys to swim in bathing suits in the bronze-cherubed fountains in front of the courthouse. When the police chased them, the boys ran dripping into the courtroom for Lindsey's protection. His tactics persuaded the city council that if the city could afford decorative fountains, it could also afford free public baths for those who needed them.

Lindsey's court uncovered a grafting scandal that extended to the board of county commissioners and involved a number of respectable businessmen. Despite considerable pressure on him to back away, he pursued the issue and obtained convictions. About the same time, Lindsey investigated conditions in the cotton mills just outside Denver. He visited factories and homes and talked with the children. Convinced that they were being illegally exploited, he saw to it the company was prosecuted and that the owner was fined. The mill moved out of town, and Lindsey thereby incurred the wrath of a portion of the business community.

In the 1904 election Lindsey had the support of the ministers, the women's clubs of Denver, and the newsboys, who openly campaigned on Denver's streets for "our little Ben." He managed to gain the nominations of both major parties, against the opposition of political bosses. Although he won, technical legal considerations forced him into another election that proved to be extraordinarily dirty. But Lindsey came out on top and subsequently battled the utilities, the railroads, the water company, and other interests. Lindsey prosecuted some of these interests through his own court. In his book *The Beast* (Lindsey & O'Higgins, 1910), he exposed other aspects of corporate corruption.

Lindsey worked to reform the election laws, "to restore the tools of democracy to the people," and in retaliation, the political bosses tried to pass bills to undercut his position as county judge. Rallying the support of newspapers, the newsboys, and the women's clubs, Lindsey beat down the attempt to get him and also managed to win some election reforms. His battles against the corporations constantly kept him in the public eye. In 1908, when the political bosses succeeded in keeping from him the nomination of both parties, he ran as an independent, gaining widespread support because of his efforts at political reform and his work on the juvenile court. He was elected with a vote almost as great as the total received by his opponents on the Democratic and Republican tickets.

Lindsey continued to win election or appointment regularly until 1924 when the state supreme court threw out the results of the election on technical grounds, thus depriving Lindsey of his position. In 1924 the KKK was riding high in Denver, and Lindsey was a prime target. He was later disbarred by that same court, so that he never again could serve the juvenile and family relations court in Denver. By 1927, his ideas on companionate marriage had made him notorious and lost him much of the support he had previously commanded among the clergy. But Lindsey had become a world-renowned figure, speaking and writing about the juvenile court. There were numerous magazine articles by him and about him in both professional and popular periodicals. The *Ladies' Home Journal, McCall's, Everyman's, Outlook,* and the *Literary Digest* regularly carried features about Lindsey. He was known and widely respected throughout the country, although one writer stated he was better thought of in the East than in the West.

Lindsey's political history helps us understand the operation of his juvenile court. His effectiveness as a juvenile court judge was due to more than his informal approach and his reliance on probation. In any case, these techniques were less successful in other courts. His effectiveness was the result of his commitment to helping the people battle the forces that were hurting them. In his political role, Lindsey gave the disenfranchised a stake in government, and he made the facilities of his court available to children so that they could obtain redress for their grievances. In electing Lindsey, the people of Denver were truly electing their own representative. Clearly, the attitude toward Lindsey as a political agent extended to Lindsey as a helping agent. If he were on the side of the people in his political attacks on their oppressors, then in his work in the court, he could not have been perceived solely as an agent of society bent on controlling and punishing the deviants. Lindsey was a person who wanted to help, and in his hands the court became a helping agency. His manner and methods on the court and his political and reform activities supplemented and validated each other.

Not only was Lindsey's court made part of the community by his political activities; it also was made part of the community by his helping methods. Lindsey related the work of the court to the work of the schools and used his knowledge of individuals and groups to foster the work of the court. Children became agents of the court, a form of delinquency control in the very institution that had formerly served only to punish and to exclude them from society.

The Effectiveness of Lindsey's Work

How does one evaluate the effectiveness of an institution? Is its humaneness as important as the concrete results it achieves? Can one evaluate the court in terms of dollars and cents, as Lindsey did in one article? Is the purpose of the juvenile court to prevent the development of adult criminal careers? And even if one could state that its purpose was to prevent delinquency or adult criminal careers, it is not clear that the relevant variables are understood sufficiently to enable a fair evaluation.

In Lindsey's day, before the juvenile court was instituted, children were thrown in jail. The effectiveness of the probation system could be tested by seeing how many children who were placed on probation were eventually sent to the state industrial school. Scattered figures are available for Lindsey's court and for several other juvenile courts of that time, but the comparability of the figures is poor. They are taken to different base lines and cover varying time periods with little or no reason offered for reporting data for five months, fifteen months, or two years. It is as if the several courts haphazardly selected figures for public reports, thus making it difficult to compare them. The criteria used in deciding when a child had committed a sufficiently serious breach of probation to warrant incarceration were never explicit. In addition, the judge's attitude was, and is, important (Polier, 1964). An extreme example is the St. Louis judge who seemed prejudiced against an early probation law governing the treatment of juveniles and who placed only one boy on probation in a year and a half (International Prison Commission, 1904, p. 163). The availability of facilities for incarceration is still another issue (Polier, 1964).

Despite all of these limitations, some of the figures from the early juvenile courts provide some basis for comparison, however poor, with current figures. First, let us offer some figures from several juvenile courts and then concentrate on the figures that Lindsey provides for the first three years of his court. The 1903 Chicago juvenile court records show that 1,301 cases were adjudicated over a nine-month period. Of these, 715 were placed on probation, but 505, nearly all repeat offenders, were sent to the John Worthy School, an institution with a reputation for strict discipline. There is no way of estimating from these data the percentage of probation violations within any one year, but one can safely assume that it is probably no lower than 33 percent. For the year 1903 in New York, 13.1 percent of children on probation were subsequently committed for violating their probation. In the Philadelphia court, for seventeen months, from June 1901 to November 1902, only 27 cases out of 1,008 placed on probation returned to appear before the court more than once. In that same court, from October 1903 to January 1904, of 367 under probation, only 22 returned to the court two or more times. In St. Louis, in a five-month period, fewer than 8 percent were returned to the court for sentencing after they had been placed on probation (International Prison Commission, 1904). As late as 1925, Charles W. Hoffman claimed that only 6 Cincinnati boys, of a population of 1,200, were in the state reform school (Hoffman, 1925, p. 259)). None of these figures gives any indication of repeated offenses committed by the children on probation, nor do they tell us anything about the number of children who later became adult criminals. For a rough comparison, we may look at the figures summarized by Sutherland and Cressey (1960) for more recent times. They cite delinquency recidivist rates of no lower than 25 percent and as high as 88 percent for a five-year period. Other sources (Children's Bureau, 1963b) suggest that about 20 percent of all children on probation violate the conditions of probation and are returned to the juvenile court. In all, then, there is some suggestion that the early courts were effective.[2]

Lindsey presented somewhat clearer data in a report he wrote with his probation staff (Lindsey, 1904). The report was prepared to answer requests for information about the juvenile court, and it was also designed for distribution at the Denver Juvenile Court Exhibit at the St. Louis World's Fair in 1904. In addition, the bulk of the document appears as part of the report of the International Prison Commission (1904) to the U.S. Congress. The fact that the report was used for such varied purposes is important to evaluating its significance. We have been unable to locate any independent statistical survey of the effectiveness of Lindsey's court however, and so it is necessary to rely on the figures he offers. We cite these as background for a description of the operation of his court and not as definitive evidence of the court's effectiveness.

During the Denver juvenile court's first two years of operation, 715 children came before it, charged with delinquency. Of this total, 554 were placed on probation, and 40 were committed to the state industrial school. (The disposition of the remainder of the cases is not clear. Some were dismissed, and others were remanded to their parents' custody.) Of the 554 on probation, only 31 committed offenses that required commitment to the state industrial school. For the two years, 1902 to 1903 (the data overlap with the material just reported), 719 cases were heard. Most of these were placed on probation and in the two years, only 23 children committed further offenses serious enough to warrant commitment. These 23 constituted 3.2 percent of the cases placed on probation. Lindsey did not say how many children appeared before the courts more than once, and, of course, there is no way of measuring hidden delinquency, but Lindsey asserted that "95 percent have not committed a second offense, after a period of one to three years" (Lindsey, 1904, p. 35). Later, he revised the estimate to 80 to 90 percent. He also noted that over 50 percent of all the boys discharged from the courts before the establishment of the juvenile court did commit further offenses within comparable period of time. A 1913 political pamphlet quotes the warden of the state penitentiary who observed that in thirteen years of the court's existence, only three who were in the state penitentiary had ever been in juvenile court.

The 1904 report offers some evidence that Lindsey's court helped reduce delinquency in the community. The matron of the police department wrote: "We do not have half as many boys placed under our care by the police as we did two years ago, and I attribute the fact to the manner in which the Juvenile Court is conducted" (Lindsey, 1904, p. 1850). Endorsements of his work by several police chiefs can also be found in later political pamphlets collated by the Denver public library.

The chief officer of the Special Secret Service of the Union Pacific Railroad Company wrote in the 1904 report:

> The railroad companies have for years been subject to depredations of boys, who have committed all sorts of depredations against the property of the company from breaking into cars and robbing them of their contents and the stealing of brass attachments off the cars. . . . [O]ur trouble with this class of offenders has been largely and almost entirely done away with. In fact, our losses from this class of thieves have been reduced at least ninety percent, and the boys who formerly were ringleaders and really caused these depredations are now seldom seen around our tracks or property of the company, and on each of the reporting days in your Court I can point out the boys who have caused all this trouble. (Lindsey, 1904, p. 185)

The railroad officers apparently were invited to appear in court on reporting days (an event described later), and a similar statement was made by the special services officer of the Colorado and Southern Railroad (Lindsey, 1904, p. 189). A political pamphlet in 1913 cites a letter from a railroad officer who stated that in a three-year follow-up of ninety-seven boys he had referred to the court because of delinquencies against the railroad, he had repeated complaints against only two. And the Denver department stores detective declared:

> It gives me unqualified satisfaction to inform you that none of your wards have left their reservation for a long time, and that the part of the city known as department stores, which until last year were their supposed privileged grounds, have not even seen a hunting party. We can now turn out our toy play sheep, tin soldiers and woolly dogs, bears, wild cats, deer, etc., without fear of loss or molestation. We even dare to leave children's marbles, spring guns, locomotives, trains of cars and candy carelessly about, and none, so far, have been missing. This line of stock, as you may imagine, has always had a peculiar fascination for some of our youngsters and I cannot quite understand the influence you must have brought to bear on them that such magnificent results have been attained. (Lindsey, 1904, p. 186)

Similar statements of unqualified support are presented in the 1904 report by a number of school superintendents, principals, and teachers, and numerous editorials are quoted from the Denver papers of 1904. The editorials of that day might be compared with the periodic outcries against the juvenile courts in today's newspapers.

Lindsey's Probation System

When a boy or girl was brought to Lindsey's court, the youth was classified as either a schoolchild or a working child. If in school, information about the child's school, teacher, neighborhood, home, and parents was obtained. If the child was working, the court determined the nature of the employment and, if unemployed, the court would make an effort to assist the youth in obtaining employment.

At the court appearance, the child was informed in firm but kindly terms of his or her obligation to obey the law and to respect authority in the home and in school. The child was told of the necessity to "overcome evil with good and make up for his delinquency by being just as decent and good as he can in the home, the neighborhood and the school." A child placed on probation was given the implicit trust and confidence of the court, and an effort was made to have the youth feel that a failure would "let the judge down."

On the other hand, the child was asked to provide evidence of being worthy of this trust and confidence.

> We impress him with the idea that we have no doubt whatever that he will keep his word, that he will respect his honor and the confidence we have put in him, and that no one will question his doing it, and that we have a record of that fact, we ask the boy himself voluntarily to get a report from his teacher every other Friday preceding the session of the Juvenile Court every other Saturday. This report is made upon a printed card, and details conduct in school and school attendance. It is graded: excellent, good, fair and poor. It is soon understood that "good" is satisfactory, "fair" is passable, "excellent" is particularly pleasing to us, and "poor" is very displeasing. Any boy that brings a poor report at the morning session is made to understand that it is "not square." (Lindsey, 1904, p. 72)

The actual operation of the Saturday morning sessions was vastly different in tone from Lindsey's description.[3] Lindsey discussed his reporting system with Lincoln Steffens.

> "What I was after," the Judge explained, "was something for which I could praise the boy in open court. Believing in approbation as an incentive, I had to have their reports for the boy to show me in order that I might have a basis for encouraging comment, or, if the reports were not up to the mark, for sympathy. It didn't matter to me very much what the reports were about. . . . But you can see that these fortnightly reports were an excuse for keeping up my friendly relationship with the boy, holding his loyalty, and maintaining our common interest in the game of correction he and I were playing together." (Steffens, 1909, p. 132)

The Saturday Morning Report Session

Fortunately, Steffens provided a detailed description of the Saturday morning report sessions. Lindsey practiced a form of group therapy in his court, in which his charges were restored to a state of grace in the community.

> The boys assemble early, two or three hundred of them, of all ages and all sorts, "small kids" and "big fellers"; well dressed "lads" and ragged "little shavers"; burglars who have entered a store, and burglars who have "robbed back" pigeons; thieves who have stolen bicycles, and thieves who have "swiped" papers; "toughs" who have "sassed" a cop or stoned a conductor, and boys who have talked bad language to little girls, or who "hate their father," or who have been backward at school and played hookey because the teacher doesn't like them. It isn't generally known, and the Judge rarely tells just what a boy has done; the deed doesn't matter, you know, only the boy, and all boys look pretty much alike to the Judge and to the boys. So they all come together there, except that the boys who work, and newsboys, when there's an extra out, are excused to come at another time. But nine o'clock Saturday morning finds most of the "fellers" in their seats, looking as clean as possible, and happy.
>
> The Judge comes in and, passing the bench, which looms up empty and useless behind him, he takes his place, leaning against the clerk's table or sitting on a camp chair.
>
> "Boys," he begins, "last time I told you about Kid Dawson and some other boys who used to be with us here and 'who made good'. Today I've got a letter from the Kid. He's in Oregon, and he's doing well. I'll read you what he says about himself and his new job."
>
> And he reads the letter which is full of details roughly set in a general feeling of encouragement and self confidence.
>
> "Fine, isn't it!" the Judge says. "Kid Dawson had a mighty hard time with himself for awhile, but you can see he's got his hand on the throttle now. Well let's see. The last time I talked about snitching, didn't I? Today, I'm going to talk about 'ditching.' " And he is off on the address, with which he opens court. His topics are always interesting to boys, for he handles his subjects boy-fashion. "Snitching," the favorite theme, deals with the difference between "snitching" which is telling on another boy to hurt him; and "snitching on the square," which is intended to help the other fellow. "Ditching" is another popular subject. "To ditch" a thing is to throw it away; and the Judge, starting off with stories of boys who have ditched their commitment papers, proceeds to tell about others who, "like Kid Dawson out there in Oregon" have "ditched" their bad habits and "get strong." I heard him on Arbor Day speak on trees; how they grow some straight, some crooked. There's always a moral in these talks, but the judge makes it plain and blunt; he doesn't "rub it in."
>
> After the address, which is never long, the boys are called up by schools. Each boy is greeted by himself, but the judge uses only his given or nickname. "The boys from the Arapahoe Street School," he calls, and as the group come forward, the Judge reaches out and seizing one by the shoulder, pulls him up to him saying: "Skinny, you've been doing fine lately; had a crackerjack report

every time."[4] He opens the report. "And you have. That's great. Shake, Skin. You're all right, you are." Skinny shines.

Pointing at another, he says: "And you, Mumps, you got only 'fair' last time. What you got this time? You promised me 'excellent', and I know you've made good." He tears open the envelope. "Sure," he says. "You've done it. Bully for you." Turning to the room, he tells "the fellers" how Mumps began playing hookey, and was so weak he simply thought he couldn't stay in school. "He blamed the teacher; said she was down on him. She wasn't at all. He was just weak, Mumps was; had no backbone at all. But look at him now. He's bracing right up. You watch Mumps. He's the 'stuff' Mumps is. Aren't you Mumps? Teacher likes you now all right, doesn't she? Yes. And she tells me she does. Go on now and keep it up, Mumps. I believe in you."

"Why, Eddie," the Judge says, as another boy came up crying. "What are you crying for? Haven't you made good?"

"No sir," Eddie says, weeping the harder.

"Well, I told you I thought you'd better go to Golden. You don't want to go, eh? Get another job, you say? But you can't keep it, Eddie. You know you can't. Give you another chance? What's the use, Eddie? You'll lose it. The best thing for you, Eddie, is Golden. They'll help you up there, make you stick to things, just make you; and so you'll get strong."

Eddie swims in tears, and it seemed to me I'd have to give that boy "another chance," but the Judge who is called "easy," was not moved at all.[5] His mind was on the good of that boy; not on his own feelings, nor yet on the boy's. "You see," said he to me, "he is hysterical, abnormal. The discipline of Golden is just what he needs." And he turned to the room full of boys.

"Boys," he said, "I'm going to send Eddie up to Golden. He hasn't done wrong; not a thing. But he's weak. He and I have tried again and again to win out down here in the city, and he wants another trial. But I think a year or so at Golden will brace Eddie right up, and make him a strong manly fellow. He's not going up there to be punished. That isn't what Eddie needs and that isn't what Golden is for. Is it, fellers?"[6]

"No, sir," the room shouted.

"It would be unjust to punish Eddie, but Eddie understands that. Don't you, Eddie?"

"Yes sir, but" (blubbering), "Judge, I think if I only had one more show I could do all right."

"Eddie, you're wrong about that. I'm sure I'm right. I'm sure that after a year or two you'll be glad I sent you to the school. And I'll be up there in a few days to see you, Eddie, myself. What's more, I know some boys up there friends of mine, that'll help you Eddie; be friends to you. They won't like a kid that cries, but I'll tell them you need friends to strengthen you, and they'll stay with you."[7]

All forenoon this goes on, the boys coming up in groups to be treated each one by himself. He is known to the Court, well known, and the Judge, his personal friend, and the officers of the court and the spectators, his fellow-clubmen, all rejoice with him, if he is "making good," and if he is doing badly, they are sorry. And in that case, he may be invited to a private talk with the Judge, a talk mind you, which has no terrors for the boy, only comfort. They often seek such interviews voluntarily. They sneak into the Judge's chambers or call at his house to "snitch up" that they are not doing very well. And the boys who sit there and see this every two weeks; or hear all about it, they not only have forgotten all their fear of the law; they go to the Court now as to a friend, they and their friends. For Judge Lindsey had not been doing "kid justice to kids" very long before all Boyville knew it. The rumor spread like wildfire. The boys "snitched" on the Judge, "snitched on the square"; they told one another that the County Judge was all right. (Steffens, 1909, pp. 132–38)

The Commitment to Help

The report sessions were only one portion of Lindsey's therapeutic program. Lindsey put himself out personally and publicly for his charges. He worked to bring about reforms that were in their interest and in the interest of their families. He made himself available any time, even to the extent of interrupting an adult trial to take care of the business of a child. He was constantly in the neighborhoods and frequently visited the boys' homes. He took them out for meals, to shows, and brought them to his home. In 1915, Lindsey was fined $500 for contempt of court because he refused to reveal information that a boy involved in a sensational murder trial had given him in confidence (Personal Glimpses, 1915). In every way he could, he showed concern and a willingness to help, and he demonstrated the principle that the court was acting in the youth's interest.

Lindsey used the court's facilities in his work with youth. The annual report of the probation office for 1903 lists 3,139 interviews between the judge and the probationers (Lindsey, 1904). Moreover, the court administered 1,150 baths during the year, many of these on report day, in the courthouse basement; provided 252 jobs; provided summer employment for 77 boys; assisted 175 children in financial need; and supplied 395 items of clothing, both new and used, to children. Lindsey organized a dancing school so that young people would have a place to go to for fun. The court had an employment service that both delinquent and nondelinquent children used (Lindsey & Evans, 1925). Some boys worked in the sugarbeet fields under the supervision of a probation officer. Twenty boys were assigned to each group, and the groups met regularly with the judge and a probation worker to discuss their work. Judge Lindsey helped the boys see that their labor was important to the whole state. An undated political pamphlet, probably prepared in 1912, states that one of Lindsey's major achievements was the development of "a system of cooperative work between the school, the neighborhood, the home, the church, the businessman, and the court in the interest of the child offender."

The court was used to redress the youths' grievances. For example, at a newsboy's request, Lindsey wrote out an "injunction" to cease and desist for a policeman who had been chasing the newsboy from his favorite corner. The boy reported that the policeman actually made friends with him after the incident. A probation officer helped a child recover a bike stolen from him by an adult. In another instance the court intervened to help a youth who complained that his employer had unfairly deducted money from his salary. In still another case, Lindsey supported the leader of a street gang who fought a saloon keeper who refused to stop selling liquor to gang members. Lindsey had the saloon keeper arrested and jailed for fifteen days.

Many children came to him themselves to seek help. Often those who sought him out were adolescent girls in sexual difficulty. Lindsey arranged for treatment for children with venereal diseases without necessarily informing their parents or anyone else. His help for pregnant young women included arranging for discreet confinement and private adoption of their babies. And all this was done in the strictest confidence. Lindsey did not approve of premarital sexual relationships, but he observed that most of the young women who engaged in premarital sexual relations eventually grew up to be good wives and mothers. It was his feeling that more harm was done by public exposure in the court than by the actual sexual experience. There is no way to know how many women Lindsey helped through the years, for he kept no public record of these cases. In fact, when he left office in 1927, he openly burned his files and records to protect the confidence of untold thousands of children and adults who had trusted him. He burned the records because he did not trust the new chief probation officer, an alleged Ku Klux Klan member (Lindsey & Borough, 1931; Lindsey & Evans, 1925).

The Reduction of Fear

Not only young women voluntarily came to Lindsey for help. Lindsey shrewdly used this desire to confess, to maintain a state of grace, and the psychology of the group in his dealings with children. He worked hard to erase the element of fear from his relationship with a child so that the youngster would be willing to tell him the truth, relieve a bad conscience, and make amends. Lindsey was not soft, and he did not condone wrongdoing. But he did not prejudge and had an unwavering acceptance and respect for the humanness of the individual. He did not become indignant or punitive about sexual or aggressive transgressions or about stealing or lying. His attempt was to understand, to accept, and to point a way that would give the individual a greater sense of self-worth.

The following excerpt is from *The Dangerous Life* (Lindsey & Borough, 1931). The reader is cautioned that although it is written as a transcript, it is not a verbatim account. The dialogue was written by Rube Borough, who had observed the judge in action from a concealed vantage point in another case, and it very likely captures the flavor of what went on:

> I recall a type of old-time policeman trained alone in the ways of violence bringing a small boy into our court. As the officer came in he glared at the little fellow as though he wanted to throw a brick.
>
> First and foremost, it was evident that no love was lost between the two. The approach in both cases was one of hostility. There is such a thing as human artistry and the greatest of all human artistry, perhaps, is the artistry of approach.
>
> ''He is a bad kid'' the policeman blurted out, ''and there are fifteen or twenty more like him down there about those tracks that I haven't been able to catch because of this kid. For every time I come in sight he is on the lookout and when

he sees me coming he yells 'Jigger the Bull!' and everybody scoots and I can't ketch 'em.''

This last with an air of offended dignity.

''And when I got this kid, I asks him the names of the kids that run away and he says, 'I dunno 'em. I never saw 'em before'.''

And then as he glared at the boy: ''The little liar, he knows every one of 'em'.''

Of course there is no artistry in that method of approach. . . .

I have always made it a rule never to call a boy or girl a liar if it can be avoided—and it generally can. And yet, in the present case, I had to retain the boy's respect for the policeman as the representative of the law as best I could.[8]

''Jimmie,'' I said with such warmth of smile and attitude as I was able to command, ''I am sure you don't understand the officer and he perhaps doesn't understand you. If you did I don't think you would have run away from him and I doubt if he would think that you were really a liar.

''Now, I don't think you are a liar but I do think that you are afraid. The best boy may be afraid when he doesn't understand things. Then when he is afraid he may say things that are not what we call true.

''But that isn't because he wants to lie. It's because he's afraid and thus he doesn't know what else to do. Now don't you think that is true in this case.?''

And presently our little prisoner half smiles through his tears and becomes as garrulous as he has been dumb under the menacing glances of his captor.

''It's jest like ye said, jedge. I ain't no liar and I don't mean to do nothing wrong but I'm just plumb scared.''

''Oh yes,'' I tell him, ''I knew you would admit that. You thought you were going to get in jail or the reform school, didn't you? Well, you are not going to get in any such place. You are just going to tell me the truth, for I know you are a truthful boy. We are going to help the policeman and the neighborhood and everybody.''[9]

''Sure,'' he says, ''I'll tell ye how it was Jedge, I live down there by the railroad tracks, I do, where those kids live. And they said there was watermelons in those boxcars but we didn't find no watermelons.

''But one of de kids sez he bets there is sumpin' good in those boxes 'cause it's got sumpin' on 'em about figs and we think it is figs.

''So we gets open a box and finds a lot of bottles with sumpin' on 'em about figs and we thinks it is sumpin' good. So we drinks the whole bottle.'' . . .

And then pausing, he blurts out through his tears: ''And I thinks we've done suffered enough.''

''Well,'' I said, ''I think so, too, for fig syrup is not recommended in bottle doses.''

The policeman doesn't see the humor in this episode that spread about my room and brought titters from the few spectators. Instead he proceeds to gloat over the boy's confession as evidence of the truth of his charge that the boy was a liar, instead of just a frightened kid.

''I told you he was a liar, Judge,'' he exploded as he turned on the tearful youngster. ''Now you know you told me you didn't know anything about any of those kids and you didn't even know their names, but now you tell the judge that you knew them all.

''Now you tell the judge their names—I want their names, see!''

And the little prisoner, hesitating, appeals from the policeman to the judge, where somehow he thinks he finds sympathy and understanding, not for his sin but for himself. (For we are dealing much more with him, with what he is than what he did and perhaps why he did it.)

I see the tears come to his eyes again, but he is backed now by a certain confidence as he stands there pleading his own case.

"Do yuh tink, Jedge, that it's square for a guy to snitch on a kid?" he asks.

In some other age of Boyville, if you told or tattled on a boy, you "squealed" and a "squealer" was ever outlawed by the gang. But now if you told on a boy, in the slang or venacular of the gang, you "snitched." And that was against the law. Not our law, but theirs. The first commandment of the gang was—and I suspect ever has been and ever will be "Thou shalt not snitch (tattle)—you will get your face smashed if you do." The human quality of loyalty was involved here.

And unless I had sympathy and understanding for their law, how could I expect them to have sympathy and understanding for mine.

But did this particular type of policeman have any respect for the fact that he was dealing with two worlds and the people that inhabited them and the laws that governed them?

And did he understand that unless he had sympathy and understanding for one he could not expect, much less exact, respect for the other?

The child world and its laws were just as real to Jimmie and as much to be observed and respected as our laws; for they were based on the finest of human qualities, loyalty and respect for each other, however much sometimes through lack of maturity they seem to us to be misdirected. If they are misdirected, the least that can be expected of us is wise direction and not hostility. . . . [10]

But to return to our little prisoner—

"Of course, I don't think you should snitch on a kid," I said, "I never asked a boy to snitch on another."

My well-meaning friend, the policeman, was unable to restrain at least a mild resentment as he noticed the triumphant look in the face of the little prisoner whose plea had won the approval of the court.

"Then how do you expect me to get them kids," the policeman inquired. "I been two weeks trying to ketch 'em and now you are going to stand by that kid in his refusal to tell me their names."

"Oh, don't worry about that," I responded. "Jimmie will do more than give us their names. He will bring the boys to court. I am perfectly safe in discharging you from the case with the assurance that we will soon dispose of the lawlessness you have a right to complain about. Isn't that so, Jimmie?"

And there came a curious beam of happy assent from his dirty, grimy little face.

"Sure," he said, through a smile and drying tears. "You betcher life I'll help yuh to stop it, Jedge, if yuh jes' tell me what to do."

"Oh that is easy, Jimmie," I explained. "You go down by the tracks and tell that gang that I have asked Officer Blank to lay off the gang, that I want them to come and see me. I know they will cut it out. After school, tomorrow, you bring them here to see me. They will tell on themselves and you won't have to tell on anybody."

"Sure, I will," said Jimmie, "if I can git 'em to come."

"Well you tell them, and, whatever happens, you just report to me."

And with one final effort to restore peace between the little prisoner and the irate officer—which eventually was to succeed—I was to behold Jimmie leaving the court in triumph, rather as its agent than its prisoner.[11]

But the next day after school when Jimmie appeared in my chambers he was alone. He was crestfallen, if not discouraged.

''Well Jimmie,'' I said, ''what's the news from the front?''

But I was not prepared for the humor of the situation. ''It's like this Jedge,'' he said. ''I tolt those kids you tolt me to come down and tell 'em to come and see you. Bud dey wouldn't nobody come. One kid said if his mudder found out he had been in dis trouble, he would get a good lickin'. 'Nother said if the cop was goin' to ketch him, he would have to ketch him. And a big guy said I snitched on him and he was goin' to beat me up when I never snitched at all.''

Noting the seriousness of his last phase of his experience, I hastened to assure him.

''Of course you didn't snitch, Jimmie,'' I explained. ''The cop said there were about 15 or 20 boys in that trouble and we thought it was square to tell how many without telling any names and you said there were only seven. Now, I'll tell you, Jimmie. I think the trouble is that they don't understand. You see, you came back. Why? Because you understood and you are not afraid. They did not come because they did not understand and they were afraid.''

''Yes,'' he hastened to inject, ''and dey said dey didn't believe that yuh tolt me what you tolt me.''

''Oh, I see, you perhaps need some evidence that is more substantial than what you say you told them.''

''Sure,'' he shot back, ''Dhat's what I needs. Will you write 'em a note?''

''Sure,'' I answered, ''I'll write them a note. What will I say?''

Jimmie edged up on the table while I took his dictation.

''Tell 'em,'' he said, ''dhat no kid has snitched.''

Now self-preservation is the first law of nature and I sympathized with his precaution. Why shouldn't I? Hadn't I been a kid once myself and who has a right to deal with kids who doesn't remember the days of his childhood?[12]

''And tell 'em,'' Jimmie went on, as he gained more confidence from this response of sympathy, ''if dhey will come to court and promise dhey will cut it out and never do it again yuh won't send 'em up this time.''

Consciously or unconsciously, he knew they were afraid and fear is the father of lies, the promotor of lack of confidence between parent and child and cold gulfs that follow and the war of the gang with the police and the war of the nations with each other. For the child's case is the case of all humanity and the cause of evil everywhere. So I wrote the note to the unknown gang.

''Dear Boys,'' it read, ''no kid has snitched. If you will come to the court and promise to cut it out and never do it again I won't send you up this time. Signed, Judge of the Juvenile Court.''

That was the warrant and Little Jimmie was the sheriff, but he didn't know it.

But our little prisoner, now our assistant departed right proudly with a warrant of his own dictation. . . .

And I was to find that Jimmie's warrant was so much more effective than that which came down to us from the wigs of the law, even though they were Blackstone, Kent and others. . . .

The next day at the time set for the return of my little friend an officer came in to announce that there was a delegation from the fourth ward waiting outside to see me.

Presently there was ushered in little Jimmie with fourteen as typical ragmuffins from down by the railroad tracks as in my slumming days I had ever gazed upon.

Jimmie was proud and smiling even if the balance of the gang showed timidity—though not without hope, for they trusted Jimmie.

"Well," I said, "Jimmie, you didn't snitch on anybody."

I was looking to his world and the protection he needed there. And it did not go without appreciation as the face of the former little prisoner lighted up and there was a swelling of his chest as he gazed upon the expectant gang.[13]

"But," I continued, "since the policeman said there were about 20 kids in this trouble you thought it would be square to tell how many there were in it really. And you said there were only seven. But you didn't tell the names of any. But, Jimmie, you got fourteen here—how do you account for that?"

"Oh," he answered, "there was only seven of us in it but the rest of dhese kids dhey runs wit' us, dhey does. And dhey got interested in that note of yours and dhey wanted to read it and I let 'em and dhey said dhey would like to come, too. And I said I didn't tink you'd care."

And it was very evident that I did not care before that conference was finished. We not only had the seven who were in it, but the seven who were not in it because they had not had a chance to get in it.

And then we had a snitching bee, which means that everybody tells on himself and not the other fellow. And since everybody agreed to that—for a sense of real justice is very pronounced in childhood—if you don't keep your word to tell on yourself after you promised to, then you give anybody leave to tell if he is asked. And while that seldom becomes necessary, nevertheless, if it ever does, it is snitching on the square, and you don't get your face smashed for that.

Of course, there was not anything that that gang had ever done that they should not have done that I did not know all about before that conference was finished. All of which proved they were not liars but very truthful little citizens.[14]

We had simply laid the spell of fear and whether in court or home when that is lifted the truth emerges triumphant. Moreover and again, I learned that fear is the father of lies.

And of course, they would cut it out, as indeed they did cut it out—as 99 out of 100 such little citizens did and do when thus approached. (Lindsey & Borough, 1931, pp. 120–31)

The Power of Trust

One of Lindsey's most remarkable achievements was his success in sending hundreds of boys and young men to the state industrial school and to prison by themselves, without escort and without losing a single prisoner.[15] Lindsey began sending boys to reform school by themselves very early in his career. Lindsey's 1904 report indicates that eighteen children had been sent to the state industrial school in their own custody. Lindsey's action had significance as both a reform of police practices and a therapeutic tactic. Lindsey believed that an individual who committed an offense was not evil but merely weak. He also believed that each individual had a potential for growth and change that could be reached.

> For I had come to believe that down in every soul, notwithstanding the effects of bad environment and the defects of heredity, there was what I then chose to call the Image of God and it was up to me to bring it out. And my way of attempting to bring it out was, through personal contact, to eliminate hate and fear, instill trust, arouse the desire to please in proving there was one to be pleased, appeal to pride, loyalty, honor. (Lindsey & Borough, 1931, p. 188)

In sending youth to the industrial school in their own custody, Lindsey was appealing to this inner strength. If a youth delivered himself, entered the reform school by virtue of his own will, and proved to himself he could overcome the strongest possible temptation, then he had already performed an act that made his time in the reform school a meaningful part of his rehabilitation. In sending a youth to reform school, Lindsey would discuss the reasons with him and help the boy see that a term in reform school was the only possible course. As indicated, this was often done in public, in court, but it was also done in private conferences with the youngsters.

In the following excerpt from his first case, Lindsey gives us a good view of his approach and of his indomitable spirit:

> I began to formulate in my own mind an entirely new strategy in dealing with delinquent youth. . . .
>
> If we had to send them to places of artificial restraint, even jails and reformatories, why not send them alone, I asked myself. What stronger proof could I offer to the world of the power of those inner restraints, those forces other than violence, in which I had come to believe?
>
> I discussed the idea with some well meaning friends but they immediately vetoed it as dangerous. Nevertheless, I was determined to put it to the test. And because that seemed to be the least difficult (I afterward found this a mistake), I decided to start with the smaller boys who had to be committed to the detention schools, and the then so-called reform schools.
>
> The very first "culprit" upon whom I tried the experiment, I remember, was a little fellow about 12 years of age. The institution to which he was being sent was located almost on the outskirts of the city. I was sure, I told him before he set forth, that he could be trusted to go alone and "on his honor."
>
> He ran away. The circumstances could not be concealed from the officer who had brought the boy to court in the first place, and there was much merriment around police headquarters at my expense.
>
> I remember the officer said to me, after he had recovered from his gloat of satisfaction over the failure of my experiment:
>
> "I'll catch that little devil and I'll see that he gets what's coming to him—if I have to break his damned neck. You bet I will—no matter what it costs him."
>
> It was not hard to understand the officer's attitude. The youth had been considered an eely little rascal and had been the source of much annoyance to the police through a rather serious type of juvenile burglaries.
>
> But I could not approve of the policeman's program. I thought for a moment and, still fighting for the truth there was in the system I had conceived, I said:

"No, you are wrong. Perhaps you will have to catch him this time. But bring him back to me and *I'll* teach him how to be trusted no matter what it costs *me*.

When the little rascal was recovered and brought back to my court by the officer, I found he was fired with hate, not for me but for the officer. His attitude toward me, at first, was one of sham and pretense.

I was learning from the boy—again and again in those days I found that the boys unconsciously taught me more than I could teach them and I often wondered why the lessons escaped most of the officers I met. Why couldn't they see how violence projected violence, hate projected hate?

. . .

"Jimmy," I said, "I am very sorry for you. The officer is sure you can't be trusted because you ran away . . . but I think you can. I know you ran away because of the bad things you let get the best of you."

And then I made it plain to Jimmy that however much I liked him I had been compelled to send him to the detention school—a mild name for the "kids' jail," as it was known to Jimmy and his harum-scarum young friends.

He was penitent and a little later when he was alone with me the explanation came out through his tears.

"You see Judge," he said, "I was going to do just what you tolt me but on the way out there I passed that big field where the kids was playing ball. You know dhat's where our gang lives and, Gee whizz, I just couldn't pass dhe kids, I just couldn't, Judge. And before I knowed it I was up on my feet and I had ditched that street car. And then, when I remembered what you tolt me, I just got scared and I run home. And then, when the old man came home, I was under the bed scared to death, knowing he'd beat me up if he found me.

"I knowed I had missed my chance but I started for the street again when that fly cop comes along and pinches me on the spot. Gee, I thinks, what a fool I has been to t'row down dhe judge![16]

"The cop gives me a cussing and tells me he'll fix me dhis time. So he t'rows me in before I knows it and then I sees I ain't got no yell."

"I am glad you learned that lesson for yourself, Jimmy," I told him, "You think the cop wants to hurt you. But I hope you know I really believe that you can be trusted—I think you were just scared."

And, as in so many other cases, with this approach of tolerance and understanding, of confidence in the good rather than the evil there was in Jimmy, he rose magnificently to the occasion.

"Gee, Judge," he begged, "can't you give me another chance? Just give me the writ and I'll show 'em—I'll show 'em, Judge. I'll show that cop I can be trusted."[17]

"Of course you will, Jimmy," I answered, "I know you will."

How little I knew that it was the beginning of great events in a work that was to be known all over the world. Consciously or subconsciously, I must have been a bit apprehensive in those early days. For I stood at the window there alone in my chambers. I saw little Jimmy descend the courthouse steps—he too, alone—with his commitment papers in his own pocket—as he emerged from the great building out into the streets of the crowded city to join me in the great adventure of the "Dangerous Life."

True to his word, he braved all the temptation. He arrived alone. His fears had been conquered, his pride had triumphed. The papers were returned with "executed" in the small handwriting of little Jimmy: "Dear Judge, I am here—

> and I comes all by myself. Be sure and tell that cop." (Lindsey & Borough, 1931, pp. 166–70)[18]

It would be a mistake to believe that Lindsey relied only on some sort of personal hypnotism to achieve his results. One time the police brought a chronic runaway into court and called the newspapers to show up the "fool judge." Lindsey still sent the boy to Golden, enlisting his aid in fooling both the police and the reporters. When the boy got to Golden, Lindsey heard that others, both grown-ups and children, had called him a "chump." Lindsey then went to Golden and told the boy's story before an assemblage of inmates and staff, praising the boy for his courage, loyalty, and self-control. Lindsey repeated this speech on innumerable occasions at Golden, in court, and in newspaper and magazine articles. He succeeded in changing the norm so that

> to "ditch your paper and run," is a disgrace in Boyville now. A boy called on the Judge one day with an offer from the gang to "lick" any kid that ditched his papers or in any other way went back on the Judge, and the Judge had some difficulty in explaining why that wasn't "square." (Steffens, 1909, p. 174)

Lindsey saw to it that the very act of going to reform school became part of the culture of the group he was trying to help. His appeal to the "spark of the divine" was made in the context of helping the individual meet the norms of his community and in the context of their interdependence. Lindsey did not work with the individual alone but with the entire community of which the boy or girl, the judge, and the court were a part.

The Institution of Human Relations

Lindsey directed some of his later work to strengthening his family relations court. Finding in many instances that the law protected property rights but did not protect human rights, Lindsey succeeded in winning a number of reforms to protect the integrity of the family in cases of nonsupport or when the parents became involved with criminal matters. Lindsey attempted to establish informal procedures in a family court that could settle cases with a minimum of legal expense to the family and a minimum use of jail penalties to enforce support orders, although he did fight for a law to make nonsupport a felony in Colorado. He also attempted to resolve family disputes and to mediate in cases in which divorce was contemplated. Frequently he met with the contending parties in his chambers, dispensing with the services of lawyers, an action that won him the enmity of some members of the bar.

Lindsey actually was campaigning for a new social institution, a house of human welfare, or an institution of human relations that would take over some of the functions of the civil, criminal, and divorce courts. He had in mind an institution to deal with illegitimacy, mental or physical disease, divorce, nonsupport, desertion, statutory rape, and similar social problems. The institution would be staffed by experts in problems of mental hygiene and by lawyers with backgrounds in biology, sociology, and psychology. It would be able to call on the schools and social agencies as needed and would handle all such cases, even having the power to grant divorces when indicated.[19]

As part of his concern for people, Lindsey also favored the repeal of anti–birth control information laws. A supporter of Margeret Sanger, he had learned through his own work that sexual ignorance created many of the problems he encountered. He thus supported a bill to enable school boards to provide courses in mental hygiene, child rearing, marital relationships, and sex. He believed such education should be mandatory for adults as well as children. His concept of prevention and treatment provides an interesting service model that has not yet been implemented. (Lindsey, 1925b; Lindsey & Borough, 1931; Lindsey & Evans, 1925).

Changes in Denver's Juvenile Court

From 1901 until the election of 1924, Lindsey was able to win election or appointment to the bench. In 1924, he won his office by only three hundred votes. In 1927, the Colorado Supreme Court found fraudulent election practices that were sufficiently serious to warrant throwing out the results of one precinct. Lindsey was then removed from the bench. He claimed that the supreme court was filled with political enemies, but even if he were right, it does not account for the very close election, when in former times he had won by massive pluralities or had been appointed to office through the force of public sentiment.

Lindsey was a man of the reform era, but the 1920s was a period of great political conservatism. In 1924, the Ku Klux Klan was at the height of its influence in Colorado as it was in a number of states. The total number of KKK members was estimated to be 4.5 million (Allen, 1959, p. 47). In Denver and throughout the state, flaming crosses were commonplace. Blacks, Jews, and Catholics were terrorized. KKK members were elected to school boards and made their way into the courts. Even a mayor of Denver and a Colorado state senator were reputed to be members. Lindsey writes of armed Klansmen entering his court while he was trying a Klan leader. Characteristically, in the 1924 campaign, Lindsey spoke out against the Klan and defended the persecuted minorities.

Lindsey also continued to speak out for his concept of companionate marriage. In urging the repeal of anti–birth control information laws and the passage of more realistic divorce laws, Lindsey won the opposition of Catholic clergy, who attacked him en masse from their pulpits. Moreover, his enemies, not satisfied with ousting him, urged that his juvenile court also be abolished. The court was not abolished, but in Lindsey's place a judge was appointed who evidently did not follow his policies. Lindsey tells of his feelings on leaving the court and at the same time gives us some insight into some of the reasons that he could not continue his work.

> In the days following the court ouster I walked the streets of Denver in a daze. At night I passed the court house, where for twenty-eight years I had carried on my work, looking up at its darkened rooms with unutterable yearning.

> In these rooms I had brought back to decency and citizenship thousands of boys and girls, saved from destruction thousands of homes, written dozens of items of legislation that, after years of struggle, had become part of the established laws not only in my own state but in many other states and in foreign countries.
>
> And as I looked I saw men, women, and children, with their deep, dark secrets and their wretchedness, again streaming up the stairs to those rooms. I know their desperate need of sympathy and understanding. . . . I heard the cry of those people for the bread of life, the artistry of love, the glowing technique of service, answered by the same cold formalism that I had driven from the court a quarter of a century before. (Lindsey & Borough, 1931, pp. 10–11)

Lindsey's institution did not survive him in the form he created, and it is important to ask why. It is clear that his loss of popularity and political defeat were part of a conservative spirit that made it all but impossible for reformers to pursue any meaningful program. Lindsey was an outspoken reformer, and his proposals for companionate marriage and an institution of human welfare were proposals for reforms in two of the basic institutions of society, the courts and marriage. The temper of the times, however, no longer permitted taking liberties with basic social institutions.

The "Routinization of Charisma"

The concept of a juvenile court spread throughout the nation and throughout the world (Abbot, 1925, pp. 267–73), but Lindsey's clinical contributions had been lost. His life's work, as he so poignantly observed, "turned into a mockery."

To some extent, Lindsey was the triumph and the failure of charisma. He could in his own person gain the affection and sympathy of thousands of youths, but he was unable to institutionalize his own practice. He left neither a trained successor nor a corps of dedicated followers who would use his approach to rehabilitation through the courts. When Lindsey left the Colorado bench, his probation officers, who were his appointees, also left the court. The system did not permit him to perpetuate his practices.

His legal works gained widespread recognition, but he left no comparable body of technical and theoretical writings that could become the basis for institutionalizing his practices. The Saturday morning report sessions for example, are described only in the 1904 report and are not mentioned to any significant extent in his 1931 book. Most of the descriptions of his work appear in popular sources, and as far as we have been able to tell, his clinical work was never regarded with any degree of seriousness by mental health professionals. Their professionalism and emphasis on psychoanalytic theory and psychotherapeutic practice may have been one reason that Lindsey's innovations were ignored. In as important a survey of clinical practices in the United States as *The Child in America* (Thomas & Thomas, 1928), Lindsey is mentioned only because he organized classes in which widowed mothers receiving pensions could discuss their children's problems.

In part, Lindsey's personality may have been responsible for his failing to win adherents. He approached many problems with a head-on assault guaranteed to create opposition. His behavior seems to reflect characteristics of his father and his grandfather, both men of integrity who did not fear standing up for what they believed. From his early days in school, Lindsey did not hesitate to speak out. Perhaps his being a Catholic in a Baptist community led to his particular style. Lindsey's upper-class background and his personal experience of poverty and hardship may have given him the range of experience that enabled him to relate to both the very poor and those in positions of wealth and power. In some respects he was reminiscent of an Old Testament prophet who spoke out against wickedness and hypocrisy and found that his message brought him only vituperation and calumny.

In some respects, however, Lindsey seemed to be an impulsive, irascible character given to high-handed actions. A newspaper carried a story of another court reversing one of his decisions in a paternity case. The judge in the appeals court felt that Lindsey had acted to promote a wedding on very uncertain grounds. In another instance, Lindsey physically attacked a man who had falsely accused him of improprieties with children. The attack took place in court during the trial of a libel suit that Lindsey won. Lindsey seemed to enjoy being the center of turmoil, and sometimes, as in an episode in which he was ejected from a church for shouting at a bishop, people questioned his lack of dignity, if not judgment.

It is true that he stirred strong feelings in people. When we visited Denver in 1966, one old-timer cursed when we inquired about Lindsey. A respected professional who knew him well said that he thought the judge had been "a little crazy." Other people remembered some of his rehabilitation schemes as half-baked and told of instances in which his efforts to use people in the community to help children backfired and ended up hurting the people involved. The same man who could write legal history could also endorse a safety razor in magazine ads and have his mother appear in widow's weeds to talk for him at a political rally that threatened to go badly. Lindsey had a complex personality that attracted some people and repelled others. To what extent his individual style, like Witmer's, caused people to take him less seriously than they might have otherwise is an open question.

Lindsey lived to see the outlines of his concept preserved in institutional form, but he could not guarantee that he himself would be reproduced. Too often in the process of institutionalizing an innovation, in "routinizing charisma," as Max Weber (1964) put it, positions in an institution are filled by those who lack the appropriate human qualities, or the positions themselves are so bureaucratized that initiative and flexibility are lost. The experiences in Chicago and Denver suggest that although the innovators were aware that certain human qualities were essential to work with youths in the juvenile court, they were powerless to guarantee that people with the necessary qualities would be attracted to the work and, even if attracted, would be permitted to function as their instincts dictated.

The activities at the heart of Lindsey's work, his involvement with the community, and his effort to make the court the agent of the youths it served would today be denied to publicly supported probation workers. Certainly, reform would never become the concern of judges who rotate through a low-prestige institution

(Polier, 1964). In the last forty years, one would be hard pressed to find examples of juvenile court personnel who challenged slumlords, who fought for recreational facilities, who attacked inequities in job opportunities, who pressed for better school facilities, who fought for the rights of families on welfare, or who attacked the political failure to provide adequate resources for dealing with human problems.

Today's mental health professionals, psychiatrists, social workers, and clinical psychologists attached to the court do not view it as part of their professional function to change the situation. The present-day professional seeks to help individuals change. The professional, leaning toward psychopathology, may not believe that each person contains a ''spark of the divine'' and so does not see the necessity of helping create the social conditions under which that person can grow. The mental health professional operates within a treatment orientation that separates both patient and professional from involvement with the broader society of which they are a part. It is paradoxical that any mental health professional who acted in the tradition of a Jane Addams or a Ben Lindsey would be viewed as acting in an untraditional manner—so far have we as mental health professionals forgotten our origins.

9

The Child Guidance Clinic: A Product of the 1920s

The community child guidance clinic was the next major service that was introduced after the juvenile court. It began after World War I, during a time when reform sentiment was dead. The child guidance clinic provides a contrast with the services established over the previous thirty years.

In 1921, the Commonwealth Fund launched a bold program for the prevention of juvenile delinquency. Demonstration child guidance clinics were opened in a number of cities, and private-sector community child guidance clinics grew out of these models. The child guidance clinic, as originally conceptualized in the Commonwealth Fund's program, was meant to be a vital force in the community, closely interrelated with other agencies and the schools and designed to influence them. In today's terms the Commonwealth program would be a community mental health program. The clinics rather quickly adopted their modern form. From agencies with a mission to modify aspects of the community and to provide a preventive service, the clinics became treatment agencies, dealing with individuals seeking help. This change from conception to practice took place in a social and professional context, and an examination of the process of change should illuminate the forces shaping mental health practices.

The 1920s

The child guidance clinic took shape in the ten years following World War I that have come to be known as the Roaring Twenties. World War I had focused America's attention on an external enemy. The problems of the cities, immigrants, poverty, social welfare, and social reform, all of which concerned the previous generation, were put aside as the nation dedicated itself to making the world ''safe for democracy.'' The reform movement never regained momentum after the war, for the mood of the nation had changed, as had the problems that occupied its attention (Allen, 1959; Goldman, 1956; Schlesinger, 1957). Despite some fairly radical shifts in the role of women and changes in sexual mores, the decade of the 1920s was basically conservative, both politically and socially. It was the age of Babbitt.

From 1920 until the Great Depression, liberals and social reformers had no power or influence. World War I promoted a belief in the rightness of the American way of life and fostered an almost fanatic demand for 100 percent Americanism. The phrase *splendid isolation* and the rejection of the League of Nations characterize the feeling of the times. The Bolshevik revolution frightened the nation, and the reaction against liberalism and foreigners was pronounced. New Year's Day, 1920, saw the Palmer raids in which six thousand "reds" were arrested. Professors suspected of radicalism were threatened with dismissal; school teachers were forced to sign loyalty oaths; and thus those with unorthodox political, social, or economic ideas learned to keep quiet. Before and during the war many social workers had flirted with socialistic concepts and had been active in pacifist and antiwar movements. When patriotic societies sprang up after the war to point out the menaces to America, liberal civic and welfare agencies therefore were favorite targets (Duffus, 1938; Holden, 1922; Linn, 1935). This oppressive atmosphere led to an unrelieved conformity of ideas that stifled liberal reform.

The demand for 100 percent Americanism was manifested in a series of restrictive immigration bills. In 1921, in 1924, and again in 1927, Congress passed laws making it increasingly difficult for immigrants from southern and eastern Europe to enter this country. National, racial, and religious prejudices and intolerance were aroused again. Some of the most vicious race riots occurred in the early 1920s, and the Ku Klux Klan found fertile soil for its growth. By 1924, its membership was estimated at 4.5 million; its political power was spread throughout the country, not only in the South, for its terror touched even New York City. To be sure, despite the business prosperity, there was still poverty with its attendant disorganization and misery, but it received little attention. Even the labor unions had taken a conservative turn (Bernstein, 1966). In such a social context, the major portion of the public listened to and supported only those who asserted that this was indeed the best of all possible worlds, and it looked with suspicion and hostility on those who thought differently. It was in this climate that the radicals Sacco and Vanzetti could be tried, executed, and forgotten.

The decade of the 1920s was the businessman's era. Beginning with President Warren Harding and continuing with Presidents Calvin Coolidge and Herbert Hoover, conservative business interests found strong support in the White House. Harding called for a return to "normalcy"; Coolidge believed that "the chief business of the American people is business." The courts, the regulatory agencies, and the legislatures all supported business, and indeed there was an almost unprecedented period of sustained economic growth and prosperity. The businessman was revered, and the millionaire was an idol and an ideal. Throughout the decade it was the individual hero, a Charles Lindbergh, a Babe Ruth, a movie star, or even an Al Capone who could capture the public's attention and admiration. The tenets of conservative Darwinism had risen again, and keen, sometimes cutthroat competition was the order of the day. If a person failed in the prosperous world of the 1920s, it was because of a lack of moral fiber. The emphasis on self-help, on the individual's achieving success as a matter of will, was symbolized in Emil Coué's aphorism "Day by day, in every way, I am getting better and better." Coué, as popular as mah jongg and the crossword puzzle, captivated audiences across the nation.

Although pressures for social and economic reform were minimal in the 1920s, reform in "manners and morals" (Allen, 1959), that is, reform in personal codes and personal values, was pronounced. Such changes were in keeping with the emphasis on self-development. Judge Ben Lindsey could talk openly about the sexual needs of youth, sex education, birth control, and companionate marriage. And it was during this period that Margaret Sanger, a proponent of birth control and sexual development, achieved great prominence and established her birth control clinics.

When the Nineteenth Amendment to the Constitution was ratified in 1920, giving women the right to vote, it was a major event, but only one manifestation of the progress that women were making in their long battle to achieve equal status with men. The role of women was undergoing a drastic change. During World War I their labor was solicited, and many continued to work even after "the boys came marching home." Appliances and canned and processed foods began to free women somewhat from kitchen drudgery; new clothing styles freed them for physical activity; and higher levels of education opened their minds to new opportunities. Concomitant with the increasing independence of women and the prosperous times was a sharp rise in the divorce rate.

The virtues of premarital chastity and marital fidelity were open to question as never before; some adulterous affairs were entered openly as a matter of living out a modern value. Lindsey's proposals (Lindsey & Evans, 1925) for companionate marriage were timely, but the violent reaction to formal proposals for a change in the marital institution suggests a strong underlying anxiety about the changes that were occurring.

The erosion of the sexual double standard was manifested by women, especially young women who flung off their corsets, put on short skirts, used cosmetics openly, and danced boldly to the wild jazz that was served with the bad liquor they drank in the speakeasies. Frank discussion of sex and free love were the order of the day. Sex movies, sex magazines, and confession magazines reached millions. Automobiles provided the young with traveling bedrooms free from parental surveillance. Prohibition did not prohibit but only added spice to the adventure of crossing forbidden limits (Allen, 1959; Lindsey & Evans, 1925; Lynd & Lynd, 1929).

The 1920s also was the decade in which psychoanalysis took hold in the United States, and psychoanalytic thinking offered an intellectual, scientific rationale for the social changes. Popular psychoanalysis, stressing the free expression of sexual impulse and the danger of "frustration," filled magazines and books and was a principal topic of conversation in some circles. It was said that almost everyone who was anyone in Greenwich Village was in analysis. Psychoanalysis as a modern science was eagerly accepted by those for whom the older religious and moral symbols had grown stale, although as a discipline, it was not yet entrenched in the medical schools (Brill, 1939; Burgess, 1939).

Given the marked changes in manners and morals, it was no wonder that parents felt uncertain about what was right. To add to their uncertainty, popular magazines, Sunday supplements, and even radio, reaching millions, brought word of the change and advice about how to live in the modern world. The advice was as conflicting then as it is now. Popular Sigmund Freud, popular John Watson (Watson, 1928), and popular Arnold Gesell (Gesell, 1928) were widely disseminated. Don't frustrate your child, train the child to be what you want; the child is in a stage of life and will grow out of it; breast-feed, don't breast-feed; punish, don't punish; sex education, no sex education; your child will grow up to be neurotic. And of course, tradition was not without its defenders. The concept of the "problem child"—an essentially normal child with difficulties—prevailed in the 1920s as well (Horn, 1989)). It is not surprising that the child guidance clinics, as specialized facilities offering expertise in an ambiguous area, met popular needs and flourished. Indeed, Cross (1934) described child guidance as one of the "well defined social movements of our time."

The 1920s, then, were a period of social and political conservatism, in which the problems stemming from the societal order were neglected or ignored. It was also a period in which the apparent opportunities for the individual were great if the individual were able to play the game. If one did not succeed in life, it was one's own fault. The attack on certain traditions and the shift in sexual attitudes and in the female role created a general uncertainty and caused people to seek new values. Science was the new god, and psychiatrists, psychologists, psychoanalysts, and social workers were its disciples. In this social context, with its emphasis on the individual, which was consistent with the psychoanalytic view of the person, the child guidance clinics were born.

The Commonwealth Fund's Program for the Prevention of Delinquency

The child guidance movement grew rapidly in the 1920s. The program initiated by the Commonwealth Fund expanded from 7 clinics in 1921, with the characteristic child guidance structure, to 102 by 1927. (Lowrey & Smith, 1933; Stevenson & Smith, 1934). There was similar growth in the number of state-supported clinics during the same period (Witmer, 1940). The Commonwealth Fund clinics were highly influential in shaping the direction of that growth. The pattern of the privately supported community child guidance clinic with the now-traditional team consisting of psychiatrist, social worker, and psychologist was set by the experience of the demonstration clinics. Two of the agencies established by the Commonwealth Fund, the Bureau of Children's Guidance (Lee & Kenworthy, 1929) and the Institute for Child Guidance (Lowrey & Smith, 1933), were the principal training centers in child psychiatry and psychiatric social work and, to a lesser extent, in clinical psychology. In addition, some of the other demonstration clinics served as training agencies from their inception. Thus, the Commonwealth Fund's program not only presented model clinics, but as training centers, the clinics helped mold thinking and practice throughout the country (Horn, 1989).

David Levy, one of the pioneers of American child psychiatry, commented in the early 1950s:

> The name[1] and the formula of "child guidance" were stamped firmly on the history of psychiatry by the philanthropic and administrative support of the Commonwealth Fund. A five-year period of demonstration clinics throughout the country was, in a way, one of the most successful enterprises in the field of philanthropy when judged by a criterion used by a number of philanthropic foundations, that a project proves its worth when it is taken over and supported by the community. In that sense, the demonstration clinics were most successful. Since they were first launched, about 30 years ago, over 200 child guidance clinics have been established and supported by local agencies in these United States. (Levy, 1952)

Because of the great influence of the Commonwealth Fund clinics, it is useful to focus on their development rather than on other contemporary clinics. Although the Commonwealth Fund obviously did not establish or dictate the pattern of service in all of the clinics, nonetheless the experience of the Commonwealth Fund was important.

The Commonwealth Fund was established in 1918 by Mrs. Stephen V. Harkness "to do something for the welfare of mankind" (Commonwealth Fund, 1963). The fund reflected the ideals and practices of the previous twenty-five years. In the latter part of the nineteenth century, men who made great fortunes from the massive industrial growth of the post–Civil War period had a humanistic concern that was manifested in their lavish contributions to social betterment. Religious leaders (Abell, 1943; Hopkins, 1940) also decried the business ethics of the period and urged a more Christian approach to human problems. Moreover, there was pressure for the relief of social problems and talk of reducing inequality through the redistribution of income. Even the shocking notion of an income tax had begun to be discussed. The Beards (1927) speculated about other motives that guided the philanthropy of the day, but beyond these, philanthropy after the 1890s turned toward preventing human misery by attacking its basic social causes. The focus of attack shifted from individual charity to massive social work, but not in the casework sense. Rather, it was in the sense of contributing to an assault on sickness and misery by supporting the humanities and the sciences and by stimulating community planning to solve the problems.

It was in such a tradition that the Commonwealth Fund initiated its first major undertaking in the field of health and welfare, the Program for the Prevention of Juvenile Delinquency, in 1921. The name and the intent to focus on delinquency reveal in the program's planning the premise that the social climate of the immediate past, with its emphasis on broad social reform, would continue in the immediate future. Some who planned the program were interested in promoting the medical model of delinquency, however, and their interests dominated practice in the decade to come (Horn, 1989).

The Commonwealth Program was a cooperative effort of the New York School of Social Work, the National Committee for Mental Hygiene's Division on the Prevention of Delinquency, and the National Committee on Visiting Teachers. In addition, the Joint Committee on Methods of Preventing Delinquency was organized specifically to act both as coordinator and publicist for this program.

The Commonwealth Fund's program, though important to promoting the growth of clinics everywhere, was not a truly innovative program. It never intentionally created a new form of help but, rather, was designed to spread existing forms of help. The New York School of Social Work was given funds for a clinic to encourage the training of psychiatric social workers, visiting teachers, and probation officers. As we have seen, probation officers were part of the juvenile court, an agency begun some twenty years earlier and past its heyday. The visiting teacher movement had also started some fifteen years earlier, and by 1921, it was well established as part of the New York public school system and others as well.

By 1914, psychological clinics were numerous enough to provide the material for a comprehensive review (Smith, 1914). Formal training in psychiatric social work was an outgrowth of the demand for help for World War I soldiers suffering from so-called shell shock. The first training center for psychiatric social workers opened at Smith College in the summer of 1918, and courses in mental hygiene and in psychiatry had been available to social work students at the New York School since 1917. By 1921, Witmer's clinic was a quarter-century old, and Healy had left the Juvenile Psychopathic Institute (by now the Institute for Juvenile Research) in Chicago to Herman Adler and had gone to Boston to establish the Judge Baker Center.

The models then, were already in existence when the Commonwealth Fund's program was established. This is not to detract from its significance but simply to reinforce the statement that the model propagated in the 1920s had been developed in the previous generation. The fund intended to add to the resources already available to the field (Commonwealth Fund, 1922).

The demonstration clinics were to be under the auspices of the National Committee for Mental Hygiene's Division on the Prevention of Delinquency. The Division on the Prevention of Delinquency was formed after a representative of the Commonwealth Fund approached the National Committee to ascertain its interest in the Commonwealth Fund's program. The National Committee showed its interest by organizing a division of delinquency in its organization (Stevenson, 1948).

The National Committee itself was a product of the earlier period of reform, the creation of Clifford Beers, founder of the mental hygiene movement. His book *A Mind That Found Itself* (Beers, 1933) alerted the public to the need to humanize the care of the mentally ill in hospitals. In 1909, with the active support of William James and many psychiatrists and philanthropists, Beers founded the National Committee for Mental Hygiene. The National Committee, at first concerned with humanizing patient care, soon turned its attention to the problem of preventing mental disorder. As a consequence of its activities, psychopathic wards were set up in city hospitals (Southard, 1914) to provide first aid for mental patients; outpatient clinics were developed to deal with incipient mental illness (Abelson, 1966); special programs were supported to train the mentally deficient for a place in society; and much political and public relations activity was designed to spread the concepts of mental hygiene (Barker, 1918; Cross, 1934).

As the concept of prevention became prevalent and the new psychology became accepted, the roots of mental disorder were traced further and further back into childhood. By 1921, psychoanalytic thinking and its derivatives had begun to pervade the fields of psychiatry, social work, and child care. It was only natural, then, that the National Committee would become interested in a program to help children, as help in childhood was seen as a primary means of preventing mental disorder and criminality in adulthood.

In January 1921, the Commonwealth Fund and the National Committee sponsored a conference in Lakewood, New Jersey, to draw up guidelines for the Commonwealth Fund's program to prevent delinquency. Representatives of psychiatry, education, and social work attended, including William Healy, Augusta Bronner, and Charles W. Hoffman, judge of Cincinnati's juvenile and domestic relations court.

The concepts of prevention guiding the demonstration clinics are explained in the following excerpts from the conference's recommendations;

> It is our opinion that lack of knowledge by teachers and school authorities of existing information regarding disorders of conduct result in many instances in the actual causation of delinquency through mismanagement of incipient disorders of this kind and, to a much greater extent, in failure to carry out preventive measures in an environment presenting many favorable opportunities. . . .
>
> It is our opinion that certain agencies created for the express purpose of protecting, training, and caring for children not only fail in many instances to prevent delinquency, on account of their poorly trained personnel and the inadequate social, psychological, and medical diagnoses on children coming under their care, but through improper methods of management and institutional administration, actually contribute to the production of delinquency in those exposed to such unfavorable conditions. . . .
>
> It is our opinion that . . . through mismanagement of their mental handicap and the lack of proper facilities for dealing with feeblemindedness in its institutional, educational, and community aspects, [many] needlessly become delinquents.
>
> It is our opinion that attainment of the general aims of the social hygiene organization will tend to reduce delinquency and that a relationship exists between sex education and delinquency which requires careful study in order to determine the efficiency of sex education and the specific methods by which it is carried on.
>
> It is our opinion that all children should have opportunities for the use of spare time in ways that, on the one hand, are not harmful to the community and, on the other, give to the children themselves opportunities for self-expression and a satisfying sense of physical and social achievement. (Stevenson, 1948, pp. 53–56)[2]

It was recommended that agencies providing suitable recreational facilities and opportunities be encouraged by generous support: "It is our opinion that the participation of children in certain industries, such as street trades, home work, and such seasonal trades as canning and berrying, increase delinquency."

The opinions and recommendations called for the expansion of the work of the juvenile and family relations courts, the further development of probation procedures, and the reform of the detention homes associated with juvenile courts and the training and reform schools.

Another passage from the conference report gives the reader an idea of its strong reform sentiment:

> While the implication of some of the foregoing conclusions is that certain agencies created and maintained for the express purpose of safeguarding childhood, and in some instances of dealing specifically with juvenile delinquency have not succeeded in making full use of their opportunities and through unfortunate conditions growing out of their work, sometimes even tend to increase delinquency, the opinions expressed are set for the purpose of directing attention to the importance of correcting specific defects and not with any idea of destructive criticism of activities which were instituted for high humanitarian purposes. . . . We believe that social agencies created for humanitarian purposes can properly be examined . . . and that the results of such examination are certain to be [as] beneficial. (Stevenson, 1948, p. 56)

The tone is not impassioned, but the message is clearly social reform to be achieved by changing existing social organizations. The report closes with a recommendation for the training of medical, psychological, and social service personnel to carry out the work of diagnosis and to disseminate modern scientific information concerning conduct disorders and the principles underlying their treatment and prevention. Those who had primary responsibility for the care of children who might become delinquent were to be trained by their participation in case studies. This was viewed as the major function of any new clinical service.

Two concepts in this report are important. First, the report adopts a situational orientation, arguing that the institution and the social setting are largely responsible for problems and that therefore the institutional or the social setting should be altered in order to permit the individual to grow. Second, in emphasizing the importance of the institutional setting, the report recommends that professional services be directed toward education and helping the institutions carry out their work. There did not seem to be any thought at this time that professional mental health workers should establish an institution that would treat patients directly. There is no statement that what was needed was a treatment facility, relatively isolated from the community. Rather, the objective was to put professional mental health workers in closer touch with the community in order to influence it. The later development of the child guidance clinic has to be considered in light of the philosophy expressed in this conference report.

Herman M. Adler, Healy's successor as director of the Institute for Juvenile Research in Chicago held views similar to those expressed by the participants in the Lakewood Conference:

> The outstanding fact, however, of this whole discussion is that the psychopathologist and the physician are coming into closer contact with social problems than ever before; and that as society is becoming more intergrated [*sic*] so the psychopathologist is becoming more specialized, and psychiatry must concern itself more and more with the social aspects as manifested in human behavior and therefore must work in greater harmony than ever before with psychologists, teachers, economists and sociologists. (Adler, 1922, pp. 19–20)

Adler's viewpoint is important because his institute was among the first centers before World War I that offered training in child psychiatry and in child guidance (Levy, 1952). Campbell (1918) made a similar statement when discussing the results of a survey of the mental health problems of Baltimore schoolchildren. Campbell said that the educational process, the schools, the visiting teacher, the psychiatrically oriented school nurse, and the school physician—not necessarily the psychiatrist—should be the prime therapeutic agents.

The thinking reflected in the Lakewood Conference statement was the thinking of the pre–World War I period. The conceptions underlying the clinics and the objectives of the clinics reflect the viewpoint of the social reformer and focus on institutional change, but the 1920s were no longer an era in which the call for institutional change would be heeded. Some of the leaders at this conference were already promoting a medical model of delinquency. Tensions between those more sociologically and those more medically oriented were already evident (Horn, 1989). But we believe that the clinics were shaped by the spirit of the1920s, which would support individual treatment and be hostile to institutional change. We shall try to trace some of the social and professional forces that resulted in changes in the objectives and practice of the child guidance clinics.

The Medical Model Versus Applied Sociology

It was obvious from the beginning that the medical model of treating the sick who appeared at the clinic was inadequate to the task of truly serving a community. Surveys and the clinics' early experience suggested the impossibility of caring for any but the tiniest fraction of cases. A survey conducted by the National Committee for Mental Hygiene in Cincinnati in 1921 found that two out of three persons who came before the juvenile court were mentally abnormal, that 13 percent of the public school population deviated from normal mental health, and that 6 percent of the public school population showed conduct disorders sufficiently severe to bring themselves repeatedly to the authorities' attention (Stevenson, 1948).

One of the first activities in 1922, of the Division on the Prevention of Delinquency was to supervise a mental hygiene clinic in Monmouth County, New Jersey. The clinic was sponsored by the Laura Spellman Rockefeller Fund. Although it was limited to diagnostic work, even this task was too large for a full-time clinic in a county of 100,000. In the second year of its operation, the clinic modified its procedures to serve three communities in the county. "Nevertheless, within three months a waiting list of 200 existed. This emphasized the immensity of the task of child guidance and the impossibility of securing coverage by direct service in any community. It showed the necessity of adapting clinical processes for indirect service" (Stevenson, 1948, p. 64).

In 1921, according to Stevenson (1948), helping a child meant that the mental health professional team had to make an accurate medical, social, and psychological diagnosis and transmit these scientific findings to personnel in the agency that referred the child. The belief was that once others had been shown the "truth" about the child and had been given recommendations about how to handle the problem, the referring agency would be served and would then, in most cases, carry out the treatment. The planning group for the Commonwealth program had recognized the institutional contribution to the problem of delinquency. The answer to the institutional problem was to change the institution by educating personnel to think in a mental health framework. The case was to be the medium of education. It was hoped that the proper training of more psychiatric social workers, probation officers, and visiting teachers would alleviate the problem, and, indeed, the Commonwealth program included the development and support of training programs. But the essence of the professional approach was to remain in close contact with the referring source in order to facilitate treatment and to educate the referring source.

By the time the program's actual plans were drawn up, about a year after the Lakewood Conference, the goals had already shifted from the reform of institutions and social agencies to a concern with prevention on a clinical level. Prevention meant the prevention of further difficulty for an individual who had already come to the attention of some helping service (Commonwealth Fund, 1922). By the time the first demonstration clinic started, the workers did not feel that reform of social agencies was their main professional concern. Rather, they felt their first professional responsibility was a thorough examination of the child and the development of a treatment plan for the agency responsible for the child (Horn, 1989).

Because the early workers felt that factual knowledge of the child would lead to better treatment by the responsible agencies, they gave little thought to gaining the cooperation of various agencies in carrying out treatment plans. By the time the clinics started to function, either by design or because the climate of the times militated against far-reaching reforms, the Commonwealth Fund's focus was on the individual child rather than on stimulating change in the agencies (Horn, 1989; E. K. Wickman, personal communication). In New Orleans, open conflict broke out between sociologically oriented social workers who wished to develop the clinic along broad preventive lines and those in charge of the demonstration clinic program for the Commonwealth Fund (Horn, 1989). This shift in viewpoint presaged an even greater emphasis on treatment centered on the confines of a therapy room, with the consequent isolation of the clinics from other agencies.

The Organization of the Early Child Guidance Clinics

Although the clinics began with a staff of a psychiatrist, a psychologist, and several psychiatric social workers, their organization and functioning in the early days differed from those of today. In this section we shall concentrate on two clinics, one in Minneapolis and one in Cleveland. They started with similar staffs, and both demonstration clinics set a pattern for service that continued when permanent clinics were established. For our purposes, it is the variety of staffing and consultation arrangements that is of interest.

The staff of the clinic in the Twin Cities (Minneapolis and St. Paul) demonstration consisted of a psychiatrist, three fellows in psychiatry, two psychologists, six social workers, and social work students. The clinic had a loaned-worker system in which thirty-nine social workers from several local agencies spent periods of up to three months as members of the clinic staff, participating under supervision, in casework. The system's purpose was to help the agency workers resolve the problems on their own and at the same time to acquaint the agencies with the services the clinic could offer (Stevenson & Smith, 1934).

Particularly close relationships developed with the Minneapolis public schools, which had a large staff of visiting teachers. When the demonstration clinic became a permanent facility, it was financed by the Minneapolis Board of Education as a result of conflicts with other agencies in the city. Its services, however, were available to the entire community. The clinic was housed in a children's hospital, which also had a school attached to it. The staff of twenty visiting teachers and ten speech teachers worked with clinic cases after they had been evaluated at the clinic. The three psychiatric social workers with the permanent clinic all had had experience as teachers, undoubtedly facilitating the relationship of the school with the clinic (Blanton, 1925; Stevenson, 1948).

With the demonstration clinic established as an educational institution, the staff participated in a variety of educational ventures through the University of Minnesota. In addition to specific training offered to psychiatrists, social workers, and psychologists, the clinic staff participated in courses for social workers and teachers and offered courses in psychiatry in the medical school and in behavior problems in the University's Department of Education. There were additional seminars and lectures in the School of Social Work. As one outgrowth of the clinic's work, the state financed a rural demonstration clinic in mental hygiene under the direction of Dr. George Stevenson (Stevenson, 1948; Stevenson & Smith, 1934).

As part of the public school system, the clinic's educational efforts continued. In an elective course in mental hygiene for high school juniors and seniors, students wrote life histories, and those who indicated a need for advice were interviewed personally (Blanton, 1925). A prevention service evolved through a behavior clinic established in the kindergartens. The kindergarten teacher filled out a behavior chart for each child. And if the charts indicated potential problems, the child's parents were invited to discuss the situation with the social worker. A course in mental hygiene was also offered for teachers and parents.

The demonstration clinic staff pioneered the open case conference as a teaching device, with the conference being continued in the permanent clinic. Staff meetings in the permanent clinic were attended by the clinic staff, the child's teacher, the school principal, the visiting teacher, and any other social worker who wished to attend. The purpose of the open case conference was as follows:

> Through discussion at this staff meeting, we hope that we may change the attitude of the teacher towards this child and towards all other children in her care, and the attitude of the principal towards her teachers and the children in the school. We hope that the teacher may stop thinking about behavior difficulties as unit characteristics, or as a moral obloquy, and think of the behavior of the child as symptomatic of some underlying cause. After the staff meeting we talk over the whole thing with the parents and try to change their attitude. We give them specific instructions about the training of the child and the social worker goes into the home from time to time to help carry out the program suggested. (Blanton, 1925, pp. 99–100)

The service's preventive nature and the participants' thinking are the following:

> In conclusion then, we believe that a mental hygiene clinic in the schools and colleges should be an integral part of the educational system, and it should not care for just those serious difficulties that might be referred by teacher or professor, but should reach all the children and students. In short, such a clinic should be a center for the education of the emotional life and for the understanding of the individual's own emotional reactions that make for success or failure. Such a clinic would be really a clinic of preventive medicine; a clinic in mental hygiene in the highest sense of the word. (Blanton, 1925, p. 101)

The Cleveland demonstration clinic began in a city where social services and some psychiatric services were available, but the clinic's explicit purpose was to build up the mental hygiene capacities of other agencies. A program of cooperative treatment was stressed. When a child was referred to the clinic by an agency, the agency social worker made a special social examination, in consultation with the clinic's psychiatric social worker, and took responsibility for carrying through the social treatment. The referring agency thus kept responsibility for treatment, and the agency workers acquired some facility in dealing with other cases requiring a similar approach. Stevenson (1948) quotes from a report describing how the agency workers used the clinic services as part of this cooperative treatment:

> All of the agencies with which we worked were most cooperative. Many of the workers fell into the habit of coming daily to the clinic to discuss various problems that had developed in their cases. For instance, here is a note in one case of a conversation between the social worker of the clinic and the probation officer: "Tommy bunked out again last night and his foster mother is terribly concerned. What do you think we'd better do? Shall I bring him in for another talk with the doctor? That last talk really impressed Tommy you know." Consequently, Tommy's case was thoroughly discussed, the foster home situation reviewed to date, and plans laid accordingly. Again quoting from notes on another case: "At last I have got Edna interested in trying to learn something, but now she is determined to be a stenographer. Do you think she's got brains enough?" The clinic worker reviewed the vocational possibilities for Edna in the light of the psychologist's report, and it was decided to try to divert her ambitions to dressmaking or millinery. (Stevenson, 1948, p. 61)

Similar consultation services were offered to the juvenile court and agencies responsible for child welfare and foster home care. The demonstration clinic supplied a psychiatrist for a summer camp for children with behavior problems and personality disturbances. In addition, camp counselors at a regular summer camp were offered an opportunity for consultation on the weekends. A number of similar services were provided to residential treatment homes, and in later years the clinic helped other agencies revamp their programs. Because of its interest in preventive work, the clinic also took a leading role in establishing the Cleveland Mental Hygiene Society. Many lectures were given to community organizations. The staff of the clinic helped train nurses and social workers and gave instruction in child guidance to prospective teachers at the local university's School of Education (Schumacher, 1948).

It is clear, then, that some of the early clinics did have close ties with the community, that they made efforts to help community agencies handle their mental hygiene problems, that there was a distinct preventive orientation, and that they saw it as part of their function to instill a mental hygiene viewpoint in those who would be responsible for caring for children.

As Schumacher (1948) points out, however, over the years the clinics saw more and more children for full study and treatment and finally decided that they could not proceed satisfactorily without full control over the case. Although cooperative services and consultative services could be helpful, the clinics felt that if the child required psychiatric treatment, they must give it (Horn, 1989). Wickman (personal communication), who worked in clinics in St. Louis, Dallas, Minneapolis, and Cleveland, notes that the practice of treating patients in the clinic increased steadily from 1922 to 1928. To some extent, in Cleveland and elsewhere, it appears that the depression of the 1930s was the death blow to such services because the agencies no longer had the funds to hire consultants (Horn, 1989). Earlier efforts to influence community agencies had been lost in the decision to take full control of a case's treatment, even though it was obvious that direct treatment could not serve social needs. What led to these changes in clinical practices?

Training and Child Guidance

Although psychiatry did provide intellectual and administrative leadership (not without considerable challenge even at the time—Horn, 1989), the field of child guidance and child guidance practices arose largely in response to the needs of the field of social work. Training was one of the first objectives of the Commonwealth Fund program, and even though the institutions were directed by psychiatrists, the training concentrated on the field of social work. The Bureau of Children's Guidance, established under Commonwealth Fund auspices in 1921, was an arm of the New York School of Social Work and had almost no role in training psychiatrists and psychologists. Its successor, the Institute for Child Guidance, was in existence from 1927 to 1933. Although only 32 psychiatrists and 15 psychologists completed a year's training, 174 social work students were educated during these years. Another 115 social work students were trained in the institute for periods ranging from one-quarter to three-quarters of a school year (Lowrey & Smith, 1933).

The Bureau of Children's Guidance and the Institute for Child Guidance still sought cooperation from outside agencies, but both decided early on to accept for treatment only those cases "in which the Bureau could carry the full responsibility for treatment and where there seemed to be at least potential cooperation from strategic persons in the environments of the children" (Lee & Kenworthy, 1929, p. 269). The focus was on the work of the child guidance clinic and only secondarily on the help that could be obtained from others who also had responsibility for the child.

The Commonwealth Fund demonstration clinics, which accepted social work trainees from their inception,[3] also emphasized a training function in the clinical setting. A significant consideration in the staffing and supervisory arrangements of these early clinics was the presence of social work trainees on the staff (Horn, 1989). The demonstration clinics in St. Louis, Norfolk, Los Angeles, and Philadelphia all had social work trainees from the beginning. Moreover, the institutions' training needs exerted pressure on case selection and on the definition of a "good case" and "adequate treatment." The ideal training situation—for trainees to see what was considered "best" in child guidance work—may have contributed to distinct constraints in the populations and the problems with which most of the later clinics worked (Stevenson, personal communication; Stevenson & Smith, 1934).

The choice to work with the individual, in contrast with an attempt to work with the larger social structure is puzzling in light of the virtual impossibility of meeting the immense demand for service by means of individual treatment. That social workers eagerly embraced the doctrines and methods of psychoanalysis is only partly explained by the power of psychoanalytic concepts, for the problems of concern to the field of social work might as easily have led to the development of an applied sociology. To illuminate the issue of why social work took the direction it did, we shall briefly examine the history of the field.

The Professionalization of Social Work

By 1899, there were two strong rival schools of thought in social work. One of these was the settlement house movement, devoted to community organization and the use of political and governmental power to gain constructive social change. The second was the charity organization society, which grew out of the custom of giving private alms. As private philanthropy grew, intensive methods of investigation were used to ensure that an unworthy recipient of charity did not mislead the donor and waste the gift. Gradually the leaders recognized the need to provide a broader helping service, when the workers were able to conceptualize the problem of poverty as more than a lack of money. What began as casework to determine eligibility for charity became a method to diagnose the social and interpersonal needs of the family as a prelude to effective treatment. The charity societies' workers generally were not concerned with the social and economic bases of poverty. Criticism of this sort would have been quite surprising at this time, for the charity societies were in part designed to show that their wealthy founders did have humanitarian concerns.

At first the casework was largely handled by untrained volunteers, but little by little the philanthropic agencies began to organize training programs for both the volunteers and the increasing number of full-time, paid workers. Experiments with summer institutes soon led to full-time social work schools; by 1919 there were seventeen such schools in the country (Cohen, 1958; Pumphrey & Pumphrey,

1961). Although the ostensible need was to produce more effective and more efficient workers, the seeds of professionalism were being sown (Lubove, 1965).

At this time when women at all levels were entering the business and industrial worlds, the best-educated ones sought new opportunities (Flexner, 1966; Klein, 1946; Smuts, 1959). The earlier volunteer social workers had been predominantly upper-class women, and the pioneers in professional social work had frequently been from the upper class as well as college educated and sought significance in their lives by engaging in humanitarian and charitable activities (Addams, 1910). The meaning of these activities was enhanced by the national and international recognition accorded to Jane Addams, Edith and Grace Abbott, Julia Lathrop, Florence Kelley, Lillian Wald, and many others.

As social work gained in prestige and the field grew, it began drawing women from a broader segment of society. The first group of settlement house workers and visiting teachers were predominantly from Wellesley, Vassar, Smith, Radcliffe, and similar institutions. The first group of seventy psychiatric social workers were largely recruited from the women's colleges that served the upper classes. The later workers, no longer of the same background, came seeking economic advantage and upward social mobility, and professional status was important to them (Davis, 1959; Jarrett, 1918; Southard & Jarrett, 1922). These changes in the background, motivation, and training of the psychiatric social worker present a set of variables demanding close attention, for the social worker is not only the gatekeeper in the child guidance clinic but also provides most of the day-to-day service.

It was in this context, a strong movement toward professional status, that Abraham Flexner, a prime mover in the reform of medical education, addressed the National Conference of Charities and Corrections in 1915 on the question of whether or not social work was a profession. He stated six criteria for a profession: ''Professions involve essentially intellectual operations with large individual responsibility; they derive their raw material from science and learning; this material they work up to a practical and definite end; they possess an educationally communicable technique; they tend to self-organization; they are becoming increasingly altruistic in motivation'' (Flexner, 1961, p. 303). He concluded that social work met the criteria of intellectuality and of altruism, but he questioned the social worker's independent responsibility and also the specific aims of the field, particularly as the divergent settings and goals influenced the ability to communicate a particular body of knowledge and technique.

Flexner's statement challenged the developing field, leading many to seek an approach that would enable social work to qualify as a profession. It was not long after Flexner's address that Mary Richmond published *Social Diagnosis*. With this text on social casework methodology, Richmond intended to formalize a communicable body of technique, applicable to the diverse settings in which social workers could be found. Her intent is clearly stated in the preface to an extraordinarily comprehensive work:

> In essentials, the methods and aims of social case work were or should be the same in every type of service, whether the subject was a homeless paralytic, the neglected boy of drunken parents, or the widowed mother of small children. . . .

> The division of social work into departments and specialties was both a convenience and a necessity; fundamental resemblances remained, however. . . . It seemed to me then, and it is still my opinion that the elements of social diagnosis, if formulated, should constitute part of the ground which all social caseworkers could occupy in common, and that it should become possible in time to take for granted, in every social practitioner, knowledge and mastery of those elements, and of the modifications in them which each decade of practice would surely bring. (Richmond, 1965, p. 5)

The availability of a work as complete and as thorough as *Social Diagnosis* was one influence leading toward an emphasis on casework with the individual. Richmond's book was important: "Clearly such evidence as is usually assembled is inadequate as a basis for diagnosis; yet, since 1917, when *Social Diagnosis* was published, social workers have had before them clear and positive statements as to the necessity of a diagnostic summary" (Additon, 1928, p. 107). Another influence came from the acceptance of psychoanalytic thinking by educated people in the United States. As Cohen put it, "The search for a method occurred just at the time the impact of psychoanalysis was being felt. Did social work, in its haste for professional status, reach out for a ready-made methodology for treating sick people, thus closing itself off from the influence of developments in the other social sciences?" (Cohen, 1958, pp. 120–21)

World War I provided still another strong impetus toward a methodology directed to treating the individual. The war had produced a large number of neuropsychiatric casualties, and to care for them, the army established a division of neuropsychiatry. In July 1918, Smith College, in cooperation with the Boston Psychopathic Hospital, announced a training program for psychiatric social workers to assist in the rehabilitation of mentally and emotionally disturbed persons.[4] E. E. Southard (1918), then chief psychiatrist at Boston Psychopathic, and Mary C. Jarrett (1918), its chief social worker, were key figures in the development of the new program (French, 1940; Lubove, 1965; Southard & Jarrett, 1922). Intensive courses were given over an eight-week summer period, and the students were sent to field placements in Boston, New York, Philadelphia, and Baltimore for six months. The following year, in keeping with their plans, Smith College expanded the course into a permanent school under the name of the Smith College Training School for Social Work.

> The purpose in view was to educate women so that they might help in getting up the social history of cases presented for diagnosis to psychiatrists, that they might be of use in the treatment of such cases, and that finally they might serve in the social readjustment of psychopathic cases discharged from hospitals. The interest of the moment was of course in mental and nervous disorders resulting from the war, but they were assured that this class of disorders was by no means confined to war conditions and that the profession for which they proposed to train would be a permanent one. (Neilson, 1919, pp. 59–60)

The trend toward work with the individual was accelerated by the fact that the alternative movement in social work, the settlement house, was declining in influence and prestige. Indeed, had they spoken out about the inadequacy of an individual, treatment-oriented approach in professional training, the settlement house workers probably would not have been heard.

The need of social work for a methodology, the development of a professional method of casework, the increasing popularity of psychoanalytic thinking—to which psychiatric social workers were exposed from the beginning[5]—work with World War I veterans, and the ultraconservative political climate of the early 1920s all led social work to focus on the individual. It took but a decade for the child guidance clinic to become an agency oriented entirely toward intramural, intrapsychic therapies and entirely away from goals of social reform or even institutional change.

Professionalization and the Process of Change

Originally, the child guidance clinics were meant to serve, and did serve, a low-income population. The demonstration clinic in St. Louis received 74 percent of its caseload from the juvenile court, but workers soon realized that children seen in the juvenile court had long been known to schools and other social agencies. It thus became clear that the clinics would be more efficient serving other settings such as the schools and social agencies, and so the referral sources increased.

E. K. Wickman (personal communication), a psychologist with the early Commonwealth clinics, considered that the opportunities for treatment provided by the juvenile court were very poor, and this, too, led to dissatisfaction with the clinics' affiliation with the court. All that could be accomplished was an evaluation that might or might not have any influence on the subsequent disposition of the case. Healy and Bronner (1948) made the same point when discussing the reasons for their move from Chicago to Boston. Stevenson (personal communication) also thought that the broadening of objectives from preventing delinquency to dealing with mental hygiene problems generally was in part a consequence of the frustration of trying to work with the court in St. Louis.

Some of the clinics did receive referrals from parents and private physicians, and these clients were more likely to be middle class. Eventually these private sources provided the majority of clinic patients. The Dallas demonstration clinic received referrals from the courts, social agencies, and schools, as well as from private physicians and parents. The complete figures are not readily available, but in Dallas perhaps 30 percent of the cases were referred from sources likely to serve middle- or upper-income populations. In the Twin Cities clinic, about 25 percent of the cases were essentially self-referred. Similarly, in Cleveland 24 percent, and in Philadelphia 33 percent were referred by parents or relatives (Stevenson & Smith, 1934). Healy and Bronner (1948) said that the same pattern held true for the Judge Baker clinic. By 1948, 60 percent of the cases were family referred.

At first, the parents' requests for help were viewed as troublesome because the staff of the child guidance clinics had to carry the entire burden of evaluation and treatment. The clinics wanted to maintain the contact with the various social agencies and the schools as a means of educating those workers, in accordance with the clinics' original purposes. Moreover, self-referred cases were expensive to handle because the clinic had to do all the work of taking histories and social investigation. However, it also became apparent that seeing self-referred cases was a way to carry the work of child guidance into other segments of the population. Finally, these cases were accepted as a means of reaching the middle classes (Horn, 1989; Stevenson, personal communication).

By 1965, public clinics funded exclusively by a government subdivision served low-income populations, whereas clinics operating under voluntary auspices, the progeny of the demonstration clinics, served middle- and upper-income populations. Self-referred cases constituted less than 3 percent of the publicly supported clinics load, whereas more than half of the cases seen in the voluntary clinics were essentially self-referred (see Furman, 1965; Furman, Sweat, & Crocetti, 1965). Similarly, in an informal survey of the population of four inner-city elementary schools, Sarason and colleagues (1966) found only a handful of cases treated or even seen by the voluntary child guidance clinics of the area. They reported a case in which a teacher attempted to arrange for the evaluation and treatment of an inner-city child. When the mother could not promise to accompany the child to the clinic visits, the clinic refused to consider the referral at all, thereby closing off that avenue of help.

When social workers influenced by psychoanalytic theory began to work with self-referred parents, they discovered that in some cases, advice to the mother about handling the child was not helpful. They thus found it necessary to study the parent–child relationship more carefully, and as they did, they found that some parents actually wanted and needed help for themselves (Allen, 1948). In a rather short time, it became a requirement for the parents to participate in the clinic's work if their children were to be accepted for service (Stevenson & Smith, 1934; Witmer, 1940). In fact, an early survey by Helen Witmer and her students (Witmer & students, 1933) showed that in some settings, guidance for children was rapidly becoming therapy for mothers. The social work literature of the late 1920s and early 1930s, in fact, focused on issues of therapy (French, 1940). This striking change took place despite clear evidence in the work of the visiting teachers, for example, and in the early work of the clinics themselves that many children could be helped by indirect treatment, by active intervention in their situations, and by active, albeit patient, pursuit of their parents. It is true that studies by Helen Witmer (1933, 1940) showed that those cases in which the parents did not cooperate were more difficult to treat or to retain in treatment, but there is no suggestion in her work that the clinics attempted to deal with parents' resistance by varying their approach.

With the new requirement of parental participation, the services provided by the guidance clinics, and also the clientele, changed. In the 1920s, social workers needed a professional method, and the method they were taught was heavily influenced by psychoanalytic and psychiatric thinking. Social workers came to depend on casework methods almost exclusively because they viewed the attainment of professional status as partly depending on a professional casework technique. In the 1920s, when people were being inundated by Freud, it is reasonable to assume that in contrast with parents who had to be pursued, the literate, articulate parent who brought a child to the clinic was likely to speak the social worker's language. Not only had that parent been exposed to the language in a narrow sense, but he or she also was likely to be from a background similar to the social worker's and to share other aspects of the social worker's life-style and values. We even might infer that the client who shared the social worker's assumptions would be seen as a more desirable client and may in fact have responded better to the social worker's techniques. It is also a reasonable hypothesis that in the attempt to enhance the professional status of social work, the workers—now an upwardly mobile rather than an upper-class group—may well have found gratification in serving a more prestigious social class (see Caplow, 1964; Hollingshead & Redlich, 1958; Walsh & Elling, 1968; Walsh & Elling, 1968; Wilensky & Lebeaux, 1965).

One of the psychiatrists at the Cleveland child guidance clinic in the 1920s pointed out that during the late 1920s, in an effort to erase the stigma associated with psychiatry and mental hygiene, clinic board members made it a practice to have their own children examined and treated (E. S. Rademacher, personal communication). Morris Viteles (personal communication) noted a similar situation in Witmer's clinic. At first, referrals were from schools and social agencies, but later, many University of Pennsylvania alumni who had heard of the clinic brought in their own children for help. It is interesting that by the 1920s, Witmer's interests had shifted from retardation to the problems of the gifted child. Lewis M. Terman's work with gifted children was just beginning, and perhaps the bright child was "in the air." For some people, it was a mark of modernity and even a status symbol to have a child in psychiatric treatment.

The early use of the clinics by upper-class members of society helped determine what clinic workers considered a desirable patient. The child guidance clinic was initially intended to treat social problems indigenous to lower-class populations, but with time the clinic changed. And as it changed, so did the clientele it served. Viewed in the broader context of the social and intellectual forces shaping the development of social work, the later focus on individual treatment, in contrast with efforts to manipulate the larger social environment, becomes more understandable.

The Fate of Cooperative Treatment

The basic concepts underlying the demonstration clinics were indirect service and cooperative treatment. The clinics were to evaluate the child, and then on the basis of that evaluation, they were to recommend treatment. The therapy was to be carried out by someone else, frequently under continued supervision and guidance from the clinic. Such a service pattern was desirable not only for its economic feasibility but also as a therapeutically preferable mode of functioning. The clinics helped workers in other community agencies deal more effectively with the mental health problems in their own settings, thereby reducing the need for outside services.

This pattern of service lasted, for the most part, for a relatively few years. By the 1930s, it was probably much more the exception than the rule, although clinics continued to maintain various consultative services (Schumacher, 1948). The Institute for Child Guidance (Lowrey & Smith, 1933) maintained a "one-day" consultation service with other agencies, for example, but its principal interest was in directly treating children and their parents.

This change in practice brought a change in the types of problems the clinics handled. Stevenson and Smith (1934) and Helen Witmer (1940) point out that for both the voluntary and the state-supported clinics there was a rough sequence of referrals. First, the clinics were sent those who were grossly mentally disabled, those who had a variety of neurological disabilities, and those delinquents who had been untouched by any other mode of help. A little later the schools sent aggressive, disruptive children. Finally, and frequently in response to the clinics' program of educating referral sources, children were referred who were shy or withdrawn or who had habit disorders or some other relatively limited psychological problem (Horn, 1989).

Some of the reasons for this change in population are related to the training needs of the mental health professions, particularly social work. In many of the early clinics and especially in the training centers, the Bureau for Children's Guidance, and the Institute for Child Guidance, cases were accepted for treatment because of their teaching potential. Both the bureau and the institute indicated that they attempted to maintain a caseload suitable for teaching purposes and that they intended to create an atmosphere in which training, research, and teaching could take place without any pressure to service the community. They both stated that their intake policies did not seriously limit the kinds of cases that came to them. But one cannot help but believe that those students who were trained in such settings must have come away with definite conceptions about what a treatable case was.

The basis on which the bureau selected cases was not made explicit. Because an important mode of teaching was casework evaluation from the point of view of the "ego–libido" method, cases that demonstrated these principles probably were selected, if not for treatment, then for teaching purposes (Lee & Kenworthy, 1929). The Institute for Child Guidance, coming a few years later, explicitly chose certain cases for treatment, and because the staff of the institute was experimenting with treatment methods for both children and parents, some of its cases probably were selected for their interest to the clinical instructors and research teams. In their discussion of the Institute of Child Guidance, Lowery and Smith (1933) observed that the institute's inception coincided with a growing interest in psychotherapeutic methods. The cases selected were those likely to fit the demands of individual therapy. By the 1930s, the "ticket of admission" to a child guidance clinic was the complaint that the child was "nervous" (Horn, 1989; Milton Senn, personal communication).

Training considerations were also a reason for the institute's dropping its practice of cooperative treatment. "Cooperative service was attempted in the earlier years, but abandoned because it did not fit into the training program; when social treatment was carried by the referring agency and psychiatric service by the Institute—the usual division of responsibility in this type of service—the training caseload was thrown out of balance" (Lowrey & Smith, 1933, p.8). Note the assumption that it was not worthwhile to teach young psychiatrists to work cooperatively with outside agency workers. Nowhere is it clear that the service was abandoned because it was found insufficiently beneficial to patients or because it was unfeasible, although the treatment's success did depend on the caliber of workers in the cooperating agency.

The clinic staffs came to feel that they were most helpful with children whose difficulties were of relatively recent origin or of lesser severity, and so they tended to concentrate on those kinds of problems. What is most paradoxical about this situation is that the mental health professional, who started out intending to help the community care for its mental health problems, ended up with the "easiest" cases and left the community agencies, presumably staffed with less well trained people, with the more difficult problems.

Many of the early clinics did continue cooperative relationships with various agencies and schools, as did most of the demonstration clinics in the beginning. In many clinics, the open case conference was an important educational tool for social workers from other agencies, physicians, teachers, and principals. These outsiders participated in discussing and planning the treatment, but although this was successful in some instances, the open conference proved to be a mixed blessing.

Participants in these conferences mentioned several problems. For one, the various agency representatives were jealous of their prerogatives and defensive of their agencies. Sometimes the agencies would fight about who had jurisdiction in a case, or the conferences would turn into a defense of a given agency's practice. In addition, the cases selected for conferences frequently were those with spectacular features, with the mental health professional attempting to dazzle the other participants. The psychiatrist often had difficulty communicating, having been trained to think and talk in the abstractions of the profession. Sometimes others interpreted the abstractions literally. As Wickman and others observed, mental health personnel tended to be rather naive because the field was poorly developed. Rademacher, for example, described how a psychiatrist would diagnose as "schizophrenic" a child who was unresponsive in the interview, because the child was showing "flattened affect." The psychiatrists came into the field without much knowledge and often without any special experience with children (Horn, 1989). And they felt forced to use an inadequate and inappropriate nomenclature because they had no other.

Both Rademacher and Wickman felt that the social workers frequently made the most important contributions to case studies because the factual and detailed histories proved helpful in understanding particular cases. In fact, psychiatry was probably as much influenced by social work as the other way around. The biologically oriented medical psychiatrist with little or no experience with children learned a great deal, and it is likely that subsequent developments in child psychiatry were shaped by this early contact with social workers (Rademacher, Stevenson, Wickman, personal communications; Stevenson, 1944).

Cooperative treatment and consultative relationships with agencies and schools ran into still another problem. Schools of social work offered training in clinical methods and in psychotherapy under the title of casework or social treatment.

While disclaiming the intent to do so, psychiatrists apparently tried to teach social workers and others to think and to act like psychiatrists. Some social workers were even taught to administer neurological tests such as the knee-jerk reflex (Stevenson, personal communication). Although there were fervent denials that the methods were designed to produce "junior psychiatrists," as far as we can determine, these negations contained an important truth. Horn (1989) noted that the increased use of psychotherapy was paralleled by increasing competition between social workers and psychiatrists about who was better qualified to do psychotherapy. In citing the apocryphal case of a young social worker who applied to the New York Board of Regents for a license to practice psychiatry, Glueck (1919) was doing more than giving voice to the attitude of "professional preciousness" (Sarason et al., 1966) detectable in the medical profession from the earliest days of psychiatric social work. Glueck, associated with the New York School of Social Work since 1917 and medical director of the Bureau of Children's Guidance, was making the more salient point that the psychiatry of that day was a medical psychiatry, not really adapted to the needs of the agency worker. The American Orthopsychiatric Association was organized in 1923 as a separate professional group, in part because of the opposition in the American Psychiatric Association to the child psychiatrist–social worker relationship (Levy, 1952; Levy, personal communication; Stevenson, personal communication).

Another problem is illustrated in Levy's (1952) discussion of the temporal course of the consultative relationship between social agencies and psychiatrists. The Kraeplinian psychiatry of the day focused on diagnosis, and many of the psychiatrists trained in mental hospitals knew nothing else. In their relationships with other groups, the psychiatrists provided them first with the labels of psychopathology. As Levy (1952) indicates, it seemed to reassure agency personnel to know they were dealing with a mentally deficient youngster or with an ambulatory schizophrenic, but after a while the agency personnel, having become adept at making their own diagnoses, would then ask for specific help in managing the patient in the community. At that point, psychiatric knowledge proved insufficient for specific working recommendations, and so the agency personnel responsible for treatment became somewhat disenchanted (Jarrett, 1918). In fact, as the social workers' psychiatric education proceeded and as psychiatrically trained workers entered the social agencies, diagnostic consultation was needed even less. Similar issues arose when psychiatrists used the technical terminology of psychoanalysis. Either the nonpsychiatric personnel learned that technical diagnoses did not readily translate into treatment programs, or they were appalled by the "down deep and dirty" interpretive concepts of an id-oriented psychoanalysis (Ridenour, 1948; Stevenson, personal communication).

It is apparent that the open case conferences, consultative services, and relationships with outside agencies were not always smooth or effective. As the psychotherapeutic model took hold, it became all too easy to justify the isolation of the child guidance clinics from the surrounding community on the grounds that the demands of psychotherapy required absolute confidentiality. By the late 1930s, open case conferences or cooperative treatment relationships were nearly extinct. In the Philadelphia child guidance clinic, for example, there were many fewer such contacts with the community agencies than at an earlier time (Allen, 1948). By the late 1930s, community agencies felt a distinct lack of communication with the clinic's psychiatric staff (Senn, personal communication).

We do not mean to imply that there were no effective consulting relationships, for obviously, there were many (Loring, 1920, is an example). The problems in the relationship, however, may have led many in the agencies and clinics to prefer to deal with their own cases, even though the leaders in the field were concerned about the isolation that such an in-clinic, intrapsychic treatment orientation would produce (Lowrey, 1948). Difficulties in managing the cooperative treatment and consulting relationships and problems in relating to the social agencies led the clinics and the training centers to give up the cooperative relationships and to focus on cases that they would treat themselves. Students in psychiatry and social work were not specifically trained to handle the cooperative supervisory or consultative relationship, and in fact there is some suggestion that such training was denigrated by the students and de-emphasized by the training centers. When psychoanalytic thinking came into the ascendency, there was real resistance to teaching community relationships or consultation as specific techniques. The general attitude of many was that anything needed along such lines would be revealed in the supervision of the individual case treatment (Lowery & Smith, 1933, Stevenson, personal communication).

In time, the clinics seemed to have selected cases amenable to the kind of treatment they wanted to offer, generally those cases offering the best prognosis for in-clinic treatment, and they left the community agencies with the more difficult problems. In this instance, as the cooperative and overlapping staffing pattern was lost, the community agencies were left without the assistance of the better-trained mental health professionals. A reflection of the totality with which the field swept toward an intrapsychic treatment orientation is the fact that there was little or no discussion of the relationship with community agencies as a technical problem of therapeutic practice. Although the initial intention was to change the agencies by means of the case study and the cooperative relationship, it was never seriously pursued.

The Child Guidance Clinics and the Schools

The demonstration clinics decided at a very early date that by the time a child came to the juvenile court, it was too late to prevent delinquency. Moreover, many of the children seen in the court had long been known to other social agencies and the schools. Child guidance clinic plans included provisions for relationships with the schools, and in fact one of the express purposes in having a psychologist on the team was the knowledge of tests that he or she could contribute to discussions with educators.[6] Because the work of the clinics centered on general problems of childhood, the clinics were necessarily concerned with the schools, and in those days, the psychologist's major therapeutic role was that of educational tutor or remedial specialist (Stevenson & Smith, 1934).

The intention of the early clinics to relate closely to the schools and the importance given to the work of the visiting teacher as an agent in delinquency prevention may be inferred from the program for the Prevention of Delinquency's plan to extend the work of the visiting teacher nationally through the schools. The Bureau of Children's Guidance, as originally established, accepted referrals from only five public schools, and the Commonwealth Fund and the Public Education Association maintained a visiting teacher to work in each of these five schools. The bureau eventually extended its services to other schools, both public and private (Lee & Kenworthy, 1929). Schools were selected on the basis of the principal's willingness to cooperate, and when this cooperation lagged, another school was substituted. The bureau had a great deal of contact with the schools. The case studies (Sayles, 1926) are replete with references to teachers' comments, visits at the school with teachers, and conferences with administrators. In individual cases, good working relationships considerably enhanced treatment. The report on the Institute for Child Guidance, however, written about five years after the bureau ceased to function, does not discuss relationships with the schools (Lowrey & Smith, 1933).

The demonstration clinic in Minneapolis kept in close touch with the public schools, maintaining cooperative relationships with the school system's visiting teachers and with workers from other agencies. In Los Angeles, the demonstration clinic worked closely with the public schools' psychological clinic, which had a staff of well-trained social workers. The clinic staff also acted as consultant to the California State Board of Education in establishing requirements for school counselors, helped develop a plan for dealing with emotionally disturbed children in three municipal school systems, and offered a summer course for teachers.

It seems clear that a primary factor in the eventual dissolution of these relationships, even when they were clinically effective, was the profound difference between the mental health workers and the teachers in their orientation toward child development and the children's behavioral and emotional problems.

The most prominent research in this area was by E. K. Wickman (1928) and was completed while he was engaged in various child guidance activities with the Commonwealth Fund's program.[7] Wickman measured teachers' attitudes toward undesirable behavior in children. He compared the weights given by teachers to several behavior problems with those given by a sample of mental health workers. The merits and deficiencies of Wickman's research are not at issue here; a critique of his studies and a review of subsequent work may be found in Beilin (1962). Suffice it to say that Wickman was careful to qualify his findings, and he discovered a major problem in his work himself that limits its interpretation: Mental hygienists were asked to rate behavior problems in terms of their significance for further adjustment, whereas teachers rated them in terms of the degree of current maladjustment.

Given these limitations, Wickman found that the teachers' ratings of seriousness fell into four groups, in decreasing order of seriousness:

1. Immoralities, dishonesties, transgressions against authority.
2. Violations of orderliness in the classroom, application to schoolwork.

3. Extravagant, aggressive personality and behavior traits.
4. Withdrawing, recessive personality and behavior traits.

The mental hygienists' ratings fell into the following four groups, in decreasing order of seriousness:

1. Withdrawing, recessive personality and behavior traits.
2. Dishonesties, cruelty, temper tantrums, truancy.
3. Immoralities, violation of schoolwork requirements, extravagant behavior traits.
4. Transgressions against authority, violations of orderliness in the classroom.

Wickman's discussion of the issue is still fresh and enlightening. He conceptualized two forms of ''evasion of social requirements,'' withdrawal and attack, postulating that these two modes of response were learned in relation to earlier social demands by parents and that they were elicited by the school's social demands. The two modes of response are explained in relation to their significance for the teacher:

> Our experimental results may be summed up in two statements: To the extent that any kind of behavior signifies attack upon the teachers and upon their professional endeavors does such behavior rise in their estimation as a serious problem. To the extent that any kind of unhealthy behavior is free from such attacking characteristics does it appear, to teachers, to be less difficult, less undesirable and less significant of child maladjustment. (Wickman, 1928, pp. 159–60)

Wickman then described the teacher's handling of behavior problems in terms of the emotional responses that the behavior elicits in the teacher:

> An inspection of the characteristic form of discipline employed for children who are disobedient, dishonest, truant, disorderly, or who offend by sexual behavior, reveals the counter-attacking nature of the teachers' behavior. Punishments in various forms and disguises are generally administered. In this connection it is essential to note that punishment is not limited to blows to the body. Wounding the child's pride, self-respect, and personal integrity may not have the sinister appearance of corporal punishment but it relieves the tension of the adult. The counter-attack may take the form of shaming the child, criticizing him before the class, exacting confessions, requiring apologies, or restrictions, impositions of tasks, negations, prohibitions, admonitions, demotions. . . . In all these methods of discipline there is a display of the aggrieved adult whose authority or personal integrity has been violated.
>
> Punishment, to be legitimate, can only be applied when it is intelligently chosen with a conception of the causes of the behavior problem clearly in mind. Intelligent punishment is precluded so long as behavior problems are evaluated in relation to their frustrating character. The requirement that the child must obey simply because ''I say so'' or because the school demands it, cannot be defended rationally. That implicit conformity to adult standards of morality, honesty, and obedience is for the good of the child is often made the justification for the methods of discipline that are employed.
>
> Insofar as the withdrawing types of behavior problems are evaluated as of least importance, we may assume that the teachers' behavior responses to these

> problems are characterized by tolerance and indulgence. . . . The show of dependency that is characteristic of withdrawing behavior makes for ease in classroom administration. It is looked upon with favor. Unsocial children often attract the favor of the teacher by applying themselves diligently to school tasks in which they find a refuge from the difficulties of social adjustment. (Wickman, 1928, pp. 162–63)

In Wickman's view, the counterattack results only in a fixation of the behavior, by either emphasizing the "bad self" or reinforcing the child's antagonism to authority. Similarly, the support that conforming behavior receives in the school setting reinforces sick behavior. From the point of view of the mental hygienist, Wickman believed that in both instances the response to a child's behavior is not dictated by an understanding of causes but, rather, by the surface manifestations of behavior and the reaction that the behavior calls forth in the teacher. In neither instance is the teacher's behavior rational.

Although Wickman recognized the school's educational purposes, both intellectual and social, he felt that the basic problem lay not in the teacher's function but in a lack of appreciation of the overt behavior pattern's significance. If attacking behavior is disruptive, blind punishment is no answer. Moreover, much of what was characterized as attacking behavior represented the normal activities of childhood. Investigatory, experimental activity and curiosity about oneself and the world, including oneself as a sexual being, are normal parts of childhood, not to be censured but to be accepted and respected as a part of development. In addition, Wickman observed that the attacking child in school did not necessarily fail in life, whereas the "good," withdrawing child was sometimes unsuccessful after leaving school.

Wickman recommended that teachers learn more about normal child behavior, including intellectual, social, and sexual development, and that they receive training in diagnosing and treating children's behavioral problems by learning to think in terms of causes and to direct their action toward the emotional and experiential factors producing the problems. He further advised that some way be found to help teachers withstand the shock of having to deal with the raw impulses of childhood so that the teachers themselves could gain better control of their own emotional states.

Wickman argued that the mental health worker had a superior, that is, a more scientific view of child development, and, moreover, that from the point of view of the mental health worker concerned with future adaptation to the world, the educator's approach was not necessarily in the child's best interests. Allen provided a sensitive appreciation of the difference between the psychiatric social-worker and the teacher, particularly in relation to the teacher's exercise of authority. Like Wickman, Allen emphasized the problems that arise because the teacher is responsible for a large group and the mental health worker tends to think in terms of the individual case, but her basic point is the same as Wickman's. When the teacher can remain objective toward individual children, avoid self-blame, and not respond on the basis of anger and frustration, then he or she can act rationally to help the problem (Allen, 1929).

Because the mental health worker, however sympathetic to the teacher's position, approached the school with the idea that the teacher must change, it would appear almost inevitable that conflict would develop. Wickman implied as much when he said that adults have difficulty changing their attitudes. It is our guess that the differences in viewpoint between teacher and mental health worker, a difference that appeared in school consultation programs of the 1960s (Grant & Stringer, 1964; Newman, 1967; Sarason et al. 1966), must have contributed to the gulf between them. Although there is no direct evidence at this point to show why the child guidance clinics retreated from their extensive contact with the schools, we believe that they retreated in part because the mental health professional did not take seriously the problem of relating to the schools. Most of the literature has been written by mental health workers, and so it would be instructive to have more data from educators. As it is, we do not know in what ways it would be desirable for mental health workers to change their concepts and approaches in order to relate more meaningfully to educators.

A Case Study

The following case study illustrates the treatment in the early child guidance clinics. Much of what follows is extracted from Sayles's 1927 volume of case reports from the Bureau of Children's Guidance. The cases, designed to show in detail how the clinics worked, were intended for an audience of agency social workers, teachers, probation officers, public health nurses, and parents, rather than for psychiatrists and psychiatric social workers. In the following case, we emphasize the various therapeutic activities undertaken on behalf of the child as well as the contacts the workers made with the home, the school, and other agencies. The reader familiar with current child guidance clinic practices might be interested in reflecting on how such activities would be received by a psychotherapeutically oriented staff.

Mildred Martin was brought to the clinic at age twelve, so unable to do schoolwork that she had been placed in a first-grade class after transferring from a parochial to a public school. "Apathetic, unsocial, sullen, she seemed to have retired into an inner world of her own to such a degree that those who observed her had become seriously alarmed about her mental state" (Sayles, 1926, p. 7). The visiting teacher who brought Mildred to the clinic had gone to the child's home to discover that Mildred's mother was devoted, sober, and industrious but that her father was an alcoholic and that Mildred had been under treatment for congenital syphilis. The visiting teacher assigned to the school took the child's history from Mildred's mother in the family's home, for the clinic procedure did not require that the parent bring the child to the clinic. Such a practice would be unusual in most clinics today. We shall omit details of the family and their history, for these are not important to our purposes. The very thorough history obtained by the visiting teacher was sufficient for the clinic's purposes, and it was not necessary for the mother to make any further visits to the clinic.

Mildred was supposed to report to a medical clinic once a week for her syphilis shots. The visiting teacher discovered she was a difficult patient, often screaming, kicking, and biting the doctors and nurses, slipping out of the clinic before receiving her shots, and pretending to her mother that she had had them. The visiting teacher treated this aspect of her problem by giving Mildred a note to the medical clinic social worker who would send back a reply. This method of checking on her progress improved both her attendance and her behavior at the clinic. The workers did not hesitate to enter into various aspects of the child's life rather than restricting their contacts to clinic visits.

Mildred was seen by a psychologist, who found her unresponsive but cooperative. He obtained an IQ of 97. The woman psychiatrist who saw Mildred did the physical examination herself.[8] The child was more responsive to her. Mildred told the psychiatrist enough for her to conclude that Mildred was probably not in an early stage of dementia praecox but, rather, that she had a deep emotional difficulty related to the problems uncovered in her school and home environments. Mildred was put into therapy, and though not resistant to treatment, she did express some unhappiness about having to come to the clinic because she was escorted there by the visiting teacher.

Following the series of examinations and the completion of Mildred's history by the visiting teacher, the clinic staff worked out a plan of treatment. One of the first elements in the plan was permitting Mildred to go to her clinic appointments unescorted by the visiting teacher. To implement the treatment plan, the visiting teacher went to Mildred's home and discussed it with the child's mother. She asked the mother to change Mildred's room because Mildred had been fighting with a sister who teased her a great deal. The social worker (it is not clear whether this was the visiting teacher or the clinic worker) continued to visit the girl's home and to work with the mother. The nature of the contact is shown in the social worker's notes:

> Explained to Mrs. Martin how very sensitive Mildred is about her school work and urged that she use every means possible to keep her sisters from taunting her about it or from comparing her progress with that of her younger sister. Suggested that they all treat Mildred just as they would any other child her age, emphasizing the fact that now she is having the better school opportunities she will learn very rapidly and will soon be in her own class. Told her that if the family encouraged her and expected her to succeed she would be happier and would make better progress. Pointed out the danger of sensitive people withdrawing into themselves, brooding over their disabilities and difficulties and developing a set feeling of inferiority. Suggested that if there was anything Mildred did more successfully than her sisters, this be picked out for special commendation and attention. Suggested that it is better for Mildred to talk things out than to shut herself up and withdraw. (Sayles, 1926, pp. 27–28)

These notes suggest that the worker was highly directive and specific in her recommendations to the mother. The worker remarked at how able the mother was to grasp and implement suggestions; there is little to indicate that the worker found it necessary to help Mrs. Martin "work through" her feelings about the child or about the suggestions.

The social worker had less luck in working with Mr. Martin. She tried repeatedly to visit him in his home and wrote to him on several occasions. She even found an employment opportunity for him, but he never responded. Despite the failure to achieve any change or, for that matter, much contact with Mr. Martin, the worker never expressed any sense of discouragement, nor did she seem to feel it futile to pursue him. The social worker was not discouraged by the mother's lack of responsibility or by the father's lack of response, nor did she hesitate to pursue the treatment plan vigorously, even though the child resisted it.

Part of the treatment program called for Mildred to engage in activities designed to give her interests outside herself and to take her out of her home for definite periods of time, for her home was described as a depressing influence. The social worker carried out the social treatment. With Mrs. Martin's consent, the social worker found a Girl Scout troop, took the child to the first meeting, and continued to encourage her to attend meetings until she was well established in the troop. The clinic gave Mildred money for her uniform and for dues when necessary. Mildred eventually became an enthusiastic scout. During the first month of treatment, the worker also took the child on an all-day outing to a museum, lunch, and the Hippodrome, and helped her become acquainted with the public library. Later in their relationship, the worker learned that the girl liked to sew and knit. She encouraged Mildred to make Christmas gifts for friends and family, taking the child to her own home for a lesson and supper.[9] The worker also found a place for Mildred in a seaside home for the summer and arranged for financial support through a relief agency that had worked with the Martin family.

From the beginning of treatment, Mildred's visiting teacher stayed in close touch with her school. The clinic social worker visited Mildred in her classroom and had frequent conferences with the visiting teacher who in turn had periodic contacts with the child. At one point, when the child, disappointed about an expected promotion, was absent, the visiting teacher went to her home to discuss the problem with her and arranged for a school supervisor to visit the child in her classroom to explain the promotion problem to her.

After one midyear promotion, when it was discovered that Mildred was not doing very well, the clinic social worker, at the suggestion of the treating psychiatrist, interviewed her new teacher and found out that the child needed additional help. A tutor was suggested, and the psychiatrist broached the subject with Mildred in her next interview, only to discover that Mildred was concerned that others would see her with the tutor if they met in the assembly where she had been tutored previously. The visiting teacher thereby arranged for after-school tutoring, away from the public assembly room. Mildred was then able to work successfully with her tutor. The treatment philosophy did not require Mildred to confront her fear of being seen with a tutor.

The psychotherapy program is not described in any great detail. Mildred had regular appointments with the psychiatrist, with the initial interview lasting half an hour, but the frequency of appointments is not indicated in the original case study, nor is there much to indicate the nature of the psychiatrist's approach. The psychiatrist acted as an understanding friend who encouraged her to express her problems, supported her in various ways, and tried to make life seem less desperate and hopeless to her.

Apparently the psychiatrist helped Mildred anticipate and work through some of her feelings about the plans made for her. Moreover, distressing matters revealed in the interviews that could be corrected in the environment were corrected following the psychiatrist's recommendations to the clinic social worker or the visiting teacher. As the child gained confidence in the psychiatrist, she apparently ventilated some of her feelings about her poor home situation. The psychiatrist did not attempt to elicit the child's feelings of hostility toward her drunken father but, on the contrary, seemed to support her in her solicitude for him.

Mildred progressed beautifully under the treatment, which continued for about two years. Three years after the end of treatment a follow-up indicated that Mildred was a happy, successful adolescent with normal heterosexual interests. The family situation had improved somewhat, although there was no important change in the father. When his children did things that enabled him to feel proud of them, he apparently responded with somewhat greater warmth and affection for the entire family, but he continued to drink heavily from time to time.

Successful cases, of course, were selected for presentation in this source. Although such therapeutic tactics were not used in the typical psychotherapeutically oriented clinics of the 1950s, in the 1960s and later some programs did use active intervention with very dysfunctional families in an attempt to keep children in their own homes. It is certainly possible that so much manipulation of a child's life might detract from a sense of responsibility and dilute the effectiveness of a psychotherapeutic relationship. On the other hand, the results in selected cases certainly do not support the view that all such treatment was superficial or inevitably ineffective, the view of extreme proponents of psychotherapeutic methods that was sometimes expressed in discussions of other approaches.

The subsequent changes in clinic practice were due not to poor results with the methods employed but to other factors. The Bureau of Children's Guidance reported success and partial-success rates of better than 90 percent with methods such as those used with Mildred (Lee & Kenworthy, 1929). It is difficult to compare the results of treatments at a later time with those of earlier cases because the standards may have differed, and cases that withdrew from treatment or that were seen only for diagnosis may have been counted differently. However, when Horn (1989) studied a sample of the records of the Philadelphia Child Guidance Clinic from 1925 to 1944, she found that about 43 percent were adjusted or partially adjusted. She noted little difference in good outcomes by method of treatment (parent therapy, parent and child, advice and instruction to parent, modification of child's environment). However, she did find that modification of the environment as a form of treatment had a lower rate of unadjusted cases and a lower rate of withdrawal from treatment. Modification of the environment did not involve the extensive social treatment described in the case of Mildred but consisted of discussions and recommendations to the parents, with the case discussion focused on specific changes that could be made. There is little evidence that the child guidance clinics did any better with psychotherapeutic methods than with social treatment, and there is evidence that the population served changed. That the clinics had huge waiting lists attests to the need for the new service, but that does not justify their shrinking away from social responsibility.

Summary

The privately supported community child guidance clinic that we know today is a mental health agency strongly influenced in its development by the experience of the Commonwealth Fund demonstration clinics. The latter were established in the 1920s as part of a program for the prevention of juvenile delinquency. The demonstration clinics in their earliest conception were designed primarily not to provide treatment for children showing psychological problems but to enable a variety of other social, child welfare, and educational organizations to handle the problems of children more effectively. The original focus on juvenile delinquency seemed to reflect a continuation of the social reform thinking prevalent in the generation preceding World War I. After the war, when the nation became conservative, individual responsibility was emphasized, and social reform was a dead issue.

In the postwar period, there were many changes in the role of women and a considerable upheaval in prevalent sexual mores. Although many personal, traditional beliefs were challenged, the postwar generation continued to believe in science as the answer to people's problems. In the human relations field, science meant psychoanalysis and psychology. Growing mass communication spread a variety of conflicting beliefs about proper child-rearing practices and caused much parental uncertainty. It was in this context of changing values, changing sex roles, and uncertainty that the child guidance clinics offering scientific expertise grew.

From the beginning it was apparent that a treatment service could never hope to meet service needs and that it would be necessary to develop techniques for educating the relevant settings to handle their own problems. Despite this, within a decade, the clinics moved in the direction of almost exclusively providing in-clinic treatment services to parents and children. Moreover, as they moved in this direction, contacts with other agencies and institutions were minimized, and the population that was served changed from predominantly lower class to predominantly middle class.

The shift toward individual treatment, in contrast with an applied sociology directed to social and institutional reform, reflected the intellectual and social spirit of the 1920s. That spirit was receptive to a belief in individual responsibility and refractory to the concept of broad social responsibility. Moreover, changing opportunities for women led many to enter social work with the hopes of achieving professional status. Considerations of training, forces leading toward professionalism, and a desire for upward social mobility led to an emphasis on casework and therapeutic interviewing with individuals. There are a variety of suggestions that relationships with other social agencies and with the schools were not always fruitful or conflict free. Instead of approaching the problem of relating to institutions and the problems of institutional change as technical issues deserving consideration in their own right, the difficulties may have encouraged the further isolation of the clinics, which were then more interested in the problem of devising methods of directly treating individual children and their parents.

The aims, the organization, and the early experience of the Commonwealth Fund demonstration clinics and the changes over time reveal the broad and complex forces that combined to shape mental health practices, forces that mental health professionals generally ignore. The approach of the early clinics in their treatment methods and their relationships to social agencies might well be reexamined in this day of concern about the adequacy of contemporary outpatient child guidance practice to meet the challenges presented by the changing urban population.

II

THE 1960s AND BEYOND: THE REDISCOVERY OF SOCIAL PROBLEMS

Part I discussed programs originating in the Progressive era, from about 1890 to the beginning of World War I in 1914, and from 1920 to 1930. At the time we published our earlier social history *(A Social History of Helping Services),* in 1970, we were writing from the perspective of the mid-1960s when the community mental health movement was just beginning and when the nation had dedicated itself to a war on poverty. Serious problems with the welfare system were coming to public attention, and the community rediscovered child abuse. In the mid-1960s, the development of the contraceptive pill ushered in the sexual revolution and, with it, birth control and abortion reform. In the 1990s, welfare, child abuse and neglect, and birth control and abortion are at the forefront of public attention.

Services in these three areas originated well before 1890, and we found that it was not feasible to incorporate discussions of those services into the original text. We decided, therefore, to place our chapters on welfare, child protection, and birth control and abortion services in a separate section, Part II. There are two links between Parts I and II. One is an examination of the development of services in other time periods that may be characterized as reform or conservative, in order to assess the hypothesis that the tenor of the times affects the nature of the services offered. The second link is the way that the mental health field, thoroughly professionalized after 1930, related to social problems emerging at different times.

10

Changes in Mental Health Services to Children and Families: 1930 to 1980

In the 1960s, mental health professionals rediscovered poverty, child abuse, birth control, and abortion—problems which remain prominent in the 1990s. To appreciate the background of this rediscovery, it is useful to review recent social history.

The Great Depression of the 1930s stimulated a great many reforms. President Franklin D. Roosevelt's New Deal was symbolic of the effort to use the resources of the national government to help revitalize the economy and to offer direct assistance to those suffering because of economic stagnation. As part of this New Deal program, the Social Security Act of 1935 initiated a major change in human welfare services that entailed increased federal financing and control of programs, and reform and change efforts that were carried out by people who worked in the federal bureaucracy and in Congress.

Although the United States did not enter World War II until December 7, 1941, it assisted the Allies against Germany from the moment the European war began in 1939. The start of the nuclear age was marked by the explosion of atomic bombs at Hiroshima and Nagasaki, and the successful wartime science–government partnership was continued after the war. This policy resulted in unprecedented growth in undergraduate and graduate education, partly as a consequence of the postwar GI Bill of Rights. The National Institute of Mental Health, established in 1946, was responsible for rapid growth in the number of professional personnel and their employment in many different settings (Levine, 1981).

After World War II the return to normalcy was basically conservative in tone. In the 1950s, the cold war with the Soviet Union and the fear of communism were reflected in McCarthyism, paralleling the "red scares" of the 1920s. The baby boom, the idealized family as represented by many popular television programs, and suburbanization dominated the public's attention from about 1950 to 1960. The Soviets' *Sputnik,* launched in 1957, was a major blow to the nation's pride and resulted in efforts to improve the schools. The U.S. Supreme Court's 1954 *Brown v. Board of Education* decision set the stage for the civil rights revolution of the 1960s.

As a nation, the United States was prospering, but not all segments of society were benefiting equally. Many African-Americans and rural whites had migrated north in response to both wartime employment opportunities and the mechanization of agriculture that reduced the need for southern farm labor. With the movement of more affluent whites to the suburbs, the character of many cities changed. For this prosperity to continue, more investment would be required in the public sector (see Galbraith, 1958), an economic policy that was behind the physical urban renewal programs that were instituted in the 1950s.

The urban renewal programs uncovered the human misery in the cities' slums. Before he was assassinated in 1963, President John F. Kennedy had proposed the Community Mental Health Centers program. After President Lyndon B. Johnson assumed office, he continued and enlarged the Kennedy policies with the Great Society program of the mid-1960s. The War on Poverty in the 1960s, the Community Mental Health Centers Act, Medicare and Medicaid, and Supplemental Security Income (SSI) all were continuations of the efforts to provide more resources to those in need. Headstart, the program of preschool education for poor children, was an important part of the War on Poverty.

The 1960s was an era of social turbulence, change, and reform. The assassinations of John F. Kennedy, Martin Luther King, Jr., and Robert F. Kennedy; the urban riots of the late 1960s; the protests on the nation's campuses against the Vietnam War; the counterculture movement; the advent of the birth control pill; and the sexual revolution all resulted in a crisis in the legitimacy of authority. Young people experimented with marijuana and psychedelic drugs; later the country awakened to the heroin epidemic and, in the late 1980s, cocaine and crack.

The civil rights movement that resulted in the desegregation of American society was successful enough that it became the model for others to seek redress through the courts, citizen participation, and citizen activism. The women's movement was one very important effort to achieve change through these means.

In the aftermath of World War II, not all women who had entered the labor force in response to wartime labor shortages returned to their homes to take care of children and families. Even so, labor shortages continued in the prosperous 1950s and provided an incentive, along with personal economic necessity for increasing numbers of women—single, married, and mothers—to enter the work force. It was in the context of the increasing numbers of women in the labor force and their higher educational attainment that the women's movement revived in the 1960s (Levine, 1977). This movement was responsible for litigation and legislation designed to obtain equality between the sexes in employment, education, and civil rights. During the mid- and late 1960s, partly in response to the general atmosphere of liberal reform, many states repealed or amended restrictive contraception and abortion laws, and eventually the U.S. Supreme Court's *Roe v. Wade* decision in 1973 made abortion more readily available nationwide. Legal reforms made divorce easier, with the result that divorce rates rose, as did the number of female-headed households and the number of complex families following remarriage.

The presidential race of 1972, with Richard M. Nixon's resounding defeat of

George McGovern, marked the beginning of the reaction to the upheavals of the previous decade. Despite presidential opposition, however, the Democratic Congress maintained funding for the reform programs initiated in the 1960s (Levine, 1981). The growing conservatism in national politics, however, culminated in Ronald Reagan's ascendancy to the presidency in the 1980 elections, thereby marking the end of the reform era of the 1960s and early 1970s.

Changing Concerns for the Mental Health Professions

After the 1930s, and especially after World War II, the number of professional mental health workers grew, as did the centralized control of resources by government. Because of those two factors, the story after the 1930s is less one of heroic individuals, such as Jane Addams, Ben Lindsey, or Margaret Sanger, and more one of government agencies, commissions, and legislation. Law became a more important part of the professional environment. Changes in practice were initiated both by legislation involving federal funds and federal control, and by litigation to define rights in health, welfare, and mental health.

The relationship among the mental health professions, social change, and participation in the amelioration of social problems is reflected in the professional literature. For a sample of that literature we shall use the *American Journal of Orthopsychiatry* (AJO), which reaches a broad spectrum of professionals. The AJO began publication in 1930, shortly after the early child guidance clinics came into being. The AJO's publisher, the American Orthopsychiatric Association (AOA), was formed in 1924 to provide an interdisciplinary home for psychiatrists, social workers, and psychologists who came together in the early clinics. Now a major and highly influential professional organization, the AOA members' interests are not restricted to problems of children and families. The AJO currently has a subscription list of fourteen thousand (Eisenberg & DeMaso, 1985) and therefore is a good source for the topics of concern to the organization's members.

In 1985, in recognition of fifty years of continuous publication, the AOA published a single volume of abstracts of the 3,224 articles that had been published between 1930 and 1980 (Flaxman & Herman, 1985). Both the volume's topical index and the abstracts provided an excellent opportunity to examine trends that attracted the attention of the mental health community during those years.

There are many drawbacks to relying on this single source: There are many more journals than the AJO in the field, and many topics are not well represented in its pages, for example, the biochemistry and pharmacology of disorders. Furthermore, a decline in publication in some areas can be attributed to the spectacular growth of the field and the proliferation of specialty journals. We shall assume, however, if articles are not published on some topics when they might be expected or if there is a change in the number of articles on some topics, then these changes, examined over five-year periods, will reflect, admittedly imperfectly, trends in research, practice, and thinking in the field.

AJO has had only five editors since 1930 and has retained its interdisciplinary membership. It publishes quantitative and qualitative research, reviews, commentary, and theoretical and clinical studies. Over the years, more space has been devoted to quantitative research and less to reviews, but clinical presentations have continued to average about a fourth of the total number of articles (Eisenberg & DeMaso, 1985).

We used the published index to count the number of articles in each indexing category and then tabulated the articles by five-year intervals. The median number of articles is about 340, with a range from 170 (from 1930 to 1935) to 399 (from 1961 to 1965). The percentage of articles published in each five-year interval varies from 5 percent between 1930 and 1935 to 12 percent between 1961 and 1965. Because of this variability, we shall cite the expected percentage of articles in parentheses (E) near the percentage of articles that was observed (O) in the specified time interval.

Changes in Services and Theory

Child guidance. The American Orthopsychiatric Association began with an interdisciplinary focus that originated with the creation of the child guidance team of social worker, psychiatrist, and psychologist. The term *child guidance* gradually lost favor, however. Of a total of thirty-seven abstracts under this heading, 73 percent (E = 35%) were published before 1951. No articles indexed under this category were published after 1970.

The decline in interest in child guidance probably reflects shifting professional identifications and concerns. With the growth in the mental health professions after 1946, stimulated by training, service, and research funds dispersed by the federal government (Levine, 1981), mental health workers were employed in many new organizational settings. They identified more with the psychiatric–medical model and less with the model for the social treatment of children (Horn, 1989). The *psychiatry* of *orthopsychiatry* refers to this shifting identification and may have promoted it as well. Before the 1960s, social workers and psychologists were willing to concede to psychiatrists' status, authority, and therapeutic power, even as they aspired to achieve comparable status. Role issues, as reflected in the publication rate, were most evident in the organization's earliest years and peaked again in the 1950s with the growth of the mental health field and the beginning of the struggle for professional autonomy in the fields of clinical psychology and psychiatric social work (Eisenberg & DeMaso, 1985).

In the original conception of the child guidance clinic, consultation with schools, courts, and social welfare agencies was to be a major means of delivering services. But consultation soon dropped off sharply in favor of directly treating the children and their parents. About 15 percent of the entries under consultation were published before 1945 (E = 25%). This topic was of greatest interest in the ten years between 1951 and 1960 (E = 20%; O = 54%) but then fell, even though the community mental health centers programs presumably emphasized consultation. As some have noted, there may have been more talk than action in this field (D'Augelli, 1982). There was a small increase in the number of articles about consultation between 1971 and 1980.

Articles indexed under prevention appeared throughout the fifty years, with little variation from the expected frequencies for each five-year interval. The finding of little variation with time is surprising because prevention was emphasized in the Community Health Centers Act of 1963.

Juvenile court. Initially, the child guidance clinics were closely affiliated with the juvenile courts. As we noted, this affiliation was lost before 1930 owing to numerous problems. The relationship with the juvenile court was never a popular AJO topic, and only sixteen articles are indexed under this heading. Just two or three articles were published in each five-year period before 1960, accounting for 81 percent of all articles under this heading in the first thirty years and only three (E= 45%; O=19%) in the twenty years between 1961 and 1980.

Juvenile delinquency. The early clinics started out with a plan to prevent juvenile delinquency; and the organizers of the AOA had a strong interest in this area, as a number of its founders were forensic psychiatrists (Eisenberg & DeMaso, 1985). There are 155 entries under delinquency in the index. Ninety percent were published before 1966 (E = 32%), with the peak coming in the five years between 1941 and 1945. Only 10 percent were published after 1966 (E = 32%).

The greatest interest in delinquency came during World War II when the field was concerned with maternal employment and paternal absence as variables in delinquency. The decline in number of publications came close to the time of *In re Gault,* decided in 1967. This decision granted due process rights to juveniles charged with being delinquent and turned juvenile court proceedings into something more closely resembling criminal trials.

The field may have become more circumspect in its claims to be able to help delinquent youth by means of psychotherapy. In addition, vocationally oriented intervention, which became somewhat more prevalent in antipoverty programs in the 1960s (Massimo & Shore, 1963), was not in keeping with the clinic-centered treatment emphasized in the medical model.

School problems. The founders of the child guidance clinics recognized the importance of having a close relationship with the schools. Psychologists thus were added to the staffs of child guidance clinics because children with emotional difficulties also tended to have learning difficulties and behavioral problems in school. In addition to ninety-two articles indexed under school mental health, forty articles were published on learning disorders, and thirty-seven were indexed as reading problems. Some articles on school mental health were published in each of the ten five-year periods; with most coming between 1956 and 1965, during a drive to improve the schools following the Soviet Union's 1957 launching of the world's first space satellite. In that decade, AJO published 44 percent (E = 24%) of the school mental health articles.

Play therapy. Interest in play therapy, which had begun with the work of David Levy in the child guidance clinics in the 1930s, appears to have diminished substantially over the years. Fifty-six articles on play therapy were published, over half of them between 1936 and 1945 (E = 20%; O = 55%). In contrast, no articles on play therapy were published between 1966 and 1975. Only three (E = 20%; O = 5%) appeared between 1976 and 1980; these described the use of play to help children hospitalized with a life-threatening illness to work through their anxiety; the use of play and dance with severely disturbed, hospitalized children; and a case report of the use of two houses in play therapy sessions to help a child work through a conflict around separation and placement in another home. Whether this decline reflects the inefficiency or the limited effectiveness of play therapy is unclear.

Behavior modification. In contrast with the declining number of articles about play therapy, thirty-three were published under the heading of behavior modification or behavior therapy. Of these, four were published between 1936 and 1940, with the other twenty-nine (E = 31%; O = 87%) between 1966 and 1980. There was little interest in behavior modification during the years that psychoanalytic theory dominated the field. This rise in the number of publications on behavior modification is paralleled by a decline in the number on psychoanalytic theory.

Group therapy. There were 121 articles indexed under group therapy, with few appearing before 1940 (E = 14%; O = 7%). Paralleling the use of group therapy in the armed forces during World War II, there was a sharp rise in the number of entries (51%) listed under group treatment beginning in 1941 and continuing through 1955 (E = 30%). Interest in group treatment leveled off after that, although there were some articles on group treatment through 1980. These later articles generally described special-purpose and support groups.

Family therapy. The 101 entries under family therapy show a different pattern. The abstracts before 1945 (about 14%) dealt largely with the relationship between casework with the parent and therapy with the child. In the 1950s, the pattern of treating the parent and child separately continued, but with more emphasis on family dynamics. Articles on family therapy as we define it today showed a sharp increment between 1961 and 1965 (E = 12%; O = 21%). This category continued to be well represented through 1980.

Psychoanalysis. The entries showing less interest in play therapy and more interest in behavior modification, group therapy, and family therapy are paralleled by the smaller number of articles (73) under the heading of psychoanalysis. More than half (E = 25%; O = 52%) were published before 1945, and only 3 percent after 1965 (E = 31%). Psychoanalytic theory as represented in such concepts as transference, resistance, defenses, and ego pervade clinical thinking and writing. We found these concepts mentioned in many abstracts that were not indexed under psychoanalysis. The distribution over time, however, did not change when we added those abstracts to the total count. From 1966 to 1980 only 5 percent of the abstracts contained such terms (E = 31%).

Summary

As reflected in the AJO, there was a rapid decline in professional interest in some services central to the work of the early clinics. With the exception of the growing emphasis on group therapy in the 1940s, there was little change in the nature of services for children between 1930 and the 1960s. In the 1960s, family therapy methods became popular, representing an extension of both the group treatment concept and the practice of treating parents and children separately but simultaneously. Behavior modification methods also expanded in the 1960s as research showed the limited effectiveness of the predominant psychoanalytically based psychotherapies and as psychologists showed the utility of behavior and learning theory for therapy.

As developed in the 1960s, family therapy theory viewed human behavior less as a matter of individual dynamics and more as a function of a family system. That theory is context oriented. Similarly, behavior modification assumes the individual's capacity to respond to environmental stimuli and contingencies and minimizes the individual's psychopathology, although behavior modifiers do use such concepts as *behavior deficits*. This emphasis on environment as a "cause" and environmental change as a "cure" parallels the orientation toward reform during the 1960s when our viewpoint suggests that environmental explanations should be prominent.

These trends—a decline in the use of play therapy, an increased interest in behavior modification, and a movement toward family group treatment—all represent progress in the mental health fields. The changes show an understanding, coming from experience and formal evaluation studies, that what we had been doing was insufficient and that other methods should be tried. The basic treatment model did not change, however. Rather, the change occurred in the "medical model" of the delivery of services by professional personnel.

The pattern of services set in the 1930s did not shift much until the 1960s. Articles on the relationship of mental health agencies to the schools, the juvenile courts, and other social welfare agencies were infrequent in those years, which suggests that by 1930, the mental health fields had become isolated from contact with social problems stemming from poverty. The theoretical orientation and the treatment models that the field had adopted contributed to that isolation. After the 1960s, different issues occupied professional attention and energy.

Renewed Attention to Social Problems

Poverty, welfare, and social problems. One could read through all the 1930s volumes of the AJO and hardly be aware of the Great Depression, for the topic is barely mentioned. There are fifty entries in the AJO's fifty-year index under the heading of poverty, with the first not occurring until the 1950s. Only 10 percent of the poverty entries appeared before 1965 (E = 68%). Sixty-two percent of the entries are dated between 1966 and 1970 (E = 11%). An additional twenty-two articles are indexed under public welfare, but there is no public welfare entry before 1946; 86 percent of the articles were published after 1960 (E = 56%). Only two entries are listed before 1960 under the heading of disadvantaged child.

The Social Security Act of 1935 was a major reform to alleviate economic distress, and it changed public welfare decisively. This act provided for payments to widowed, indigent mothers to care for dependent children at home; later amendments funded maternal and child care programs. Nonetheless, despite the importance of this act, the AJO had no article on it until 1947, and there are few after that date.

It is not that contributors to the AJO were isolated from society. There are a number of articles in the AJO on World War II's effect on civilian life, and some theoretical articles speculating on the aggressive drive as manifested in war. Eisenberg and DeMaso (1985) point with justifiable pride to articles in the AJO in the 1950s that discussed the deleterious psychological consequences of McCarthyism. It took courage to criticize openly the "red hunts" then taking place. There were many publications on delinquency, a problem prominent on the public agenda during and directly after World War II. But much of the research and writing dealt with the effects on delinquency of maternal employment and paternal absence. The articles tended to stress the psychodynamics of delinquency and reported on efforts to treat delinquency with individual and group psychotherapy. Although Eisenberg and DeMaso (1985) correctly state that in comparison with many other professional organizations, the American Orthopsychiatric Association took an active stance on social issues, many social problems were not reflected in the AJO until the 1960s and later.

We shall turn now to an examination of abstracts on social problems that appeared in the AJO in the 1960s and 1970s. Articles about welfare, child protection, abortion, and birth control now appear.

Out-of-wedlock pregnancy. It is difficult to explain why topics that came to be of central concern in later years were not discussed earlier. Out-of-wedlock pregnancy is scarcely a new phenomenon; Juvenile Court Judge Ben Lindsey helped many pregnant adolescents (Lindsey & Evans, 1925). In the past, an unwanted pregnancy caused severe distress for many young women because of the strong degree of social opprobrium. Thirty-six articles on out-of-wedlock pregnancy were indexed over the fifty years, with 80 percent published after 1960 (E = 44%) and nearly 40 percent between 1976 and 1980 (E = 10%).

Why did the mental health field pay so little attention to this phenomenon? Surely it must have been more difficult, psychologically speaking, to have an out-of-wedlock pregnancy in the years before the sexual revolution of the 1960s than afterward. In many states, out-of-wedlock mothers were not eligible for welfare, which must have added to their plight. Did pregnant unmarried women, who may well have been desperate to the point of suicide, not turn to psychologically oriented clinics for help? Lancaster and Hamburg (1986) assert that adolescent pregnancies were rare in the distant past and frequently resulted in marriage. But we found little discussion in the AJO of the consequences for marriage and child rearing of early, forced marriages. Was it truly not a problem at all, or was it not a problem that came to the attention of mental health personnel?

Birth control. Beginning in the 1870s, laws were passed nationwide that restricted the advertisement and distribution of contraceptive information and devices. Margaret Sanger's struggle to disseminate information about birth control and contraceptive devices began in the early 1900s, and by the 1930s, her fight had been largely successful. Between 1930 and 1980, the AJO published fifteen articles on birth control and contraception. Eighty-seven percent were published after 1961, and 40 percent of the total were published between 1976 and 1980 (E = 10%).

The increase in articles on birth control emerged after the U.S. Supreme Court decided *Griswold v. Connecticut* (1965). The privacy right in sexual matters discussed in *Griswold* took full force in *Roe v. Wade* (1973). In *Griswold,* the Court struck down "an uncommonly silly" Connecticut law that not only forbade giving contraceptive information but also prohibited the use of contraceptives even by married couples. Restrictions on the distribution of contraceptive devices to minors were struck down in *Carey v. Population Services* (1977). Granted that it was illegal in most states to disseminate contraceptive devices or information to minors before that date, there are no case studies and no discussions in the AJO of the place of contraception in the lives of young people. There are only eight articles on sex education in the AJO, and interestingly, five of them were published before 1945. There is little in the AJO on sex education after that date.

Abortion. No articles on abortion were published in the AJO between 1930 and 1968. Of the twenty-two articles that were published on this subject, nineteen appeared after 1972. Before 1968, there was not even a single case study of the psychological consequences of abortion, even though the AJO published many case studies on other topics. Before *Roe v. Wade,* abortion was available in some states if recommended after a psychiatric evaluation. Articles discussing the psychological characteristics of women seeking abortions and their reactions to abortions were published in psychiatric and medical journals. But such articles did not begin to appear in the AJO until abortion reform was occurring in many states in the mid-1960s. The vast majority appeared after 1973 when *Roe v. Wade* struck down state laws that unduly restricted abortions.

Legal change has become an influential factor in determining what problems come to our attention and are studied. After the *Roe v. Wade* decision, the advent of community-based pregnancy clinics and outpatient abortion procedures made it possible to study large numbers of women who were concerned about problem pregnancies, whether they kept their babies, put them up for adoption, or terminated their pregnancies.

Child abuse and neglect. Western society is no stranger to child abuse. The topic received renewed attention in the mid-1960s with the identification of the battered child syndrome (Kempe et al., 1962), and by 1967 all the states had passed legislation that made it mandatory for professionals to report child abuse (Meriwether, 1986). We found thirty-two articles on child abuse and neglect in the AJO, nearly 80 percent of which had been published after 1971 (E = 21%). Once again, legal change put the mental health professions in touch with a problem.

Child sex abuse and incest. Articles dealing with sex abuse and incest show a pattern similar to that of articles on child abuse. Of the twenty articles indexed under these two headings, seven (E = 25%; O = 35%) were published before 1945, three of them describing the sex offender. One dealt with incest, and two were studies of sixteen children who were referred by a juvenile court because the children had had sex with adults (Bender & Blau, 1937; Bender & Grugett, 1952). One article on incest was published in 1954, the others considered sexual delinquents. There were no articles or case studies on sex abuse and incest between 1956 and 1976, although a few articles on other aspects of sexual behavior were published in those years. Forty percent of all articles (E = 10%) in the AJO on incest and sex abuse were published between 1976 and 1980, after reporting laws came into effect, thereby forcing the public's attention to the problem.

Divorce and child custody. Although we shall not discuss divorce and child custody in this book, these issues represent another instance in which changes in the law resulted in changes in clinical practices and research interests. Before 1976, there was an almost total absence of discussion in the AJO of mental health problems related to divorce and child custody. No article on divorce or child custody appeared before 1947; only one was published in that year and another in 1968. No article was indexed again under this heading until the five years between 1976 and 1980, when sixteen articles appeared. Similarly, there were no articles on custody disputes in divorce before 1972. It was not that there were no problems. Lawyers recognized that custody agreements frequently were difficult and often were the rock on which settlement efforts foundered. Before the early 1970s, however, custody disputes rarely reached the courts because of the "tender years" presumption then existing in favor of giving the custody of young children to the mother. The few articles listed under custody before the early 1970s examined custody changes related to child protection.

Articles on child custody reflect the fact that the states were engaged in divorce reform in the early and mid-1970s, amidst a rising divorce rate (Wadlington, 1984). Moreover, the women's movement attacking gender discrimination in many areas of law resulted in changes in statutory language, so as to make the laws appear gender neutral. This change in language eliminated built-in preferences for maternal custody (see *Watts v. Watts,* 1973; *Johnson v. Johnson,* 1977). The statutory changes encouraged lawyers to support fathers who sought custody, because it would no longer be a foregone conclusion that the mother of young children would be awarded custody as a matter of law. These legal changes thus brought new problems to the attention of mental health workers, even though divorce must have been as difficult psychologically before 1972 as it was afterward, and custody problems must have arisen in earlier times as well.

Summary

We have argued that various aspects of social change and of professional sociodynamics affect both the problems that come to our attention and the ideology and methods that we use to work with those problems. Now we can add legislation and litigation, reflecting social change, to the list of variables that define clinical and research problems.

Our review of the AJO abstracts shows that the problems of welfare, child protection, birth control, and abortion did not become salient for the mental health professions until the 1960s and later, with the onset of the War on Poverty, renewed attention to the mental health aspects of social problems, and the increasing influence of changes in laws on clinical practice and research. These three problems are interrelated and continue to be relevant today. The history of services and interventions in relation to welfare, child protection, birth control, and abortion is presented in the following chapters.

11

Aid to Dependent Children

Children from poverty-stricken, disorganized families; children born out of wedlock; those born physically or mentally disabled; and children who are neglected, abused, or abandoned are considered at risk for social and psychological disorders. Children have always existed in such circumstances and in larger numbers than we like to admit. The United States' welfare services, and by extension most human services for children and families, originated with efforts to aid the poor. With the development of child guidance clinics and the emphasis on psychotherapy based on a psychoanalytic model, the problems of poverty that occupied our professional forebears in the early part of the twentieth century did not really come to the attention of the mental health professions again until the 1960s.

In this section, we will describe a key child welfare program, Aid to Families of Dependent Children (AFDC). The program's origins can be traced to colonial America, but modern-day programs arose first in connection with massive immigration, beginning in the 1840s, and its attendant poverty. Welfare programs, which primarily provide support for children, can be traced to mid-nineteenth-century private philanthropy and to the New Deal reforms in the 1930s.

Welfare in Early America

Colonial ideas about appropriate care for the dependent were based on the English poor laws of 1601 (Coll, 1969). These laws were enacted because England was struggling with the problems created by the breakdown of the feudal system of mutual obligation. Under feudalism, each person had a place in society and obligations of care, however limited, extended to those who could not care for themselves. Under a wage system, however, the obligation of an employer to an employee was limited largely to paying wages, and if the worker could not work, the employer had no further obligation. Families were responsible for their own members. If the families could not take care of them, the members were left to their own devices or to whatever charity was available through the churches.

The poor laws formalized the concept of local community responsibility through local government for those unable to provide for themselves. The laws did not differentiate among categories of the dependent, except between the unemployed and the unemployable. People were financially responsible for their kin; each parish was to provide an overseer of the poor; and the law empowered the parish to commit to institutions those who could not be cared for by either themselves or their families. Most poor families received some "outdoor" relief in their own homes (Katz, 1986). Neglected or dependent children worked as apprentices or "indentured servants" if they could, but many were placed in almshouses or poorhouses. Some were placed with families under an agreement to care for them in return for a sum of money.

The colonies followed the English pattern. Urban communities created undifferentiated almshouses and workhouses that cared for the dependent and the deviant (Bremner, 1971; Rothman, 1971). From the early 1800s to about 1875, institutions predominated. Orphan asylums were opened by private charitable or religious groups. By the 1830s, specialized institutions for deaf, blind, and mentally disabled children prevailed. Houses of refuge for delinquent youth, organized by private charities, appeared in the 1820s. By the mid-nineteenth century, state reform schools emerged. In those days, children who were difficult to manage and homeless children who made their livelihood on the streets by their own devices were sent to such schools (Folks, 1902; Lundberg, 1947; Rothman, 1971). Even with these institutions, however, the need for other forms of care grew.

The Children's Aid Society

By the 1840s, prolonged economic depression and increasing numbers of immigrants seeking work in the new land compounded the problems of urban poverty. Tens of thousands of Irish and then German immigrants came to the United States. The problems intensified after the Civil War when the immigration resumed. Most immigrants did not go west but settled in the cities where employment was uncertain and marginal and life in the tenements was hard, particularly for those coming from rural areas.

Charles Loring Brace (1880), a minister and an astute observer of the culture of poverty, described bands of homeless children, deserted by their families, who lived where and how they could. In the late nineteenth century, homeless children were called *street arabs, gutter snipes,* or *waifs*. Street arabs were nomadic, traveled in bands, were predatory, and had a loose government, their own dialect, customs, and traditions. Gutter snipes were weaker children, less able to keep up with the street arabs, and reduced to scavenging in garbage heaps for subsistence. Waif was an English legal term meaning "goods found of which the owner is not known." The term applied to children who had been abandoned by their parents or were held by them in a state of cruel bondage. Observers agreed that in any given year in the mid-nineteenth century there were as many as thirty thousand children roaming the streets of New York City (Needham, 1887). Brace worked with the children of Irish and German immigrants; later, the children of poverty came from southern and eastern European immigrant families.[1] Describing them as the "dangerous classes" because he saw in them the potential for social breakdown if not revolution, Brace was a leader in the "child-saving" movement designed to rescue children from vice and degradation.

A new period in child welfare was marked by the formation of the Children's

Aid Society of New York in 1856, with Brace as its executive director. Earlier, many abandoned or orphaned children had been placed in almshouses, and such infants had been turned over to foundling homes. During the 1860s, states established welfare boards to supervise the private charitable agencies, which received some public funding. The state welfare boards made efforts to remove children from almshouses, and some states passed laws barring the placement of children in almshouses and requiring their removal from them (Folks, 1902). But removing children from almshouses also meant breaking up families. In that day, "child saving" often meant saving children from their poor parents (Katz, 1986).

The child-saving movement provided a partial alternative to long-term institutional commitment. By law, children could be committed by the courts to child-saving societies and then placed in facilities temporarily until they were later boarded out or indentured to families rather than placed in institutions for prolonged periods.[2] Institutional placement, however, continued to be an important disposition, and in 1910, 108,000 children were in specialized institutions rather than in almshouses (Warner, 1919). The supply of homeless children was indeed large.

The Children's Aid Society, developed in the spirit of child saving, created many useful local community services similar to those provided by the settlement houses in a later day. Brace tried first to attract homeless and poor children to boys' meetings, or evangelical prayer meetings. These initially proved a failure because of the rowdiness of the youths they attracted. The inadequacies of the prayer meetings soon led Brace and his colleagues to develop other, more directly beneficial services, including daytime "industrial schools," night schools, reading rooms, and lodgings for homeless newsboys, for which the newsboys paid a fee.

The industrial schools taught girls work skills and provided food and clothing to the "well behaved." "Prominent ladies from all the leading sects" volunteered to work with the street girls: "a flock of the most ill-clad and wildest little street-girls that could be collected anywhere in New York. They flew over the benches, they swore and fought with one another, they bandied vile language, and could hardly be tamed down sufficiently to allow the school to be opened" (Brace, 1880, p. 138).

The prominent ladies taught sewing, knitting, crocheting, and the like. Brace believed that the caring ladies and the children developed a relationship that was instrumental in the children's change. "They [the children] felt a new impulse—to be worthy of their noble friends" (Brace, 1880, p. 139). Once trained, the street girls were often placed in domestic service. Brace claims that many who grew to have a new sense of self-esteem also married "above their stations," thus improving their social and economic conditions. Needham (1887) describes similar programs and success stories. Citing official records, Brace claimed a 90 percent decline in the imprisonment of female vagrants where his industrial schools opened (Brace, 1880, p. 142).

Emigration as a Solution

Despite the good results claimed for these community-based services, Brace believed that the best solution for abandoned, urban children was to place them with farm families in the developing West, where there was a labor shortage (Wiens, 1984). The plan took advantage of the growth of the West and the need for labor in farm families. He used the laws that allowed the courts to commit neglected, destitute, or abused children, or those who begged, to child protection or other charitable societies or to "some respectable family" (Needham, 1887).

The idea of emigration may have come from the solution to the poverty problem proposed for adults by William Booth, the founder of the Salvation Army. In England, a plan was put into effect to send children to Canada. The children were sent first to training schools where they received medical examinations, baths, and clean, comfortable clothing. They were also given some education, including religious lessons and instruction in farm work, before they were sent off to be placed with families in the New World (Needham, 1887).

Brace's plan worked a little differently. Beginning in the 1850s, he placed advertisements in city and rural weekly newspapers and accepted applications from families who wanted to receive children. Not surprisingly, each applicant wanted a "perfect child," pretty, good-tempered girls, or boys of good stock who would never steal apples and would use language of "perfect propriety." The objective of meeting each applicant's specifications soon proved impossible. Brace then followed a second procedure:

> We formed little companies of emigrants, and, after thoroughly cleaning and clothing them, put them under a competent agent, and, first selecting a village where there was a call or opening for such a party, we dispatched them to the place.
>
> The farming community having been duly notified, there was usually a dense crowd of people at the station, awaiting the arrival of the youthful travelers. The sight of the little company of children of misfortune always touched the hearts of a population naturally generous. They were soon billeted around among the citizens, and the following day a public meeting was called in the church or town-hall, and a committee appointed of leading citizens. The agent then addressed the assembly, stating the benevolent objects of the Society, and something of the history of the children. The sight of their worn faces was a most pathetic enforcement of his arguments. People who were childless came forward to adopt children; others who had not intended to take any into their families, were induced to apply for them; and many who really wanted the children's labor pressed forward to obtain it. (Brace, 1880, pp. 231–32)

The "orphan train" project was not without its critics. Brace's version represents the perspective of the author of the plan. Some claimed that Catholic children were made into Protestants, that because the children were renamed, brothers and sisters might marry, that the children sometimes were sold into slavery, or that the agents made money unscrupulously on the transactions. Some considered that Brace's agents were overly aggressive in sweeping up children to send west (Thomas, 1972; Vandenpol, 1982). Poor, urban parents were often distressed when faced with the prospect of giving up their children, even to a good home in the West.

Brace asserted that he placed as many as 24,000 children with families in the West, at a much lower cost than institutional care, which he claimed would have been prohibitive for that many children. Needham wrote that from 1853 to 1882, 67,287 children were placed in western homes. He noted that the New York Juvenile Asylum also transplanted children from "the slums of the city to the homes of the prairies" (Needham, 1887, p. 328). Brace did some follow-up studies, which he admitted were inadequate, but he believed that if the children were under fourteen when sent out, most would soon be indistinguishable from the neighboring children in the same communities. The older children were not indentured but were free to leave if ill treated, or the farmers could send them away if dissatisfied. Brace's evaluation of the orphan train project's success may have been overly optimistic, however, as he overlooked many children who had multiple placements or who returned to the city and rejoined their families (Katz, 1986). As a measure of his various programs' success, he cited additional figures from New York City records showing drastic declines in arrests for vagrancy and for many different crimes among youths (Brace 1880, chap. 36).

The Charity Organization Society

In the late 1800s, the needs of the urban poor stimulated the development of private and church-related charitable enterprises, in addition to personal philanthropy. For example, Mrs. Etta Wheeler, who rescued Mary Ellen (whom we will discuss later), thus initiating child protection services, was a church worker in St. Luke's Methodist Mission (Lazoritz, 1990). Early in his work, Brace (1880) was already concerned about the duplication of services and the consequences of "pauperization" for those who became enmeshed in the service system.

> The number of poor people who enjoy a comfortable living, derived from long study and experience with these various agencies of benevolence, would be incredible to any one not familiar with the facts. They pass from one to the other; knowing exactly their conditions of assistance, and meeting their requirements, and live thus by a science of alms. (Brace, 1880, p. 384)

The proliferation of private charitable services led to the development of the Charity Organization Society (COS), with the first one established in Buffalo, New York, in 1877. Within five years, 18 more had been founded, and twenty-five years later, 170 were operating (Seidl, 1989). The COSs marked the beginning of professionalization in social work and an attempt to base charity on science. The societies maintained central registries and used them to attempt to eliminate duplicated services to the same family and to promote cooperation among agencies. The COSs also devised methods to investigate families not only to prevent fraud but also to tailor assistance to the family's needs.[3] Mary Richmond's (1965) text on casework methods was an outgrowth of this work. In addition, the societies either provided immediate emergency relief or served as referral sources to send anyone in need to the appropriate agency. They developed the volunteer, "friendly visitor" to establish personal relationships between the poor and the well-to-do. The societies served as employment agencies for the disabled or for those who through "some perversion of character or defect of mind or body cannot fit themselves into the industries of the time'' (Warner, 1919, p. 469). They provided education and recreation, activities not dissimilar to those carried out by the settlement houses ten years later. The COS also undertook to disseminate information to the public at large about charitable work and the prevention of dependence (Warner, 1919).

The mix of evangelical-influenced help, supervision of the home, and investigation, with the police sometimes used as the investigators, was criticized by labor unions and settlement house workers. Katz quotes a poet, John Boyle O'Reilly, who wrote:

> That Organized Charity, scrimped and iced,
> In the name of a cautious, statistical Christ.
> (Katz, 1986, p. 82)

The poor were dependent on the goodwill of their voluntary benefactors. And although their benefactors were ostensibly private agencies providing charity in fulfillment of a charitable impulse, the agencies received a substantial portion of their budgets from public funds. Brace (1880) states that fully 50 percent of his money came from public funds in payment for services. In those years, there was no statutory entitlement to welfare payments outside institutions, even though the funds came from the public treasury.

Mothers' Pensions

Institutional care of children came under critical scrutiny by state boards of charities that were organized in the mid-1860s to supervise public and private charitable enterprises. In exercising supervision, members of the state boards became concerned about the inadequate, if not scandalous, care of children in almshouses. In some states, separate orphanages for children had already been established. Foundling homes for abandoned infants were also troubling to authorities because of their high mortality rates. Such institutions in general came under criticism from welfare authorities:

> The care of destitute infants (children under two years of age) is sharply distinguished from the care of older dependent children [in almshouses]. Among the former the death-rate is the principal index of success or failure, while among the latter the death-rate is always low and the attention must be given to evidences of right or wrong development afforded by the character and subsequent careers of the children. Formerly, in a majority of cases it mattered but little to the individual infant whether it was murdered out-right or was placed in a foundling hospital—death came only a little sooner in one case than the other. . . . A death-rate of 97 per cent for children under three years of age was once not at all uncommon. (Warner, 1919, p. 251)

By 1875, state laws had been passed prohibiting the commitment of children to almshouses and requiring their removal (Katz, 1986). Many children were then boarded out in private homes, with the state paying the family a fee for their care (Folks, 1902).

Some of the most desperate circumstances were presented by infants whose mothers, many of them unmarried, had neglected or abandoned them. Unwanted infants who were not murdered were often placed in foundling homes.[4] Brace (1880) also cited studies showing that mortality rates among infants in some poorly managed foundling institutions rose above 90 percent and rarely were below 50 percent. These shocking figures led to some reform and efforts to board infants with individual families rather than place them in large institutions. With boarding, mortality rates dropped to around 30 percent, the norm for the time (Brace, 1880), and later dropped still further to 10 to 12 percent (Warner, 1919, p. 255).

The movement to provide pensions for indigent or widowed mothers to care for their children in their own homes was tied to a political and ideological struggle about whether institutions were preferable to "outdoor relief" (i.e., relief given in one's own home in the community). Opponents of outdoor relief argued that relief in the community in one's own home would result in pauperization. Private charities, which often benefited from public funds supporting children in church-related institutions, competed with public welfare agencies, with the private charities arguing against public welfare (Folks, 1902). In time however, outdoor relief won out over the proponents of institutional care, although the institutions never disappeared but coexisted with other means of providing assistance to those unable to care for themselves.

After the 1870s, surveys showed that a large number of children were taken from widowed mothers who worked at bare subsistence wages or who had no source of income, without other evidence that the child was exposed to immoral conditions or was exhibiting difficult behavior. Reformers pointed out that a destitute child could be removed from its mother, placed in an institution, and then boarded out to some other woman to take care of the child in her home at state expense. Because of legal restrictions, in many states no funds could be given directly to a poverty-stricken mother to maintain her child at home. By the early 1900s, a number of states had changed their laws to provide mothers' aid, pensions, or allowances to support children in their own homes. Often these allowances were granted by the juvenile or family courts (Lubove, 1968; Lundberg, 1947; Tiffin, 1982; Vandenpol, 1982).

The movement toward mothers' pensions administered by juvenile or family courts was a step toward the shift from private charity to public entitlement. In 1909, the White House Conference on Dependent Children recommended that aid be given directly to mothers to prevent the removal of their children from their homes. The conference report recommended private assistance, but it soon became apparent that private philanthropy could not cope with the task. Women's organizations such as the General Federation of Women's Clubs, the National Congress of Mothers, parent and teacher associations, and settlement house workers lobbied for public support for indigent mothers. Proponents of mothers' pensions argued that children could be maintained in their own homes at half the institutional cost (Tiffin, 1982).

Private charities and the Charity Organization Societies argued vehemently against public assistance. Public relief officials, they said, were poorly paid and were untrained, with the consequence that public relief efforts were poorly administered and subject to political manipulation. Aid viewed as a right only encouraged permanent dependency and undercut the social values of noblesse oblige, thereby increasing the distance between the rich and the poor. The opponents predicted that if aid were provided by the government, it would soon cost huge sums. Another argument against public relief was that no rehabilitation effort was associated with it. Simply giving away money was viewed as an inferior form of social work, and to social workers striving for professional recognition, the denial or de-emphasis of higher forms of casework undermined their claims to professional status (Lubove, 1968; Tiffin, 1982).

Mothers' aid, as mothers' pensions or mothers' allowances, although a form of relief, was usually presented as a child welfare measure. These allowances were written into the legislation of almost all states between 1911 and 1935 and were successful despite vigorous lobbying by private charities (Lubove, 1968; Tiffin, 1982; Vandenpol, 1982). The provision of mothers' aid, originally supervised for a short period through the juvenile courts (Ben Lindsey was active in campaigning for mothers' pensions), was eventually centralized in state agencies which also were responsible for supervising the activities of local public welfare authorities.

Many of the state laws had restrictions severely limiting who could receive aid and when. The restrictions were designed in part to save money and in part to make it difficult for fathers to avoid responsibility for their families. Some states would not permit deserted mothers to apply for relief until one year after the desertion, and then the woman had to agree to cooperate in prosecuting the deserting father for nonsupport. In some places, divorced women were ineligible for aid. Eligible women also had to be "proper," in morals, in physical health, and in mind. A man living in the home thus could make a woman ineligible for a pension. The use of tobacco and the failure to provide religious education or to learn English were grounds for refusing to grant support. Many of the laws made no provision for money to support the mothers in addition to an allowance for the children, and most states would not provide for expectant or unwed mothers. The allowances were often determined by a budget for each family, a practice stemming from the desire to have charity administered on a scientific basis (Tiffin, 1982).

Mothers' pensions were viewed not as a right but as a response to an abnormal condition of dependency. There was no recognition of the necessity to provide economic support as an end in itself in a wage economy in which in contrast with the mutual obligations of feudal social organization, the only tie of social responsibility between employer and employee was the payment of wages in compensation for labor. There was no recognition that wages were uncertain in different phases of the economy and that some people were simply unable to save enough to tide them over difficult times. Because the built-in structural characteristics of poverty were not recognized, those who accepted pensions continued to be stigmatized (Lubove, 1968; Piven & Cloward, 1971; Tiffin, 1982).

The Social Security Act of 1935

The Social Security Act was one of the major reforms instituted as a response to the Great Depression of the 1930s. Upon entering office in 1933, President Franklin D. Roosevelt's administration initiated many proposals to cope with the depressed economy and the misery it spawned. The Social Security Act was in part an answer to the popular Townsend movement in the early 1930s when hundreds of thousands of older people demanded pensions of $200 a month for everyone over age sixty. The Social Security Act, providing for relief for older Americans, also was a response to revelations of dismal conditions in the poorhouses and almshouses that housed many of the indigent elderly.

The act established the principle that everyone who had worked was entitled, as a matter of right and not as a matter of charity, to some minimum pension when they grew old and unable to work or when they were disabled. If death took the family breadwinner, then the widow and children of the covered worker were to be protected. Because the act provided for income to survivors of workers who were covered by Social Security, many fatherless children who otherwise would have been on welfare were covered by this insurance. Retired persons, widows of workers, their children, and disabled workers moved from a deviant to a more socially acceptable social position, permitting greater self-esteem as well. No longer were those unable to care for themselves to be stigmatized; they were to be supported through the social insurance to which they or their parents had contributed. Thus the onus of being poor or dependent was removed, and a structure was created to deal with a recurring and enduring problem.

The Welfare Program: Aid to Dependent Children

As a result of the efforts of settlement workers, the U.S. Children's Bureau was established in 1912 with former settlement house workers Julia Lathrop and Grace Abbott as its first and second chiefs. The Children's Bureau participated in the development of the Social Security Act in 1935. Aid for children was included in the social security relief package (Altmeyer, 1968; Lubove, 1968; Piven & Cloward, 1971; Witte, 1963).

The Social Security Act provided for aid to dependent children in their own homes, for local services for the care of neglected, homeless, and delinquent children, for maternal and child health services, and for assistance to disabled children.[5] Two categories of recipients were created. One consisted of workers and their families, covered as a result of the social security tax on their earnings, and the other included people who were dependent but had not paid into the social security fund, with the bulk of these being children.

The two categories of people were treated differently. Once eligibility was established, the workers' families received a regular check in the mail and were not subject to supervision. But the second category of people were subject to means tests for eligibility and were supervised thereafter by government welfare departments. This aspect of the Social Security Act (Aid to Dependent Children, or ADC, later named Aid to Families with Dependent Children, AFDC) established the basis for the current welfare programs. This reform did not reduce the stigma of dependency, although the welfare program was structured as a temporary expedient for those who could not work for a short period of time. Welfare was not conceptualized as an earned right in the same nonstigmatizing way that we have come to think about social security pensions, disability, or payments to widows and children of covered workers.

The Social Security Act was a decisive event in American public welfare. A welfare structure directed by the federal government was put into place in every state and county in the nation. Most welfare functions were transferred from the private to the public sector, and control of the programs shifted from the local and state level to the federal level (Lubove, 1968). The Social Security Act conditioned federal assistance to the states on the passage of state legislation that made aid mandatory in every local unit of the state and that established a state welfare agency to supervise the program.[6] This system of federal reimbursement in exchange for compliance with federal rules resulted in broader categories and less rigid criteria to determine eligibility for support. Many more states granted aid to children of unmarried mothers, to children whose fathers were absent or had deserted the family, and to children who lived with relatives other than parents. Successive amendments to the Social Security Act increased the allowances.

The Aid to Families of Dependent Children (AFDC) program grew slowly, and it was not until 1945 that almost every state and all the territories had accepted the federal plan. During the war years, the requests for aid were low:

> A decline occurred during the period when mobilization of workers for war industries offered many inducements to women and children to secure employment. Undoubtedly many mothers who had been maintaining their families on grants which were often inadequate for their needs chose to make other arrangements for the care of their children and to take advantage of the comparatively high wages they were able to earn. (Lundberg, 1947, p. 178)

After World War II, the AFDC program became the fastest-rising expenditure for most local communities, obligated as they were by federal and state law to share in paying welfare costs. At that time, the more affluent whites were moving to the suburbs, and the cities were filling with the poor, many of them African-Americans who had come north in response to wartime employment opportunities. Some had come because the mechanization of agriculture had driven them off the land. Black workers were paid lower wages than whites were and more often were unemployed. By 1960, 58 percent of all AFDC families lived in large cities, most of them in the inner cities. By 1961, although whites continued to make up the majority of all recipients of welfare, compared with their numbers in the population, blacks were overrepresented on the welfare rolls.

The combination of rising welfare costs and the visible numbers of blacks on the welfare rolls created resentment in some segments of the public and led to legislative investigations. Many on the AFDC rolls were out-of-wedlock children whose fathers had deserted the families. In some cases, unemployed fathers left their homes so that the mothers could apply for assistance. Other investigations showed that many on the rolls did not meet some of the rigid eligibility requirements adopted by many states and therefore could be considered welfare cheats. For example, in some states if a woman was able to work, she was not eligible for relief, whether or not jobs were available. Some of the harsher regulations were struck down in the courts in response to legal challenges.[7]

The process of applying for welfare was difficult not only for the client but also for the caseworker.[8] Clients at public welfare centers often had to wait for long periods before being called. On days when the intake workers' calendars were full, prospective clients were asked to return another day. Sometimes clients confronted impolite or indifferent caseworkers who had been instructed to give information sparingly. Workers also asked clients personal questions about sexual behavior, and if they made home visits, the caseworkers inspected the clients' personal possessions and interrogated their children about any male visitors to the home. Investigations for eligibility required contact with relatives, in the hopes that others would be shamed into taking responsibility for the prospective welfare client. These attitudes resulted in some people's turning away from assistance in dismay; almost everyone was enraged at the process. Sometimes applicants who persisted through the intake process were turned down arbitrarily, and sometimes those who did obtain welfare benefits had their benefits terminated arbitrarily (May, 1964; Piven & Cloward, 1971).[9]

After World War II, when caseloads grew rapidly, caseworkers had little or no training for their jobs. A bachelor's degree in any field was all that was required, and in some states less education was acceptable. In 1960, only 1 percent of welfare caseworkers had a two-year master's degree in social work (May, 1964). The job was frustrating and low paid, caseloads were high, and the paperwork was overwhelming. Workers had as much trouble seeking reimbursement for expenses they incurred (e.g., using a car to make a home visit) as clients did in getting welfare payments. The workload was such that department regulations often could not be met, and workers could not provide much by way of services. Rules prevented them from providing birth control information, even though many clients had out-of-wedlock children. It was no wonder that surveys showed that worker morale was generally low or that turnover rates were high (e.g., reaching 50 percent in Erie County, New York, in 1960).

Piven and Cloward (1971) argue that all relief measures are designed as temporary cushions when the labor market cannot absorb workers and that they are often sops to prevent civil disorder. They contend further that aid is set at a level low enough to encourage workers to reenter the labor market when the need for labor increases. In order to keep the welfare rolls within limits, the process of receiving aid is purposely made difficult and demeaning. In their view, eligibility requirements and application procedures were designed to keep people off the welfare rolls.

> These practices are not merely the consequences of carelessness or inaccuracies inherent in a cumbersome bureaucracy. Rather, secrecy, intimidation, and red tape are adaptive patterns designed to inhibit completion of the application pro-

> cess and facilitate arbitrary rejections and terminations. Agency procedures are designed to make it easy and safe to reject or terminate cases, but complicated to accept them. (Piven & Cloward, 1971, p. 160)

AFDC payments were set at a minimum subsistence level, and then the basic payment was supplemented by allowances for special needs or extraordinary circumstances. Observers claimed that workers often did not tell clients about the amounts to which they were entitled or that workers sometimes actively discouraged clients from requesting the extras for which they were eligible. Based on the age and sex of each child, the caseworker had to work out a budget down to the penny for each family. The smallest change in the family's circumstances, such as a child's leaving the home to visit with relatives for a summer vacation, required that a new budget be worked out. The client bore the burden of proof that the extras were necessary, with the result that clients were in the position of begging for items that could justifiably be called necessities. Such extras included items of clothing or school supplies for children, furnishings, household appliances, day care, telephone, transportation, education or employment training expenses, extra or special food allowances, and money for utilities, heat, and rent.

Although the system was cumbersome—one worker noted that he had to complete twenty-four separate forms to get the first check to a client (May, 1964)—a client finally on the welfare rolls was given little help in overcoming problems and getting off the rolls. Supervision of most clients was more likely to be perfunctory than close, but the "helping relationship" was coercive; the helping person was a caseworker who also controlled the purse strings (Handler & Hollingsworth, 1971).

National Welfare Rights Organization (NWRO)[10]

The difficulties that clients experienced in applying for welfare, accepting supervision, and, especially, obtaining the welfare allowances to which they were entitled were grievances that became the core of a protest movement. In the 1960s, the time was right for poor people to organize on their own behalf. The Economic Opportunity Act of 1964, the basis for the Kennedy–Johnson War on Poverty, with its concept of maximum feasible participation (Moynihan, 1969), provided the authority and the funds for the establishment of federally funded community action programs with neighborhood service centers (Sarason et al., 1966). Workers in many of these centers began to act as advocates for people in their neighborhoods, helping them obtain welfare, determine underbudgeting, and fight for increases in their allowances.

The staff of Mobilization for Youth (MFY) in New York City, one of the early antipoverty agencies, recognized that it was a losing battle to take on the welfare system one client at a time. That is, winning benefits for one client by badgering the welfare department did nothing for the next client.[11] The MFY staff then decided to organize the AFDC mothers to press for their allowances through a form of collective bargaining. Accordingly, mothers in the MFY neighborhood prepared requests for winter clothing allowances for their children. Written requests went to the local welfare office and then to the New York City commissioner; when they received no reply, the group threatened to picket. The commissioner responded to the threat by agreeing that all who were entitled to receive grants would receive them. He also agreed to meet with the recipients to work out grievance procedures.

Word of the victory spread quickly. In New York City alone, thousands of welfare families in other neighborhoods organized and were assisted in the same way. Similar organizations of welfare mothers sprang up in many other cities in the country. Bailis (1974) notes that in Boston, the first local group was organized in 1963, in part under the influence of Students for a Democratic Society (SDS). Similar groups were organized in Minneapolis and St. Paul as mother's clubs, with the assistance of social workers based in community agencies (Hertz, 1981). In 1966, Cloward and Piven circulated a draft of an article arguing that by encouraging as many people as possible to enroll and by taking full advantage of all welfare entitlements, the welfare program could be brought into a state of crisis, and reformers could press for a guaranteed annual income. The welfare rolls increased sharply, but it is debatable whether these organizing efforts were largely responsible.

At about that same time, Dr. George Wiley organized the Poverty Rights Action Center in Washington, D.C., drawing on his experience as an associate director of the Congress of Racial Equality (CORE), one of the civil rights organizations fighting for desegregation and other civil rights. The center encouraged welfare rights demonstrations in a number of major cities, and many but not all were successful. In August 1967, delegates from sixty-seven local welfare rights organizations met in Washington to form the National Welfare Rights Organization (NWRO).

The NWRO rejected the idea of flooding the rolls with new recipients of welfare and chose instead to work with those already on the rolls to obtain more benefits. The NWRO's membership grew rapidly, although the exact numbers were never clear. It had 22,000 dues-paying members, but at its peak of influence it may have reached as many as 75,000 to 100,000 people if one included family members. Its membership was concentrated in a few large states and in the large cities.

In addition to using indigenous leaders, the local NWRO-affiliated organizations used neighborhood legal service lawyers and VISTA volunteers (Volunteers in Service to America was a domestic Peace Corps program). The NWRO supported professional organizers who went to communities to organize local groups and statewide coalitions. The organizers, trained in the Alinsky tradition of helping groups help themselves (Alinsky, 1971), taught individual welfare recipients how to get what was due them from the welfare program. Often they insisted, as a price for their help, that women join the local group and participate in its activities. The local groups agitated and sat in at welfare offices, and they may have influenced some welfare workers to make more liberal decisions about benefits. In some areas, the groups were so influential that welfare authorities even consulted with the NWRO representatives about welfare problems. Despite these successes, however, the seeds for conflict were sown because many local people felt that the organizers were manipulative, employing the rhetoric of participation, but keeping control in their own hands.

At the national level, the NWRO leaders lobbied Congress, consulted with the secretary of health, education and welfare and the White House staff. The NWRO won church and foundation grants and a contract to monitor a major government welfare program.

At the height of the NWRO's success, however, with one blow, welfare authorities undercut the major basis for organizing welfare recipients. The welfare authorities reacted to the special-needs grant problem by simply eliminating it and going to a flat-grant system.[12] The organizers soon learned that poor people were willing to participate as long as concrete benefits were quickly forthcoming but that it was difficult to maintain a large organization of poor people when their participation produced no immediate, tangible results.

By the early 1970s, internal conflicts had factionalized many of the welfare rights organizations and coalitions. In addition, the political climate had changed. Government and citizens in general were less receptive to welfare recipients' claims for social justice; not only were they seen as the cause of the increasing local tax burdens, but in the late 1960s, many minority poor also were rioting in the inner cities. Negative public opinion hurt the NWRO's fund raising: fewer resources supported fewer organizers; and so many local groups simply ceased to function.

The internal conflicts centered on many issues: the division between centrally supported organizers, who were most often white, and the indigenous leadership, who were most often black; between white liberals and blacks who, in a ''black power'' era, were increasingly insistent on controlling their own organizations; between males who dominated the paid positions and female members of the organization who believed they should be entitled to powerful, paid positions in the organization; and between those who saw an organization built around welfare as unpopular and stigmatizing and those who wished to continue to pursue policies to improve the lot of those receiving welfare.

The movement's fatal flaw was that it was never financially self-sustaining. The NWRO had prospered because of Wiley's abilities to convince church and private foundations to give his organization sizable grants. But the organization was never able to secure sufficient funds from its membership. Its black power rhetoric also alienated white liberal friends who contributed some funds but who were offended when the increasingly black-dominated organizations made it clear that all they wanted from whites was money, without participation and without accountability. Demands to churches for $500 million in ''reparations'' reached unreceptive ears and were viewed as unreasonable. After a few years, therefore, churches and foundations began looking for other recipients for their money. In the mid-1970s the Protestant churches, which had been losing members, began looking with favor on the support of evangelical efforts rather than social action programs. When Wiley left, or was pushed out of the NWRO, he took with him his personal ties to funding sources. None of the later leaders of the NWRO had links to such sources, whose interests had shifted, in any event.

In the NWRO's later days, black women who had been welfare recipients became increasingly vocal and insistent on moving into positions of power in the organization. They began to see welfare not only as a poor people's issue but also as a woman's issue. Some of the NWRO's leaders believed that all mothers, married or not, should be counted as working because they were caring for children and so should be given economic credit and recognition for their work. The women who moved into leadership wanted to press for a guaranteed annual income, for welfare programs that would allow them to stay home with their children while receiving support—in a manner that did not undermine their self-esteem—and for programs that would offer opportunities for real training and real jobs. This program was articulated late in the game, however, after the peak influence of the NWRO had passed and when the body politic was no longer amenable to reform programs.

A few statewide and national groups, some of them offshoots of welfare organizing, were able to continue some of their efforts, but as the political climate changed, the influence of local leaders and groups diminished. By the mid-1970s, the NWRO had nearly faded from the scene, and if some statewide organizations remained, their activities were much reduced. The NWRO finally closed its doors in March 1975.

Bailis (1974) states that it is difficult to assess the full effects of the welfare rights movement. In Massachusetts, a small number of welfare recipients used the movement as a means for personal social mobility. Some took positions in government, and others continued to function on advisory bodies and related groups and organizations. Community organizers learned that it was indeed possible to organize the poor along racial and ethnic lines if the issues were highly visible, highly salient to the members, and promised immediate payoffs. But they also learned that it was difficult to retain the poor in social action organizations once the short-term objectives had been met. It also proved to be easy for the authorities to undercut the movement's base by changing the rules of the game. Poor people simply did not organize around abstract issues.

An Intractable Social Problem?

In 1988, about 11 million people were on the AFDC rolls. Of those who received aid, 7.3 million were children. Many of the recipients are children born out of wedlock. More than half of welfare expenditures go to women who had their first child before they were twenty years old (Klerman, 1986). The availability of welfare probably does not increase pregnancy rates, but it might induce women to keep their children (Chilman, 1986). Welfare payments also may encourage young women to establish independent households, and living apart from the influence of parents or grandparents may not be best for an adolescent trying to complete school or plan for the future (Klerman, 1986).

With sporadic efforts to redesign the system, the welfare program has improved only a little since the late 1960s (Jansson, 1988). Indeed, a political administration hostile to welfare programs can develop creative methods for harassing welfare clients. During the Reagan presidency, a practice was instituted called *churning*. Using regulations that required periodic recertification, local welfare agencies sent letters to recipients containing questionnaires to be returned by a specified date. Thousands were cut off each time a recertification letter went out because the questionnaires were filled out improperly or returned after the deadline. An internal audit of the New York City welfare agency determined that 54 percent of the cutoffs could be characterized as inappropriate. More than two-thirds of those who were cut off returned to the welfare rolls within six months (Funicello & Schram, 1989).

From time to time, different programs are tried, to provide services or to help welfare recipients leave the welfare rolls. Workfare is currently popular, although the success of experimental workfare programs has been shown to be marginal at best.[13] Successive changes in law have authorized various services: to support children and families in the home by paying for outpatient child guidance services through Medicaid, homemaker services, day care, and educational or vocational training (Kadushin & Martin, 1988; Katz, 1986).

Mental health workers have been peripherally involved with the welfare system. Traditional mental health services may be unsuitable for most welfare recipients, but since the 1960s with the advent of the community mental health centers movement, mental health workers have had more interaction with welfare clients.

Welfare is still considered a program to aid people in temporary need. As a nation with Puritan forebears, we still believe that there should be little or no need for a welfare program, that everyone should be self-supporting, and that we should assist only those who are deserving. We have not yet squarely faced the problems created by the changing nature of work. There now are fewer and fewer positions paying even at the official poverty level that can be filled by those with minimal reading and mathematical skills. Basic medical care is not available to many workers earning marginal incomes when they do not receive adequate medical insurance through their jobs and are not eligible for Medicaid. Many people in low-income jobs are priced out of the child care market. The housing markets, driven in part by the gentrification of the cities, have priced the poor out of their homes and have helped create the widespread conditions of homelessness. Too many children are being reared in welfare hotels and are undersocialized to the society's dominant values. The rampages of youths in ghettos who sometimes go on destructive "wildings" and the gang warfare to gain hegemony in local drug markets can easily lead us to remember Brace's description of the "dangerous classes":

> It has been common, since the recent terrible Communistic outbreak in Paris, to assume that France alone is exposed to such horrors; but in the judgment of one who has been familiar with our "dangerous classes" for twenty years, there are just the same explosive elements beneath the surface of New York as of Paris.
>
> There are thousands on thousands in New York who have no assignable home, and "flit" from a attic to attic, and cellar to cellar. There are thousands more or less connected with criminal enterprises; and still other tens of thousands, poor, hard-pressed, and depending for daily bread on the day's earnings, swarming in tenement-houses, who behold the gilded rewards of toil all about them, but are never permitted to touch them.

> All these great masses of destitute, miserable and criminal persons believe that for the ages the rich have had all the good things of life, while to them have been left the evil things. Capital to them is the great tyrant.
>
> Let but Law lift its hands from them for a season, or let the civilizing influences of American life fail to reach them, and, if the opportunity offered, we should see an explosion from this class which might leave this city in ashes and blood. (Brace, 1880, p. 29)

Perhaps Charles Loring Brace's chilling rhetoric was designed merely to stimulate private contributions to charity, but we heard similar predictions before our ghettos exploded in the late 1960s. We cannot expect to eradicate poverty or to cope with all of the evils at once, but we do need to renew our efforts. As Sarason (1978) noted, when we deal with intractable social problems, there are no right answers:

> The scientist who enters the arena of social action would do well to be guided by the values he attaches to the facts of living. . . . [T]here will be no final solutions, only a constantly upsetting imbalance between values and action; the internal conflict will not be in the form of "Do I have the right answer?" but rather of "Am I being consistent with what I believe?"; satisfaction will come not from colleagues' consensus that your procedures, facts, and conclusions are independent of your feelings and values, but from your own conviction that you tried to be true to your values; you will fight to win not in order to establish the superiority of your procedures or the validity of your scientific facts, concepts and theories but because you want to live with yourself and others in certain ways. (Sarason, 1978, p. 379)

Today, we are trying to come to terms with built-in problems. The problems may be deeper and may involve more difficult issues—drugs and the ready availability of deadly weapons; large numbers of unwed teenage parents; children who are not prepared for school and schools not prepared for the children; crack babies and AIDS babies; a job market that cannot absorb unskilled workers; homelessness, including many homeless children; and a housing stock that needs to be thoughtfully refurbished so as to avoid the worst failings of public housing in the past and to enable poor people to live in decent accommodations. It may be that today's problems of poverty are as much women's problems as anything else (Harrington, 1984), but this diagnosis does not in itself point to a solution.

The problems have not gone away during the "benign neglect" years of conservative political administrations. If anything, they have grown worse. Charles Loring Brace's "solution" of sending poor children out West is not available to us today. The social security system is stretching to meet its obligations. A new War on Poverty may not be an answer, either. It is not a solution to blame the victim and to point to the psychological or mental limitations of the population at risk as the cause of their problems. It is clear that we ignore contemporary social problems at our peril. We may be dealing with intractable problems, and there may not be optimal solutions, but none of that excuses a lack of effort, not only because we feel imperiled by the state of our "underclass," but also because we owe one another obligations of care simply because we all are human.

Summary

We have always had a problem in deciding what to do about those among us who are dependent. Private charity was helpful to many, but it also was demoralizing and stigmatizing. Charles Loring Brace, responding to the poverty of the time, created programs to provide opportunities for abandoned and neglected youth. His major accomplishment in placing thousands with farm families in the West was based on the assumption that the youths would respond to normal conditions and develop normally in good home environments. His was an ecologically oriented theory that given proper resources, youth would blossom. The later efforts in the 1870s to develop a scientific charity were designed to separate the worthy from the unworthy and required treatment of the unworthy to help make them self-sufficient. In this more conservative period, when the ideas of social Darwinism prevailed, the focus of efforts was more on the individual's pathology and less on his or her conditions of living.

Although there were efforts to provide mothers' pensions before the Social Security Act of 1935, poverty was stigmatizing. The Social Security Act, cast as insurance, thus immediately de-stigmatized all those who were eligible for it by virtue of having worked or being dependent on someone who worked. The social security laws produced a massive and profound change in the way we looked at certain problems. After World War II, a welfare program built into the Social Security Act—meant as temporary assistance for those who would reenter the work force after a temporary inability to work—grew to include large numbers of women and children who would stay on welfare for a long period of time. Because we continued to view poverty as stigmatizing for those who had not earned their security and because we wanted to contain expenditures, the process of applying for welfare and staying on it was made difficult and demeaning.

In the 1960s, there was a brief effort to organize the poor to act on their own behalf. This experiment in empowerment was effective for a short while, but it proved to be too little, too late. The effort at organizing the poor to act in their own interests also proceeded on the assumption that given the proper encouragement and the proper organization, poor people would not need therapy but could act in their own interest and thus cure themselves.

The problem of providing for those unable to care for themselves may be intractable and will not give way to a scientific solution. There may not be an optimal solution, but only one that allows us to feel we are living up to our own values.

12

Child Protective Services

Early History

Until comparatively recently, children had few legal rights. Parents were protected by law from assault by their children, but children had little protection against their parents, who in ancient times could sell children into slavery, mutilate or kill them, or abandon them as infants (Brace, 1880; Johnson, 1984; Mason, 1972). Parents had minimal duties to educate and maintain their children, and there were few restrictions on the discipline they could impose (Mason, 1972; Wadlington, Whitebread, & Davis, 1983). In colonial days, the system of committing destitute children to almshouses was supplemented by the use of apprenticeships and indenturing. Statutes passed as early as 1735 permitted neglected or abandoned children to be taken by the community and "bound out" to other families. The purpose was probably less to protect children than to minimize the burden on the public of children who could not care for themselves or who might become criminals.

Before the nineteenth century, only destitute children or children engaged in wrongdoing became the object of public attention and care. Parental wrongdoing was by and large ignored (Mason, 1972). But, as early as 1825, as the cities grew and people lived in closer proximity, statutes began to appear emphasizing the right and the duty of the public to intervene in cases of parental cruelty or gross neglect seriously endangering the health, morals, or elementary education of children. By 1833, New York City had a law that permitted municipal officials to commit abandoned, suffering, or neglected children, or those whose mothers were "immoral," to an almshouse or other means of care (Folks, 1902). Some of the later laws included abandonment or cruelty to children (as defined in the penal code) (*Matter of Knowack,* 1899).

The existence of the laws indicate that authorities were well aware of acts of abuse, cruelty, and neglect of children by their own parents. Homeless and abandoned children were the most visible in the mid-nineteenth century, but the standard of care in the children's own overcrowded tenements was also poor. Brace (1880) described many examples of children who were maltreated, abused, and beaten, in addition to being forced into street trades (rag pickers, street musicians, flower girls, match girls, bootblacks) to earn some income for the family. Sex abuse and incest were not uncommon:

> Living, sleeping and doing her work in the same apartment with men and boys of various ages, it is well-nigh impossible for her to retain any feminine reserve, and she passes almost unconsciously the line of purity at a very early age. In these dens of crowded humanity, too, other and more unnatural crimes are committed among those of the same blood and family. (Brace, 1880, p. 55)

The Society for the Prevention of Cruelty to Children[1]

Child protection developed in the 1870s out of a growing concern with "saving" children from their poor parents. The theory was that if such children could be removed from their parents' influence and brought up properly in an institution, they might be prevented from entering a life of crime (Katz, 1986). The earlier laws to protect children from cruelty and abuse had rarely been enforced. No agency had the jurisdiction to investigate instances of abuse of children or to enforce existing laws without a complaint to the police by the child against a caretaker or a parent, an unlikely occurrence. In any event, courts were reluctant to convict parents who had been charged under criminal statutes involving assault, because it seemed difficult to draw the line between punishment and cruelty (Mason, 1972; Paulsen, 1966).

In 1867, Henry Bergh, a rich philanthropist, organized the Society for the Prevention of Cruelty to Animals (SPCA) in the United States, based on a model he had observed in England. The SPCA preceded by several years the first society to protect children (McCrea, 1910). In 1874, a dying woman told a mission worker, Mrs. Etta Wheeler of the St. Luke's Methodist Mission, that a child named Mary Ellen who lived next door was regularly beaten and mistreated. The woman said she could not die happy until she had revealed the child's mistreatment. Mrs. Wheeler, with no authority to enter the child's home to gather evidence of the child's mistreatment, tried to get the police and several charitable institutions to act, but she found none who could or would, until she appealed to Henry Bergh. As president of the SPCA, he conducted an investigation and then went to court with a petition alleging cruelty to the child. Mary Ellen bore the marks of brutality with bruises on her arms and legs, and a cut near her eye made by a scissor point. He requested that she be taken away from her custodians, even though he recognized that the matter was not within the SPCA's jurisdiction. His attorney, Elbridge T. Gerry, acknowledged the lack of jurisdiction and attributed the participation of the SPCA's president to his "feelings and duties as a humane citizen."

Based on Gerry's creative interpretation of an existing law, the court issued a special warrant directing that the child be taken into custody and brought before the court (Mason, 1972). In court, it was revealed that as an infant Mary Ellen had been abandoned by her mother after her father died in the Civil War. A charity organization had placed Mary Ellen in the care of the couple now charged with felonious assault. The child testified at the hearing and later at the criminal trial of her foster mother; the jury found her guardian guilty, and the judge sentenced her to one year in prison.

Mary Ellen was placed in the "Sheltering Arms" asylum, a home for delinquent girls. Lazoritz (1990) notes that not only was this the beginning of child protective services, but "ironically it was also the first inappropriate placement resulting from such a case" (Lazoritz, 1990, p. 144). Her savior, Mrs. Wheeler, followed up, went to the judge, and received custody of the child. Mary Ellen was raised by Mrs. Wheeler's family in upstate New York. She married and had two daughters who said that she did not talk much about her own childhood, although she still carried some of the physical scars. Her daughters described her as "a real sweet lady" who would occasionally spank her daughters "as any good parent would do." Mary Ellen died in 1956 at the age of ninety-two.

Mary Ellen's case was notorious in its day, generating a great deal of publicity, including songs lamenting her plight and praising her courage. In December 1874, her case and revelations about others led to the organization of the New York Society for the Prevention of Cruelty to Children (NYSPCC). Elbridge T. Gerry and Henry Bergh were key figures in the organization of the new society (McCrea, 1910), and Gerry later became its long-term president. A new society had been necessary because the existing social agencies provided care for children only after they had been legally placed in their custody. No society had the authority "to seek out and rescue from the dens and slums of the City the little unfortunates whose lives were rendered miserable by the system of cruelty and abuse which was constantly practiced upon them by the human brutes who happened to possess the custody or control of them" (Bremner, 1971, p. 190).

A listing of the cases that came before the NYSPCC and similar societies organized in other states would be familiar to those currently working in child protection: a twelve-year-old victim of incest with her father; a nine-year-old, badly scarred from whippings, whose stepmother had placed her on a hot stove and left her naked in a cold room in the winter; a fifteen-year-old struck on the neck with a hot poker; three children aged two to five found drunk along with their intoxicated mother; three children, aged three to thirteen, found dirty, ragged, almost naked, their mother dead, and their inebriated father apparently uncaring; four children ages five months to nine years left at home alone for weeks at a time by their mother (Bremner, 1971, pp. 203–4).

The NYSPCC was organized under the New York laws of 1875 which specifically provided for the incorporation of societies for the prevention of cruelty to children in each community. The societies had the authority to bring before the courts complaints of any violation of laws pertaining to children, and they were empowered to assist in the prosecution of those complaints. The police were directed to assist the agents of the societies in carrying out their duties. The societies' agents were empowered to take temporary custody of children who were charged with the commission of crimes, who were victims of violence, who were being held as witnesses to crimes, and who were found in a state of destitution (*People of the State of New York,* 1900, pp. 241–42).

Within five years, similar societies were organized in New York State in Buffalo, Rochester, Brooklyn, and Richmond County, and in Portsmouth, New Hampshire; Wilmington, Delaware; San Francisco; Philadelphia; Boston; and Baltimore. By 1910, almost all states had some version of the New York law, and the movement eventually became worldwide (Lundberg, 1947; McCrea, 1910).

When the laws giving the state authority to intervene to protect children were challenged in the courts, the laws were generally upheld as a valid exercise of the state's *parens patriae* powers (Bureau of Education, 1880; Mason, 1972). By 1882, the NYSPCC was well established, and its powers and jurisdiction had expanded to include children whose parents allowed them to be "exhibited" or to participate in circuses, to beg, peddle, or play musical instruments in a "wandering occupation," or in any other way exposed the children to danger or involved them in "any indecent or immoral exhibition or practice" (Laws of New York, 1881).

An Agent of Law Enforcement

Under Elbridge Gerry's leadership, the early SPCCs made it clear that they were arms of law enforcement rather than welfare or charitable agencies. Gerry believed that scientific charity was expanding into realms about which its practitioners had no expertise. In 1908, in an address to the annual convention of the Societies for the Prevention of Cruelty to Children, he stated:

> There is nothing to-day which scientific charity does not seek to appropriate to itself, and when it cannot absorb collateral work it endeavors to obtain possession of the subject of that work and utilize it for its own ends. Our workers should be careful to remember the copy book axiom and mind their own business, politely suggesting the like course to outsiders who endeavor to improve upon it. (McCrea, 1910, p. 140)

The NYSPCC was unwilling to accept cases on the basis of investigations made by other agencies, and for a long time, it refused to cooperate with other social agencies, even to the extent of referring cases to them (McCrea, 1910). Many of the statutes passed to protect children following the formation of the SPCCS were introduced into the state's penal codes (Paulsen, 1966). The SPCCS were chartered as law enforcement agencies; indeed, the statutes in most states gave the societies police power (McCrea, 1910). They were not defined as charities and were not subject to the supervision of the state board of charities, even though they received public money (*People of the State of New York,* 1900). Their badge-wearing agents in the magistrates' courts investigated complaints and had the authority to take the child from the home (Folks, 1902). In addition, the New York society's agents functioned alongside the district attorneys to prosecute the cases.[2]

Some agents were quite aggressive in searching out abused children, not always following the niceties of the law in entering homes to seek children they believed needed protection. Gordon (1988) described the way that agents of the Massachusetts Society for the Prevention of Cruelty to Children worked. The society's agents did not make arrests themselves but called on the police for help. The agents roamed the streets looking for children who appeared neglected or abused or who were out of school. They then imperiously summoned people to their offices, threatened to arrest families or to remove their children, and interviewed neighbors and relatives to collect incriminating evidence. And when they entered homes, they sometimes seized whiskey bottles or even told people they found objectionable to leave the premises.

As one observer later put it:

> Child protection is a specialized service in the field of child welfare. . . . It aims to obtain results through advice, persuasion and parental education, but, when necessary its agencies are equipped for the effective use of compulsion, discipline or punishment through a personnel trained in the use of the law and legal machinery for a social purpose. . . . The child protective agency stands as "the friend, protector and sometimes the avenger" of helpless, oppressed children, the innocent victims of brutality and crime, of vicious and degrading surroundings, and of abuse and neglect of every kind. (White House Conference, 1933, pp. 354–55)

The mandate given to the SPCCs represented a shift in public attitudes toward the rights of government vis à vis the rights of parent. Previously, child savers had accepted abandoned children. Now they developed an ideology to support intervening in an intact family, although the SPCCs asserted that they were intervening only in the most serious cases.

> The state should interfere as little as possible with the economy of the family and the liberty of the individual to pursue his own happiness in his own way. And as a general rule, parents are the best guardians of their children . . . but when parental love fails, and the offspring is either abandoned or educated in vice, the state may rightfully intervene. Its right is derived from its duty to protect itself and to protect all of its people. (D. D. Field, quoted in Bremner, 1971, pp. 205–7)

Florence Kelley, a settlement house worker, regarded the legislation permitting intervention in the family as the assertion of "the superiority of the moral over the legal qualification of the home in securing the child's welfare" (Kelley, 1882, p. 96).

Criticism of the Child Protective Societies

Given their broad charge and their police powers, the SPCCs inevitably received criticism. Many parents feared their arbitrary authority, which seemed to be exercised against the poor and the immigrants but rarely if ever against the wealthier families. Although not the SPCCs' responsibility, false reports were often filed, some maliciously, as a way of harassing disliked neighbors (Gordon, 1988). Foreshadowing a conflict that would arise again and again later, some middle- and upper-class patriarchs opposed any effort to extend child protective services because they believed such an extension would interfere with parental rights to use corporal punishment (Nelson, 1984). The NYSPCC also had to fend off charges that it was designed to favor Protestant organizations over Roman Catholic ones. Although the complaints reflected concerns about childrens' religious upbringing, institutions received allowances for caring for children, and to some extent, religiously operated agencies wanted to receive their fair share of the welfare pie.

The SPCCs however, claimed they made efforts to place children in institutions operated by churches of their religion (Folks, 1902).[3]

Vigorous enforcement of the new and existing laws increased the number of children coming into public care. By 1890, the NYSPCC controlled 15,000 children and had an annual budget of $1.5 million, which came from both public and private sources (Folks, 1902). In general, the SPCCs tended to favor institutional rather than foster placement. For example, in 1900, the NYSPCC placed 2,407 children in institutions and 6 in home care.

Although the SPCCS had statutory authority to inspect institutions, critics complained that they rarely did so. The critics further claimed that SPCC officials were more concerned with placing the children in institutions than in restoring them to their families after an emergency had passed or in supporting adoptions of children who had been removed from their homes. The practices varied in different locations. The NYSPCC placed a far higher percentage of its children in foundling homes and institutions than did its counterpart agencies in Massachusetts and Pennsylvania. The agencies in these latter two states, although maintaining their separateness, were more willing to cooperate with other social welfare agencies in making plans to care for the children in their own homes. Gordon (1988) states that the Boston SPCC was often used by its clients for a range of general welfare services. In Pennsylvania, the Children's Bureau, a centralized clearinghouse, was used to find the agency best suited to provide the services that each child needed. But the SPCCs in most other states followed the New York model (McCrea, 1910).

Laws governing guardianship, custody, adoption, and the termination of parental rights were unclear and far from standardized. The legal status of the abandoned children whom Brace placed with families in the West, the children who were committed to institutions, boarded out, or were removed from their homes was unsettled (Lundberg, 1947, pp. 358–72). In theory, the institution to which a child was committed controlled the child's discharge back to the parent. In practice, the institutions deferred to the SPCCs when it came to discharging children. Parents had little protection, and lower-class parents in particular came to fear the arbitrary seeming authority of the SPCCs, whom they sometimes called "the Cruelty" (Bremner, 1971, pp. 210–211; Folks, 1902; Gordon, 1988). The NYSPCC, viewing parents as difficult to deal with, was usually reluctant to return the children to homes that its workers felt were inadequate.[4] As one observer put it, they were dealing with parents "who are indifferent, stubborn, often vicious and resentful of interference" (White House Conference, 1933, p. 360). Modern-day critics however, believe that the SPCCs, whose members were drawn most often from the upper classes, had a vision of the patriarchal family and so were especially hard on single mothers (Gordon, 1988).

Not all parents were unable to change. In one case, Charles Knowack and his wife Johanna petitioned the New York court to return their four children, who had been removed from their home in June 1895. The children had been committed to the society because they were neglected, and their parents were indigent and intemperate, although it was not clear that they were alcoholics or merely drank liquor, a moral offense in the eyes of temperance advocates. Two years later, when the parents asked for the return of their children, the NYSPCC refused. The NYSPCC did not dispute that the parents were sober, industrious, and employed, had cooperated with the society in its plans, were free of debt, and had a bank account and that the children wanted to return to their parents. In the case at hand, the NYSPCC challenged the authority of the court to return the child to its parents. The court found that it had ample powers, whether sitting as a court of equity or using appropriate statutes, to return the children over the society's objections (*Matter of Knowack,* 1899).

Sex Abuse and the Juvenile Protective Association

The sexual exploitation of children has always been a difficult problem and was difficult for the child protection agencies as well. When a specialized juvenile protective association was organized in Chicago in the early 1900s in connection with Hull House and the juvenile court, in addition to its purpose to attack conditions associated with juvenile delinquency, the association tried to rescue young women prostitutes. Some had been forced into prostitution by "white slavers," but others found the life more attractive than working as domestics or chambermaids at paltry wages (Addams, 1912). Juvenile protective association workers monitored playgrounds, parks, dance halls, and other places where youths were to be found. They attempted to reach young women who were in moral difficulties or who were about to become unmarried mothers. Prosecuting the men who had sex with underage women did not prove useful as a protective action. When adolescent prostitutes who might have been complainants appeared in court, they were asked harassing questions by the police or the city attorneys. The legal rights of such young women were often trampled. As Jane Addams put it, they "belonged to a class of women who are regarded as no longer entitled to legal protection" (Addams, 1912, p. 195). The consequences for the women were often worse than those for the alleged perpetrators.

Other sex abuse cases presented equally difficult problems. Separate juvenile protection associations were organized in some communities because child protective workers became aware of "newer menaces" to children:

> The most flagrant and most complicated cases in child protection are doubtless those involving assault or rape of young girls and sex practices of or upon young children of either sex. . . . The children's protective work of this nature is on the whole badly done, and mostly for the reason that the punishment of the offender has been the principal consideration and not the safety and well-being of the child.
>
> A few communities of the United States have equipped themselves with suitable police and court personnel for these cases, and besides are having their laws so wisely administered that not only the culprit's rights are safeguarded but also the young prosecuting witnesses are protected from humiliation and degradation. (Carstens, 1924, quoted in Bremner, 1971, p. 220)

The problems of sex abuse of both boys and girls were well known to authorities. Gordon (1988) found that incest and sex abuse were far from rare in those early days of child protection and, indeed, later on as well. About 10 percent of the Boston Society for the Prevention of Cruelty to Children's cases involved sex abuse or incest. The 1930 White House Conference report observed:

> Children and particularly girls, need a vigorous agency in every community for their protection from early sexual irregularities. Prosecuting attornies in many communities attempt to render this service with varying success. In many places the prosecution for sexual abuses of girls below the age of consent are apt to be futile because the testimony is poorly used, or brutal and demoralizing to the girl witness. Court procedure should give more adequate protection to the girl without removing reasonable safeguards for the accused. (White House Conference, 1933, p. 358)

Given the awareness of sexual abuse going back to the earliest days of child protection, it is remarkable that child sexual abuse as a social problem disappeared as a cause for concern, not to reemerge until the late 1970s and 1980s. Apparently the issues in investigating and prosecuting the cases all were well recognized, and the need for specially trained police and child protection workers was well known. But somehow that wisdom disappeared as the problem of sexual abuse became professionally invisible, and so much had to be relearned in the contemporary period (Goodman, 1984; Haugaard & Reppucci, 1988).

Public-Sector Control of Child Protective Services

Child protective services originated in the private sector, and the SPCCs that were organized on the model of the New York society were privately sponsored. Eventually, child protective services moved from the private to the public sector. Forces for centralizing such services existed at both the federal and the state levels.

At the federal level, the Children's Bureau, organized in 1912, had jurisdiction in child protection, but judging from its in-house publications before 1959, child abuse and neglect had a much lower priority than did other issues related to infant and maternal health.[5] Indeed, a summary of early twentieth-century federal programs affecting children and youth made no mention of child protection or child abuse (Interdepartmental Committee on Children and Youth, 1951). Instead, the Children's Bureau emphasized the important aim of reducing the infant mortality rate, and it advocated the Shepherd–Towner Act of 1921, which laid the foundation for maternal and child health programs under the later Social Security Act. The Shepherd–Towner Act encouraged each state to develop maternal and child health care programs through the state health departments. Because by then institutional care had given way to foster care and to adoption programs for neglected, abused, abandoned, delinquent, or illegitimate children, the bureau stressed those rather than the child-rescuing efforts related to child protective services.

The Social Security Act of 1935, requiring centralization of welfare services and uniform services throughout a state, built on a trend that had started earlier. Beginning in 1867, New York State had created a state board of charities with the authority to visit and inspect all charitable institutions in the state. The NYSPCC, however, had avoided state supervision on the grounds that it was not a charitable but a law enforcement agency (*People of the State of New York,* 1900). In 1912, jealously guarding its independence, the NYSPCC opposed the formation of the federal Children's Bureau on the grounds that creation of the bureau was not within Congress's express powers and would be an unconstitutional infringement on the states' power to legislate in the area of child welfare (see Bremner, 1971, pp. 771–73).

At the state level, centralization of welfare proceeded because a large percentage of the funds to support children's services was already coming from state and local governments. For example, of twenty-eight institutions for children in New York City, twenty-five received public funds, and seventeen of them received more than half of their money from public sources (Bremner, 1971, pp. 280–81). With funds came the power to regulate. Professional considerations also led to the consolidation, if not the centralization, of services. Some SPCCs recognized the necessity of supplementing child protective work with preventive and rehabilitative work with the family.[6] Some private child welfare organizations, which had not provided such services before, took on child protective functions. Children's aid societies and SPCCs sometimes merged, offered new services, and eventually many became general social service agencies.

By the 1920s, the desire for state-supported and state-administered child protective agencies was evident. In some communities, public welfare agencies were given the authority to provide general social services. Some child protective work was accomplished by city welfare boards, although they were not very vigorous in their pursuit of child abuse and neglect. Often public and private agencies existed side by side. In 1930, the White House Conference concluded:

> Child protection, although a public duty and a proper governmental function, has been left largely to private initiative and support, with the result that large areas of the country, especially rural sections where it is more frequently needed, are without child protection services of any kind, while in many places where it exists, it is poorly done, without proper standards of service and trained personnel. (White House Conference, 1933, p. 388)

The White House Conference report advocated that child protection be established within counties, under state supervision, as a government function. That recommendation was eventually reflected in the Social Security Act of 1935, which included federal funds for a wide range of child welfare services, but not for institutional or foster care or services provided by private agencies. The main purpose of the Social Security Act was the reform of social agencies and the social welfare system; using the spending power leverage, the Social Security Act conditioned federal funds on the development of central state welfare authorities and local community services. Despite the subsequent development of public welfare services, however, often there was no specifically identified child protective service within the public welfare agencies (Kadushin & Martin, 1988).

The Rediscovery of Child Abuse

The rediscovery of child abuse owes a great deal to the work of the Children's Bureau, starting in the 1950s. During World War II, the Children's Bureau had priorities other than child abuse, including day care for children of mothers who had entered the labor force in response to the demands of war, and maternity care for military wives. Immediately following the war the bureau focused on problems of juvenile delinquency.

Some popular interest in child abuse was aroused by medical reports in the scientific literature dating back to 1946 when several publications, based on studies using newer radiological techniques, reported multiple, healed fractures in children (Paulsen, 1966; Pfohl, 1977, Sussman & Cohen, 1975). The Children's Bureau had been actively collecting information on child abuse since 1955 and, despite its major attention to other issues, had been encouraging some research on the subject (Nelson, 1984). Beginning in about 1959, the Children's Bureau noted an increasing number of reports of physical abuse from hospitals where children were brought for treatment of their injuries.

California had a child abuse reporting law in place by 1962. About that time, the Children's Bureau issued guidelines and suggested language for a model child abuse reporting law (Children's Bureau, 1963a; Paulsen, 1966, 1967; Mason, 1972). Over the next few years, the American Humane Society, the American Medical Association, and the American Academy of Pediatrics also issued guidelines for legislation (Sussman & Cohen, 1975). The support of these prestigious groups was important to the quick passage of state reporting laws, although the bureau's guidelines provided the model in most cases (Paulsen, 1966, 1967).

In 1962, a Colorado group of radiologists, pediatricians, and psychiatrists conducted a survey of hospitals and district attorneys who reported a large number of cases of children with serious physical injuries, many followed by severe brain damage and death. Based on the survey and their own clinical experience, the group labeled the clinical condition the battered child syndrome. The Colorado group's findings about the frequency and severity of injuries to children inflicted by parents or foster parents, and their comments that the conditions were inadequately handled by physicians were published in the *Journal of the American Medical Association,* thus reaching the general medical community (Kempe et al., 1962). The article was accompanied by an editorial endorsing medical responsibility for dealing with child abuse. Nelson (1984) makes the interesting point that "medicalizing" child abuse removed it from the social context of poverty, in which child abuse was often found. Rather than view child abuse and, later, neglect, as a problem with complex causes in the social environment, it came to be seen as a matter of individual deviance.

After the American Academy of Pediatrics held a symposium on the battered child in October 1961, the Children's Bureau convened a group of consultants to advise it of possible actions and issued a set of guidelines for emergency social services for the care and protection of children (Mason, 1972; Paulsen, 1966, 1967).

The Children's Bureau and its successor agencies continued to support research on child abuse, and the reports of research in professional journals were picked up by the popular media. The combination of professional articles and legislative development fueled the widespread media attention to child abuse. Not only were there numerous reports of child abuse in the newspapers in all states, but also the *Saturday Evening Post, Life,* and *Good Housekeeping* ran child abuse stories, and popular medically oriented television soap operas such as "Ben Casey," "Dr. Kildare," and "The Nurses" featured episodes on child abuse (Pfohl, 1977). Popular attention in turn persuaded legislators to continue to deal with the problem of child abuse (Nelson, 1984).

Reporting Laws

Before 1962, emergency room physicians could scarcely avoid seeing signs of abuse, but their duties were not clear. Some may have been psychologically unable to recognize that parents would inflict such serious injuries on their own children. Confidentiality as an ethical and legal obligation may have prevented many physicians from initiating complaints against their patients, and others may have been reluctant to become involved with criminal prosecutions of offenders. One solution was the reporting statute.

The widely disseminated, rapidly increasing information coming from radiological and related studies and the support of diverse voluntary and professional associations provided the background for the passage of corrective legislation. Academic pediatricians working with other civic leaders pressed for reporting legislation (Pfohl, 1977). The promotion of legislation making child abuse reporting mandatory succeeded with remarkable speed in all states. Within five years, every state had some form of legislation requiring that physicians report suspected child abuse to authorities.

Five years is an unusually rapid period for legislative innovations to spread nationwide. Nelson (1984) states that it usually takes about twenty-five years from the time the first state passes legislation until most states have a similar law. Reporting legislation was easy to pass: There was no opposition; legislators were voting for the protection of children; and initially the legislation cost nothing.

Legislators did not anticipate the impact of the reporting legislation. In Florida, for example, there were seventeen reports in 1970 before reporting legislation, and over nineteen thousand in 1971 after legislation (Nelson, 1984). Initially, the increased number of reports was not accompanied by sufficient appropriations for investigations or for services to families (Paulsen, 1967). Eventually, however, the flood of reports uncovered the necessity for remedial and preventive services as well as investigative ones.

The early legislation directed reports to agencies that would initiate criminal prosecutions. Those agencies were not very successful in achieving cooperation with pediatricians and social workers who believed that parents should be treated rather than prosecuted (Nelson, 1984). As a result, later legislation mandated that reports be made to child protective authorities in addition to, or instead of, the police. The list of mandated reporters grew, and the definitions of what was to be reported soon included neglect, sexual, emotional, and physical abuse. The inclusion of neglect drew attention to the plight of families in poverty. The failure of a mandated reporter to carry out the duty to report could result in criminal penalties, and such a failure could provide a cause of action for malpractice suits.[7]

Sexual abuse came to the attention of mandated reporters as they began to examine children more closely. In 1978, Kempe published an article in which he pointed to the sexual abuse of children and adolescents as a hidden and neglected area, adding to the mental health professionals' interest in that subject.

The volume of reports increased steadily with the advent of reporting laws (Eckenrode et al., 1988). However, even though the laws were designed to make it easier for professionals to make reports about their patients and clients, there may well be underreporting of incidents. In order to protect the families, private physicians serving a more well-to-do clientele may underreport to a greater extent than do physicians in public agencies who serve poorer people. Some professionals are reluctant to report because they fear that any action would interfere with the therapeutic alliance (Watson & Levine, 1989). Some believe that the child protective response may be more destructive than the child's current situation, and some are reluctant to become involved because a significant percentage of such cases may be disputed and may require court appearances and a loss of professional time. The situation may be changing, however, as the public becomes more familiar with and more accepting of reporting laws (Sussman & Cohen, 1975).

The Renewed Federal Role

The growth of laws requiring child abuse reports was accompanied by legislation authorizing services to cope with the human problems uncovered by the reports. The 1960 White House Conference recommended that the states authorize every community to designate a specific agency for child protective work. Federal legislation taking effect in 1962 authorized reimbursement to the states if they provided foster care for children receiving AFDC funds who lived in their own homes but who were in need of protection. The states could receive the funds if they used public welfare workers to provide the services. The federal legislation tied income maintenance and child protection closer together and encouraged the movement to place child protective services within public welfare agencies. By 1966, almost every state had publicly administered child protective services, but many communities failed to establish a unit designated as a specialized child protective service. Senator (and later Vice-President) Walter Mondale, and Congresswoman Patricia Schroeder led the effort for the federal Child Abuse Prevention and Treatment Act of 1974, which required separate services and also mandated child abuse reporting laws. Title XX of the Social Security Act mandated protective services. By 1978 all the states had such services (Kadushin & Martin, 1988).

Nelson (1984) observed that the federal act was shaped to avoid political opposition from President Nixon. The legislation emphasized abuse rather than neglect, which would have raised questions about poverty, a subject then being quietly ignored after political opposition grew to the Kennedy–Johnson War on Poverty. In order to minimize opposition from conservatives concerned about protecting family integrity from government intrusion, the legislation singled out "excessive" corporal punishment, implying the legitimacy of some corporal punishment. States could give authority to child protective workers, physicians, and police to take temporary protective custody of abused children. In addition to creating questions about intruding on family prerogatives and privacy, removing children from their homes raised questions about possible psychological harm as well. Goldstein, Freud, and Solnit (1979), who had earlier championed the primacy of psychological over biological parenthood (Goldstein, Freud, & Solnit, 1973), now raised questions about the harm in removing children from their homes. They opposed such removal except under the most extreme circumstances. The removal of children from their homes placed responsibility for them in the state's hands and increased the need to find suitable substitute care and foster homes.

Although federal legislation did not require that the states establish central registers, the states have been doing so since the 1960s. Each state manages its central register somewhat differently. In general, a reporter notifies either a "hot line" or the local agency, which then reports to the central register, which determines whether previous reports have been made about the same child or about the person who allegedly committed the act. These registers raise the same problems of privacy and the protection of confidentiality that other data banks raise. States have responded with various devices to protect confidentiality, to limit access to the information, to notify people whose names are included in the register, and to give them the opportunity to challenge the findings and have the records expunged if the information alleging abuse cannot be substantiated (Sussman & Cohen, 1975).

Parents Anonymous and the Self-Help Movement[8]

In 1969, there was little help available for parents who were abusing their children or feared they were about to harm them. One such woman was Jolly K., who founded Parents Anonymous (PA), along with Leonard Lieber, a professional social worker. The organization had its beginning when Jolly K., seeking help at a local child guidance clinic because she feared she would harm her children, was told there would be a six-month wait. Jolly K. objected vigorously and was then referred to Lieber for an emergency evaluation for possible admission to a state mental hospital. Her emotional outburst resulted in a call to child protective services as well. Lieber began to see Jolly K. in twice-weekly psychotherapy, but it did not go well. According to their account, the idea for a new organization emerged during a psychotherapy session and progressed as Lieber encouraged Jolly K. and even referred a second woman from his own caseload to her.

The new group began under the name of Mothers Anonymous, and little by little, its membership and the concept of the program grew. The name was changed to Parents Anonymous (PA) in recognition that child abuse is a problem of fathers as well as mothers. PA advertised in the papers and eventually drew the media's attention; a segment showing a Mothers Anonymous meeting played on local TV news in Los Angeles and was picked up on network television. PA extended its program to the prison system and began a regular meeting at a women's correctional institution. By 1971, PA was given space on the Child Welfare League's regional conference, and the organization gained legitimacy when it received prominent mention by Kempe and Helfer (1972).

Parents Anonymous never assumed an antiprofessional stance. Indeed, its rules required a professional sponsor and a lay leader for each group, and it attracted some notable figures in mental health to its board, including Dr. Roland Summit. It followed the Alcoholics Anonymous model and did not engage in social action to ameliorate conditions related to child maltreatment but, rather, focused on changes in the members' own feelings and behavior. Over the next two years, the organization grew steadily, with new chapters starting when individuals learned about PA from various sources, including the media, or when former members moved to different parts of the country. By mid-1973, PA claimed forty-seven chapters in thirty-five states. Jolly K. was invited to address child abuse advocacy groups interested in sponsoring new programs, and in 1973 she testified before Senator Walter Mondale's committee. As a result of her testimony, the program received further national attention.

Sensing a need for services on a national level to help found and support fledgling chapters, Jolly K. and Leonard Lieber began to seek other funds. They obtained a small grant from a private foundation and then a federal demonstration grant to support organizing activities on a national level, including the establishment of a national hotline. By 1975, with 150 chapters, PA was able to organize a national conference on child abuse. With the growth of reporting laws and professional and public attention to the problem, by the end of 1976, PA grew to 450 chapters in the United States and Canada. By 1979, it had representation in every state and the District of Columbia, and many chapters were started in foreign countries, especially Great Britain.

The organizing of chapters was not as smooth a process as this brief chronicle suggests. They grew slowly and required great persistence by local leaders to keep the effort going. Sometimes chapters would fail when a local leader moved or "burned out." But the work was continued by remarkably dedicated individuals who found their own salvation in helping others cope with the problem of child abuse. Wheat and Lieber (1979) describe many moving personal stories of the local leaders whose commitment and dedication contributed immeasurably to the organization's growth.

With the extension of reporting laws, the rapid growth in number of reports, the overwhelming demand for foster care, and the growing recognition that alternatives to removing children from their homes were scarce, Parents Anonymous became an important resource. The national PA office helped write manuals for organizing and running chapters, recruited and appointed regional coordinators to support local chapters, and struggled to develop policies for problems that the groups encountered. The manuals included a statement of PA's program, essentially its ideology, definitions of child abuse, rules for maintaining anonymity and mutual respect, and other guidelines for achieving its goals of controlling child abuse.

Parents Anonymous has recognized that it is difficult for its members to deal forthrightly with problems of sexual abuse. It has tried to provide some guidance in that difficult area but without a great deal of success. Its policy has been to refer people with sex abuse problems to other agencies. PA professionals and lay members also obey reporting laws; difficult as it is, their policy is to protect the child by reporting child maltreatment, even at the risk of losing members. PA accepts court-referred clients but provides reports only on attendance and not on the substance of the discussions in PA groups.

An evaluation of PA was conducted in 1975–76 as part of a demonstration project. A survey revealed that 83 percent of the members were female; about two-thirds were white, with moderate to low incomes; and about two-thirds were married. The respondents expressed a great deal of satisfaction with the program, and they reported reductions in abusive behavior toward their children. Satisfaction with the program, reduced abuse, increased knowledge of child behavior and child development, wider social contacts, and higher levels of self-esteem were correlated with length of time in the program. More rigorous experimental evaluations to establish causal relationships are not possible. The survey provides no information about the outcomes of those who drop out of the program, and PA acknowledges a dropout rate of 20 to 30 percent within the first few weeks of attendance.

Parents Anonymous is an interesting example of the growth in self-help groups and of the positive relationship possible between professional and laypersons. PA appeared at a time when professional mental health services were unable to provide much assistance, particularly on the scale that was required after reporting laws threatened to swamp the child protection system, and when research and clinical developments in treating parents with child abuse problems were wanting. Its growth and the loyalty of many of its members indicates that PA plays an important role in many people's lives, even though it may not be a definitive answer to a problem and even though we cannot demonstrate to our scientific satisfaction that the PA program is the "active ingredient" in helping parents overcome a child maltreatment problem.

The Mental Health and Social Science Communities

Based on the scarcity of articles in the *American Journal of Orthopsychiatry* (AJO), it seems that before the 1960s the mental health professions were not deeply involved in treating child abuse. But the AJO was not alone in its neglect of the problem. In 1978, Kalisch published an annotated bibliography containing over two thousand citations of the child abuse literature, and most had been published after 1970. Nelson (1984) found a similar distribution, over time, of publications on child abuse in both the professional and the popular media. The professional literature has burgeoned, however, since a National Center for Child Abuse and Neglect was established under the Child Abuse Prevention and Treatment Act of 1974.

The mental health profession's lack of participation in diagnosing and treating child abuse reflects the neglect of the problem for decades by all the professions, including physicians who encountered many indisputable cases of severe physical injury and deaths resulting from child abuse. Mental health professionals, treating children in their offices, rarely saw the physical signs of abuse. Before the 1970s, no one used the laws to pursue emotional abuse (Garbarino, Guttman, & Seeley, 1986).

In the early days of the dominance of psychoanalytic theory in the helping professions, the denial of sexual abuse was so strong that Summit (1988) characterized it as a "shared negative hallucination." Summit believes that the taboo is so strong because adults simply do not wish to think about the sexual abuse of children, and in the past those who did call attention to the existence of sexual abuse were often attacked by their peers. Freud at first believed that many of his patients had been abused sexually, but he recanted and called the reports fantasies after others criticized his conclusions. Mental health workers influenced by psychoanalysis probably did not look for evidence of sex abuse. Indeed, given professional beliefs that incest occurred only in very sick families and that discussions of incest with patients was psychologically dangerous for patients except under the special circumstances of psychoanalytic treatment, the topic might have been carefully avoided (Haugaard & Reppucci, 1988).

In recent years, mental health professionals have been much more involved in clinical work and research in child maltreatment and sexual abuse (Haugaard & Reppucci, 1988), a result of the mandated reporting laws affecting every profession dealing with children. Mental health professionals act as reporters, and they are called as witnesses in maltreatment actions in family and juvenile court and in criminal actions as well. Client–patient confidentiality often cannot be preserved because the privileges that protect confidences revealed in treatment from exposure in legal proceedings have been withdrawn by statute when child maltreatment proceedings come into the family courts. New problems arise in this context. The insistence by many mental health professionals and sometimes by the courts, that alleged abusers acknowledge responsibility for the abuse raises Fifth Amendment self-incrimination problems under some circumstances, because the individual may be coerced into making an admission that can have drastic legal consequences (Levine & Doherty, 1991). In a significant minority of cases, a mandated report can have an adverse affect on psychotherapy (Watson & Levine, 1989).

Increasingly, mental health professionals are seen as a resource for the courts. Referrals to specialized sex abuse evaluation and treatment units take place not only to help the child or the family deal with the trauma of the abuse but also to help the child cope with the problem of going to court and giving testimony. The testimony of mental health specialists can provide critical corroborating evidence both when a young child is not qualified to testify in civil or criminal court proceedings and when the child is not called to testify in family court in order to be protected from psychological stress. Under these circumstances, hearsay testimony is allowed as evidence (Levine & Battistoni, 1991). Referrals under court order are common for the treatment of sexually abused children and families. Mental health workers also participate in other programs such as those designed to improve parenting skills in an effort to keep children in their homes or to return them to their parents if the children have been removed from their homes.

At another level, social science research workers are assessing the capacities of children as witnesses and their credibility to adults. Their efforts may contribute to the development of policies to protect children and create conditions that will elicit the best testimony that children are capable of giving, while also protecting the rights of the defendants. Goodman (1984) has been a leader in such work, and her research, along with that of others, was cited in recent U.S. Supreme Court cases, *Coy v. Iowa* (1988) and *Maryland v. Craig* (1990). In *Craig,* the U.S.

Supreme Court interpreted the Sixth Amendment confrontation clause to permit a vulnerable child to testify using one-way video.

There seems to be no decline in the number of cases of child abuse and child sexual abuse coming to the attention of child protective authorities and the courts. Mental health workers are trying to devise methods to identify the characteristics of sexual and other abusing parents, to understand the circumstances under which abuse takes place, and to diagnose and treat parents and children (Wolfe, 1987). Frequently, mental health workers are asked to assist the courts in determining whether a child has been sexually abused and, sometimes in family court cases, to identify the abuser (Bulkley, 1989). Observers have also noted a rise in complaints about sex abuse in divorce and custody disputes, leading to new questions about the rates of, and the potential for detecting, false allegations (Nicholson & Bulkley, 1988).

The demand for assistance from the courts has far outrun the research community's capacity to provide instruments and observations reliable enough for valid testimony. At present, due to the rising tide of complaints and the public's desire to protect the child, weak evidence is often admitted and used as the basis for adjudications in family and juvenile courts (Levine & Battistoni, 1991). There is some justification for this policy. In many cases, the child's parent is the abuser, and often the other parent will not provide emotional support for a child who has revealed abuse (Everson et al., 1989). Thus the child is deprived of the care of its natural protectors, and it makes sense that society should tip the scales in favor of protecting the child rather than the parents.

Whether this desire to protect the child leads to miscarriages of justice is not yet clear. Several well-publicized cases of alleged mass child sexual abuse included sensational allegations, and battles of experts who sharply disagreed with one another's methods and conclusions (Corwin et al., 1987, Farber, 1990). Some actions of overly zealous prosecutors and child protective workers, following the lead of physicians relying on ambiguous physical signs of sex abuse, have led to renewed attention to the necessity for care in prosecuting allegations of child abuse (Haslam, 1989; Van de Kamp, 1986). A national citizens' group has also emerged, called VOCAL, or Victims of Child Abuse Laws, whose purpose is to correct the current system's deficiencies, especially in regard to accused parents (Hechler, 1988).

Some critics charge that the entire child protection and foster care system is overwhelmed by numbers and in serious disarray. There are a number of lawsuits against social service agencies, in an attempt to force local and state governments to fulfill their duties to serve children (Barden, 1991). Critics level serious charges against what they characterize as arrogant and mindless "child savers," successors to the agents of "The Cruelty," who exercise arbitrary power, under the guise of "helping," to the detriment of children and families (Wexler, 1990). We can expect some change in the system as its inadequacies are revealed and new solutions are tried.

Today, the successors to the early SPCCs are public agencies whose workers are valiantly trying to do the best they can with limited resources. Judging by the evidence of history, child abuse is an intractable problem, but mental health workers are in the forefront of those seeking solutions. We may be approaching some of the problems of abuse more forthrightly than in the past, and in this openness, there may be solutions that will cause more good than harm.

Summary

Child protective services began in the 1870s, when a conservative social Darwinism was dominant. Designed to protect and to save children, the services assumed that parents were bad, and so children were frequently moved to inadequate institutions. Child protective services were not concerned with the social conditions under which parents and children lived. The parents were thought to be responsible for all inadequacies, and their children were taken away from them by quasi policemen. Neglectful or abusing parents were symbolic of the deficiencies of the lower classes, and child saving was an emblem of the nobility of the child savers.

In the early 1960s, child abuse was medicalized. Children were to be rescued from parents who were at first believed to be psychiatrically disturbed. The medicalization of child abuse drew attention away from the correlation between child maltreatment and poverty. In the early 1970s, in order to obtain federal legislation, the correlation between child maltreatment and poverty was hidden for political reasons. Reporting laws are another case in point. These laws were designed to protect children, but the resources necessary to enforce them simply were not sufficient. Reporting laws were perceived to be an inexpensive means of showing concern for children. The creation of Parents Anonymous in the late 1960s and early 1970s reflected the participatory ethos of those years. PA was formed in collaboration with a professional social worker, and after a few years the organization decided to follow the route of accepting federal money. PA also serves local social service offices, often on contract. PA may do excellent work with those who seek out its help, and our comments are not meant to denigrate its work. We merely want to point out that it would be possible for this group to assume other functions and to work for preventive programs and improvements in welfare, housing, day care, and employment, all factors related to child maltreatment. That it does not may reflect something of its origins and the context in which it has chosen to work.

There is no question about the necessity of protecting and providing for children who cannot rely on those who should be their natural protectors. The child-saving movement undoubtedly has helped many children who were being treated inhumanely. At present, the many cases coming to the public's attention has stimulated much research on parenting, personality development, the treatment of parents and children, the detection and prevention of child maltreatment, and children as witnesses when such cases are prosecuted. There already have been important effects on the service system, and recently the research has entered into a U.S. Supreme Court decision defining the constitutional right of confrontation in a criminal trial (*Maryland v. Craig,* 1990). The reverberating effects of the Mary Ellen case are still being felt throughout the child welfare, the child mental health, and the legal systems.

13

Birth Control and Abortion

Birth control and abortion services are important to the prevention of many social problems related to mental health. In 1978, the President's Commission on Mental Health Task Panel on Prevention identified "the national epidemic of teenage pregnancies, unwanted births, premature parenthood" as a major target for preventive intervention. In 1977, a government-sponsored conference on preventive programs listed family planning for younger, unmarried women as a priority need (DHEW, 1979). Although the problems may be more related to poverty and a lack of prenatal care than to the age of the mother, the President's Commission report noted that pregnancies among teenagers are associated with both physical and psychological effects on the young women and on their children. They are subject to higher-than-average maternal and infant mortality and morbidity, and their pregnancies are more likely to be unwanted[1] (Garn, Pesick, & Petzold, 1986). Unwed teenagers who carry their pregnancies to term are prone to problems in that they have higher rates of drug and alcohol use as well as low achievement in school, and they may more often be depressed than nonmothers are (Hamburg, 1986).

Out-of-wedlock teenage pregnancy has educational and economic consequences for the teenage mother and for society in general. The teenage mother is less likely to complete high school, less likely to enter the labor force, and more likely to be dependent on welfare than are other teenagers or mothers who have their first child after their teen years. About 60 percent of welfare costs are attributed to women who have conceived before age eighteen (Klerman, 1986). Out-of-wedlock pregnant adolescents are more likely to have premature and low-birth-weight infants than are older mothers. Those factors are in turn associated with higher rates of neurological impairment in the infants and consequent school and behavioral problems in those children who survive to school age (DHEW, 1979). There is some reason to believe that children of teenage mothers are at higher risk for child abuse, although the data are far from clear on this point (Gelles, 1986). Children of teenage mothers have poorer adaptations, on the average, later in life than do children of women who are past their teens when they have their first babies (Furstenberg, Brooks-Gunn, S. Morgan, 1987).

More than half of all out-of-wedlock teenage pregnancies end in abortions, and abortions are primarily responsible for the declining teenage birthrate, despite a pregnancy rate that has remained constant or has increased in recent years (Konner & Shostack, 1986). About 25 percent of all abortions are performed on teenagers (Henshaw & Silverman, 1988). Compared with those who carry their pregnancy to term, teenagers who choose abortion are more likely to come from families of higher socioeconomic status, are likely to be doing well in school, are more likely to be employed, and are more likely to have a supportive relationship with the putative father (Chilman, 1986). Although abortion decisions are not made lightly, we have no reason to believe there are serious emotional or physical sequelae to undergoing an abortion, although controlled studies involving long-term follow-up are not generally available (Adler et al., 1990).

Today, a number of service facilities provide counseling for family planning, birth control, and abortion. The clinics serve an important preventive function, for many of their clients are adolescents. As part of their counseling function, the clinics administer pregnancy tests and offer information and counseling about alternatives such as carrying the fetus to term, keeping the infant, giving the child up for adoption, or terminating the pregnancy. (Since 1988, federally funded family planning clinics have been restricted from providing counseling or referral for abortions. This regulation was upheld by the U.S. Supreme Court in May, 1991: *Rust v. Sullivan,* 1991). Clinics may also provide prescriptions for birth control pills, diaphragms, condoms, and intrauterine devices. Planned Parenthood has the largest network of facilities, with some two hundred affiliates and two million clients annually. In addition, similar services are offered through state, county, and municipal health departments (Joffe, 1986). In recent years, about one hundred school-based or school-linked clinics have opened. These clinics, often the subject of controversy in the community,[2] offer general medical services, including reproductive health and sometimes prenatal care as well (Glasow, 1988).

Since *Roe v. Wade* (1973), freestanding abortion clinics, both nonprofit and for profit, have opened around the country. Many of their clients are unwed and adolescent. The clinics provide free pregnancy testing as a means of attracting clients, and many clinics offer counseling on birth control, sexuality, and issues related to abortion or continuing a pregnancy to term. In recent years, personnel working in the clinics have been subject to direct pressure from antiabortion forces, in the form of picketing, demonstrations, and illegal blocking of clinic entrances, attempts to stop women from keeping appointments to have abortions, picketing of the homes of abortion clinic staff members, and other harassments. In a few instances, clinics have been bombed or vandalized. Most of the employees of the freestanding clinics persist in their activities despite the pressures, in part because they have an ideological commitment to freedom of choice and some identification with the women's movement (Joffe, 1986).[3] In addition, the antiabortion activists exert indirect pressure through political persuasion, letters to the media, threats of boycotts, and the picketing and harassment of landlords who lease space to clinics.

The struggle between prochoice and antiabortion or right-to-life movement activists represents an intense political and social conflict. That bitter struggle is a reminder that it has been less than twenty years since the *Roe v. Wade* decision reduced the level of government intervention in private abortion and contraception decisions to a point closer to what it had been from colonial days until the post–Civil War period (Mohr, 1978).

Although one of the current conflicts is about providing contraceptives to high school students, most people are not aware of the struggle to make contraceptive information and devices freely available and how recently this struggle was won. Until the early 1930s, it was illegal to send contraceptive information or devices through the mails (Kennedy, 1970). Although not enforced against physicians, the ban on importing contraceptive devices was not removed from federal law until 1971, after four years of effort by a New York congressman who acted on a constituent's complaint that a customs agent made her throw her diaphragm into New York Harbor before allowing her to reenter this country (Reed, 1978). *Carey v. Population Services International,* striking down a ban on distributing contraceptives to minors, was not decided until 1977.

The rights to contraception and abortion are linked, although historically, supporters of one form of family limitation did not necessarily support the other. The dissemination of contraceptive information was not regulated in the United States in the eighteenth and early nineteenth centuries. Under the common law, abortion before quickening was never a crime. States did not begin regulating abortions until well into the nineteenth century, with the early laws designed to protect women from dangerous medical procedures carried out by untrained practitioners. A combination of increased political and social conservatism, a fear of domination by foreign immigrants, conflicts within the feminist movement, and developing professionalism in medicine resulted in restrictive laws, most passed between 1870 and 1900, which persisted until the late 1960s and 1970s (Mohr, 1978).

The Regulation of Abortion

In the late nineteenth century, laws regulating abortion and contraception were viewed as solutions to a set of problems that had been troubling some descendants of Puritan America for quite a long time. During the previous half-century the United States had grown and transformed politically, economically, and demographically. Andrew Jackson's election to the presidency in 1829 symbolized a shift in political leadership from the gentry whose position derived from landownership to a new class whose wealth and status derived from industry and commerce. Many believed that a slower-paced, more graceful way of life in America had given way to the unbridled pursuit of money and power (Tucker, 1855).

As the workplace moved from the home to the factory, many middle-class women could afford to stay home, and the ideal married life centered more and more on the nuclear family (Reed, 1978). The fertility rate of native-born, white Protestant American women had been declining since the early 1800s. On the average, American women had 7.04 children in 1800, 5.21 in 1860, and 3.6 in 1900, reflecting the economic changes in American life (Reed, 1978). Children were not an asset in an industrial or a commercial economy, except among the very poor who did not educate their children but sent them out to work or to engage in street trades such as peddling or rag picking (Brace, 1880). The motivation for many to limit the size of their families is generally attributed to these economic considerations.

Methods of limiting family size or getting rid of unwanted children have always included murdering them, abandoning them, selling them, and giving them up for adoption (Petchesky, 1984).[4] These methods aside, if families were to limit family size through fertility control, they had to abstain from sexual relations or use contraceptives and abortion. The known contraceptive methods were far from perfect, but they certainly were not totally ineffective. By 1865, various physicians had endorsed coitus interruptus, spermicidal douches, vaginal diaphragms, vaginal douches, and rubber condoms that were manufactured after the 1846 invention of the rubber vulcanization process (Reed, 1978). Popular home medical manuals contained contraceptive information and described methods for inducing abortions. Physicians and midwives performed abortions, evidently with a reasonable degree of safety, especially if done early in the pregnancy (Mohr, 1978).

The declining birthrate of native-born white Protestants in the mid-1800s occurred simultaneously with the social problems accompanying the upsurge in immigration and the growing fear that the ''native stock'' would be overwhelmed by (inferior) foreigners who were breeding much more rapidly than were the native born (Kennedy, 1970; Mohr, 1978). Reed (1978) and Mohr (1978) argue that the ''regular'' medical men, who were then still drawn from the middle and upper classes and who were among the few educated people, shared the popular xenophobia. Mohr (1978) states that some doctors began to withhold contraceptive assistance from patients on the grounds that as physicians, they knew best who should use contraception and that most middle-class women should have babies. Despite the physicians' attempts to limit the supply of information, contraceptive materials, and abortions, the demand for them remained high. Without an enforceable medical monopoly provided by state licensing, many suppliers entered the field. Given the rampant commercialism of the day, daily and weekly papers contained thinly disguised ads for contraceptive devices, for medications of dubious effectiveness and safety to induce abortions, and for abortions using surgical means.

Laws that regulated abortions had appeared between the 1820s and 1840s. Before 1840, antiabortion legislation was directed against quacks and apothecaries and did not carry criminal penalties for the young unmarried women who had most of the abortions. The rate of abortions rose sharply between 1840 and 1870 when abortions were sought by married women seeking to limit their family's size (Mohr, 1978).[5] The first laws addressed unsafe conditions, but twenty years later the regulation of abortion became part of the medical field's drive for professionalization. In the 1800s, physicians' abilities were not well regarded because they really had few effective remedies, and many practiced medicine with little or no formal training. Proprietary medical schools, dependent on student fees, were often little more than diploma mills. Although the formally trained ''regular'' physicians sought licensing laws at a fairly early date, state legislatures were generally unresponsive to their pleas (Mohr, 1978). (The famous Flexner report of 1910, designed to remedy deplorable practices in medical education, was the culmination of fifty years of work toward professional status.) Abortion was one of the areas that attracted a great many nonmedical or poorly trained practitioners. By eliminating abortion, the "regular" physicians hoped to reduce competition and clear their ranks of poorly trained "irregular" medical practitioners.

The medical attack against abortion continued through the middle years of the nineteenth century. The American Medical Association (AMA) had organized in 1847, and state and local medical societies had become more active as well. Eager to have control over health policy physicians wrote medical texts and tracts arguing against abortion. As they became aware of the continuity of fetal development, quickening became more controversial as a demarcation point for the beginning of life. The *New York Times* and other newspapers began to print sensational stories about deaths following abortion, and eventually a small handful of clergy picked up the cudgels as well (Mohr, 1978).

More restrictive antiabortion legislation was introduced into many states between 1860 and 1880. Provisions for licensing physicians, and state boards of examiners for physicians, were often linked to antiabortion legislation. By this time, most physicians were affiliated with organized medicine, and thus sanctions by local or state medical societies had more force. Because there were questions about the ethics of abortion practice, there was little research on abortion. The practice was driven underground, despite acknowledgment that with advances in knowledge of antiseptic practices and surgical technique, abortion could now be a safe procedure. After 1880, evidentiary standards and burdens of proof shifted to make it easier to criminally prosecute abortionists.

By the 1880s, the pattern of abortion changed again. Middle-class women were increasingly using more effective means of contraception, and illegal abortions were sought mainly by the poor or the unmarried. Middle- and upper-class women continued to have access to abortions through private physicians, but poor women were driven to seek assistance from nonmedical abortionists who functioned at the borderline of legality and sometimes crossed that line (Mohr, 1978; Reed, 1978). This pattern continued until the 1960s, when some states began to reform their abortion laws.

The Comstock Laws

After the 1870s, antiobscenity crusaders entered the picture, targeting those who provided or advertised abortions and contraceptive information, and drawing no distinction between pornography and fertility control. By the late 1870s, public opinion had apparently turned against abortion and the open advertising of those services, in part as a reaction to the ethos of the immediate post–Civil War period. Broun and Leech (1927) described the ten years following that war as

> a period of melodramatic crime, of corporate theft, of extravagance, [and] of specious prosperity. . . . Social behavior, notoriously lax in the years succeeding the war dissolved in the easy warmth of plentiful money. . . . Then in September of 1873, the country's bric a brac prosperity crashed. . . . With the collapse came panic, followed by an era of business depression. . . . A powerful social reaction set in. . . . The voice of the reformer was heard in the land. The stage

was set for a stern and rigorous revival of the spirit of the Puritan forefathers. (Broun & Leech, 1927, pp. 75–76)

The targets of the Puritan reformers were "Billiard Saloons, Theaters, Gambling Hells, Porter Houses and Bar-Rooms, Houses of Prostitution and Assignation, and Concert or Pretty Waiter Girl Saloons" (Broun & Leech, 1927, p. 77). Obscene books and papers, they claimed, were sold on the streets "as freely as roasted chestnuts." Feminists opposed prostitution and pornography as degrading to women. Others feared that if men used prostitutes, in part as a reaction to abstinence as a form of birth control, venereal disease would be brought back into the home. The separation of the amative from the propagative aspects of reproduction was seen as especially degrading; that is, contraceptive use was equated with prostitution (Reed, 1978). It was in this atmosphere that antiobscenity laws were passed at the federal as well as the state or local levels of government.

Anthony Comstock, the chief enforcer of the antiobscenity laws, was a fundamentalist of Puritan descent, born in Connecticut in 1844. In his book *Traps for the Young* he stated: "I believe that there is a devil. Those who disagree with me in this may translate my language. All I ask is that they admit the vital truth on which I insist. Let my language be considered symbolical, provided the evils I denounce are regarded as diabolical" (Comstock, 1967, p. 239).

Comstock served in the Civil War, but apparently he was not popular with his army mates, probably because of his disapproval of their flirtations with the devil and his tendency to press his religious convictions on others.

After the Civil War, Comstock worked in New York City as a dry goods salesman, married in 1871, taught Sunday school, and visited hospitals and jails with missionaries, but according to his diaries, he did not feel fulfilled. In 1871, outraged by the fact that two saloons were open, despite Sunday closing laws, and goaded by the lack of interest the police showed in his complaint, he made a citizen's arrest himself. Although there were some laws restricting the sale of pornography, these too were not enforced. Comstock started buying what he deemed pornographic works and then asking the police to prosecute. Generally they refused. But, when Comstock, who was achieving some notoriety, wrote to the board of the YMCA expressing his concerns, they quickly spun off a Committee for the Suppression of Vice in 1872, with Comstock at its head. The committee eventually grew into the New York Society for the Suppression of Vice. After 1875, agents of the society were granted police powers under New York state law (Ch. 205, Laws of the State of New York, 1875). Using that office, Comstock launched a one-man crusade against dealers in obscenity and, to some extent, against the publishers as well.

With some local victories in hand, backed by influential board members, and using his store of pornographic and obscene materials, Comstock influenced Congress to pass a statute in 1873 (amended in 1876 to close its loopholes) to prohibit the mailing "of every obscene, lewd, or lascivious book, pamphlet, picture, paper, writing, print, or other publication of an indecent character, and every article or thing designed or intended for the prevention of contraception or procuring of abortion." The statute also prohibited the importation of contraceptive materials by anyone, with no exception for physicians. The statute,[6] popularly known as the Comstock law, attached criminal penalties for its violation.

Comstock himself was appointed as an unpaid postal inspector in charge of enforcing the law, and armed with the statute and his commission, he swooped down with a vengeance on pornographers and polluters of the moral environment. One of his major triumphs—at least as judged by the publicity it garnered—was his 1878 arrest of Mme. Restell, a very rich abortionist with an elite clientele. Although she had been arrested several times before, after this arrest Mme. Restell committed suicide, an act that may have reflected a turning point in social attitudes toward abortion (Broun & Leech, 1927; Mohr, 1978).

Comstock also warred against "evil reading," which included light literature, newspapers, and weeklies whose stories acquainted their readers with the details of lurid crimes, "half dime novels and story papers." He saw the devil's hand in theaters and low plays that were to him nothing more than recruiting stations for brothels and drinking saloons. He opposed lotteries and other gambling dens, including pool halls with their inevitable hustlers. What Comstock could not attack as a postal agent—because his jurisdiction was confined to materials sent through the mail—he prosecuted through his role as executive director of the New York Society for the Suppression of Vice using state and local laws and the police powers granted to officers of the society (Comstock, 1967).

Comstock's activities brought him into conflict with many prominent figures, including not only George Bernard Shaw (who coined the term *comstockery* to refer to the mindless censorship of art, literature, and theater), but also Bernarr MacFadden, a health and fitness enthusiast who arranged exhibitions of the human body fully clad in union suits, sometimes with red ribbons draped over them, as well as the New York Art Students' League where nude models posed. Comstock never quite attacked nude paintings exhibited in museums, but he argued that when a nude painting appeared on a picture postcard, it became lewd. He pursued those who advocated free love and obtained an indictment against a man who had published Walt Whitman's "To a Common Prostitute" and "A Woman Waits for Me."

Although Comstock was frequently the subject of satirical cartoons and derision for his zeal in stamping out the obscene and pornographic, he pressed forth, undaunted in pursuit of his mission. His method of attack included direct purchases of questionable materials by undercover agents and answering ads by mail, tactics still used by the postal authorities. He would send money and request the vendor to mail the materials or the publications to an innocuous name and address. Upon receipt of the materials, he might arrest the culprit himself (Bennett, 1878). He screened bulk mailings of publications and seized those as well, although he claimed that he never opened sealed mail (Comstock, 1882, reprinted in Bremner, 1971, pp. 222–26).

A lone zealot could not have succeeded as well as Comstock did. Despite his critics, his views were shared by many in the larger society. State legislatures and the U.S. Congress had passed the laws against pornography, obscenity, contraception, and abortion. Time and again the courts upheld Comstock's methods and condoned his seizure of materials. Even as enlightened a person as Jane Addams shared some of Comstock's views: "We are authoritatively told that the physical difficulties [of coping with puberty and the developing sex drive] are enormously increased by uncontrolled or perverted imaginations, and all sound advice emphasizes a clean mind, exhorts an imagination kept free from sensuality and insists on days filled with wholesome athletic interests" (Addams, 1912, pp. 105–6). She went on to discuss the stage with its lurid advertisements, novels, and dancing in dance halls where "improprieties are deliberately fostered. The waltzes and two-steps are purposely slow, the couples leaning heavily on each other barely moving across the floor, all the jollity and bracing exercise of the peasant dance is eliminated, as is all the careful decorum of the formal dance" (Addams, 1912, p. 106).

Comstock succeeded in driving pornography underground. He was responsible for the arrest of thousands and for interrupting the flow of a great deal of material. Within a year, he claimed to have seized and destroyed 134,000 pounds of books, 194,000 obscene pictures and photos, 60,300 rubber articles, 5,500 indecent playing cards, and 3,150 boxes of pills and powders. Apparently, he kept enough of the pornographic materials to show to Congress in 1877 when he beseeched them to close loopholes in the law (Broun & Leech, 1927). He also succeeded in making advertisers more responsible for what they printed. The pages of ads for abortions, contraceptive devices, and dubious medications disappeared from newspapers and magazines because Comstock had the power to keep publications that ran such ads from using the mails. His unremitting enforcement resulted not only in suppressing the distribution of pornographic and obscene materials but also in suppressing the dissemination of scientific information about abortion and contraception. Indirectly, his efforts probably limited research on those subjects as well.

Anthony Comstock's attitude toward these matters is revealed in his explanation of why he opposed contraception. He asked, "Are we to have homes or brothels?" (Broun & Leech, 1927, p. 249). Premarital sex, of course, was viewed as an evil:

> When you hear one declare that no unmarried man can live a continent life, and that in fact all young men have sexual intercourse, occasionally before marriage, you may set that man down as an impure man. . . . He is a liar and libels thousands of pure men who would sooner pluck out the right eye than defile themselves by illicit intercourse. (Hunter, 1900, p. 146)

Birth control devices made possible the separation of the amative from the propagative aspects of sexuality, and Comstock and like-minded others opposed contraception because of its implied approval of sexuality for its own sake.

It was Comstock's power to ban publications from the mails that led to his battles with Margaret Sanger.

Margaret Sanger and Her Birth Control Clinics

Margaret Sanger began her work in the early 1900s. She not only promoted birth control and established the first clinics in the United States, but she also played an important role in the development of the birth control pill. The very existence of Planned Parenthood birth control clinics and similar public clinics supported by local and state governments owes a great deal to Margaret Sanger's work. As one of her biographers said about her: "Through these achievements she had a greater impact on the world than any other American woman" (Reed, 1978, p. 69).

Margaret Higgins Sanger (1884–1966) was born into a poor family in Corning, New York.[7] Her father was an improvident stonecutter, who was also a freethinker. Very early, she came to associate large families with poverty, unemployment, drunkenness, cruelty, fighting, and jails. Her sickly mother died after bearing eleven children, and after that time, her father apparently became quite tyrannical. Originally she wanted to become a physician, but receiving no encouragement in that direction, she went to a small college and then trained as a nurse. In college, she became interested in women's rights, and while in nurse's training, she met and married William Sanger, an architect.

The Sanger's New York apartment became a salon for radicals, labor leaders, anarchists, communists, socialists, IWW (Industrial Workers of the World) leaders, and socialites, some committed to working-class causes, and some taken with "radical chic." Margaret Sanger was not a salon radical. On several occasions, she participated in strikes in Hazelton, Pennsylvania, at which she was arrested, and in Lawrence, Massachusetts, and in Patterson, New Jersey. She evidently was inclined toward activism and was not afraid of taking risks. Influenced by Freudian thought, by Havelock Ellis's writings, and by Ellen Key, a Swedish writer, Sanger began to differ from other feminists of her day in advocating sexual love as a means of developing the inner self.

While living in New York, she worked as a private-duty nurse specializing in the care of postpartum women. Many of the mothers, especially poor women on New York's Lower East Side, asked her how they could avoid further pregnancies. At that time, she could only recommend abstinence and coitus interruptus, both unrealistic methods because they depended on the male's self-restraint. She often saw fifty to one hundred women standing in line, their shawls covering their heads, waiting for the local five-dollar abortionist. She heard about and witnessed deaths due to botched abortions and knew many women who were worn out and ill because they had borne too many children spaced too close together. She heard of women who had committed suicide when they learned they were pregnant once again, and she often wrote and spoke of the death of one of her patients from an infection following a self-induced abortion. She lectured to groups of working women on sexual hygiene and wrote articles on the subject for radical publications.

Sanger began to read about contraception in the New York Public Library. She found little on the subject, although contemporary writers claim there was information available at the time (Kennedy, 1970; Reed, 1978). In 1913, she and her husband went to Europe where he painted and she studied European birth control methods. After a few months, she left her husband in Europe and returned to New York with their three children (a few years later she and her husband were divorced).[8]

Upon her return to the United States in early 1914, carrying samples of contraceptives, Sanger made plans to publish her own magazine, *The Woman Rebel.* The magazine advocated radical causes, but it focused on rousing working-women's consciousness about the right to control their own bodies. To exercise that right, they needed to know how to prevent conception and disease. Comstock's agents quickly discovered her journal. An issue with an article entitled "What Every Girl Should Know," about venereal disease and feminine hygiene, was banned from the mails. Other issues of *The Woman Rebel* were also banned, but Sanger persisted anyway. She had not yct published anything explicitly on contraception, although articles advocating sexual liberation had appeared, and by then she had decided on the term *birth control.* She also wrote a pamphlet, entitled *Family Imitation,* on methods of contraception. It was only with great difficulty that she found a publisher willing to print 100,000 copies, which he did clandestinely.[9]

In 1914, Sanger was indicted on nine counts of having broken the federal Comstock law. Her lawyer urged her to plead guilty and pay a fine, but she was insistent that she would not do it, that she had done nothing wrong, and that she wanted to test the law. Because she did not feel ready to go to trial and did not want to be lost to her cause by sitting in prison, she adopted a pseudonym and fled to Europe via Canada, leaving her children behind in private schools. Before departing, she sent copies of *Family Limitation* to labor leaders in many parts of the country, with instructions to distribute the pamphlet in her absence.

In Europe, she met with Malthusians who were concerned about population control and with other British intellectuals who told her what they knew about contraception. She learned of British freethinkers who had been prosecuted for disseminating birth control information but who had won their cases on appeal. When she met Havelock Ellis in 1914, he instantly became a friend and tutor, helping her learn and putting her in touch with people who might help her.[10] Even though World War I had begun, she traveled on the Continent, and when she visited birth control clinics in the Netherlands, she was impressed. Education was not enough, she concluded, it was also necessary to have clinics to provide personal instruction if contraceptive methods were to be effective.

Anthony Comstock intruded on Margaret Sanger's life again in 1914 while she was still a fugitive from justice and subject to extradition. One of Comstock's agents persuaded her husband William to give him a copy of *Family Limitation.* A few days later, Comstock himself arrested William. Comstock wanted to find out where Margaret was hiding, but he got no help from William, who admitted he gave out the pamphlet but refused to plead guilty, refused to pay a fine, and refused to tell where Margaret was in Europe. He was sent to jail, and the entire affair received a great deal of publicity.

Margaret returned to New York in 1915. In her absence, the National Birth Control League had been organized, but the league did not want to be associated with her because its aim was to change the law through persuasion and lobbying rather than activism. With her case still pending, her friends and the distinguished constitutional law authority Samuel Untermeyer urged her to plead guilty in order to avoid going to jail. Sanger lost her attorney when she decided to present her case in her own way, to arouse the public by arguments in court.[11] Aware that public opinion seemed to be on her side, on the night before her trial she addressed a large rally organized by influential friends. At first, the government tried to postpone the trial and then, a few weeks later, apparently embarrassed by the case, withdrew prosecution on the two-year-old indictment. Sanger was annoyed that the law had not been tested, but as a result of the publicity, she went on a national speaking tour to promote the idea of establishing birth control clinics.

Although Anthony Comstock died in 1915, his spirit remained very much alive, and Sanger was faced with opposition in many cities. She noted that Hull House, for reasons that are unclear, prevented her from speaking in the stockyard district of Chicago. Jane Addams was deeply involved in the fight against prostitution, and her colleagues may have seen birth control as a means of promoting the practice. In St. Louis, Catholic groups threatened to boycott the theater that had been hired for her talk. Judge Ben Lindsey, however, supported her when she spoke in Denver.

In the fall of 1916, when Sanger returned to New York, she decided to open a birth control clinic to challenge the New York law that made it illegal to give contraceptive information or advice, for any reason. Although no physician was willing to help her test the law, she went ahead and opened a clinic in Brownsville, a Jewish and Italian section of Brooklyn. To be sure that the district attorney knew what she was doing, she wrote to him about her intention to dispense contraceptive information and gave him the clinic's address. The police did not disappoint her!

The clinic, staffed by Sanger, her sister Ethel who was a nurse, and another assistant, opened on October 16, 1916. Based on the advertising she had circulated in the neighborhood, some of it in Yiddish, the clinic did a booming business. She saw 464 clients in just a few days. Margaret Sanger not only wanted to assist people to achieve better contraception; she also wanted to do research on the effectiveness of various contraceptive methods. Toward that end, she intended to keep careful clinical records. A few days after she opened the clinic, a female police undercover agent came asking for assistance. A day later, the clinic was raided, the 464 case records were confiscated, and Sanger was jailed. Within hours of her release on the next day she reopened the clinic and was soon arrested again.

Her sister Ethel was tried first.[12] After being convicted, Ethel went to jail and engaged in a well-publicized hunger strike. When Margaret went on trial, she was unsuccessful in making a constitutional argument against the statute. The court was willing to allow her to remain free pending appeal, provided that she would agree not to break the law again. Much to the chagrin of her attorney, she refused and was sentenced to thirty days in the workhouse which she served amidst great publicity. It is a mark of her courage and her commitment that asserting she was not a criminal, she refused to allow herself to be fingerprinted when she went into the workhouse and again upon her release, although the issue threatened to delay her release.

Sanger eventually lost the appeal of her case, but she won a victory for her cause. The appeals court interpreted the New York law as permitting a licensed physician to provide birth control information for the cure or prevention of disease, and the court went on to define disease very broadly. This judicial interpretation made it possible for any physician to provide birth control information and birth control devices without fear of prosecution. Thus, she had won an important point through her litigative strategy.

Upon her release from prison, Sanger continued her educational efforts. She distributed her publications by means of street sales and street-corner lectures, techniques she had learned from British suffragettes. Convinced that the war had dampened interest in her cause, she went to England in 1920, renewed her ties with British intellectuals, including H. G. Wells and her good friend Havelock Ellis, and continued to gather birth control information.

In the 1920s, Sanger's interests shifted. Realizing that middle-class women were her chief constituency and that it was difficult to engage working-class women in a movement, she became less involved with working-class issues. She flirted with eugenic arguments for using birth control, even though the arguments had racist undertones, especially when recommended to control the reproduction of biologically inferior people. Eugenics was popular in the 1920s among native-born Americans concerned about the masses of immigrants in their midst and fearful of the spread of the Russian revolution. In addition to lecturing, Sanger wrote a book, *Women and the New Race,* which sold 250,000 copies. By 1921, she was well enough known to receive an invitation to lecture in Japan, in a lecture series at which she was one of four speakers, the other three being Albert Einstein, H. G. Wells, and Bertrand Russell.

Margaret Sanger continued to be interested not only in promoting birth control but also in doing research on birth control methods. She hired a physician to run a clinic whose dual purposes were to provide contraception and to conduct research, something the medical establishment was not yet doing.[13] She founded a journal, the *Birth Control Review,* which eventually became a respected outlet for scientific publications on birth control. She also encouraged reputable manufacturers to make high-quality diaphragms and condoms. Sanger organized national and international conferences on birth control and spent many years lobbying unsuccessfully to change state and federal statutes. Those years were not without strife between her and physicians wary of working with lay people. In addition, personality conflicts and competing advocacy groups with different birth control agendas led to disharmony within the birth control movement.

In 1929, Sanger's clinic was raided once again by police seeking evidence that her physician was prescribing birth control without adequate evidence of disease. The case was dismissed after a number of physicians testified to the validity of a broad definition of disease. For all practical purposes, by the mid-1930s birth control was legal in that the physicians prescribing it were not prosecuted. The one remaining restricted area for physicians was the importation of contraceptive devices. The *One Package* case (*United States v. One Package,* 1936) finally removed that roadblock as well. Judge Learned Hand ruled that it was irrational to bar physicians from importing birth control materials that were intended for the prevention and treatment of disease. Once again, progress came by the litigative rather than the legislative route.

By 1938, birth control was recognized by the federal government. Title V of the Social Security Act of 1938 provided financial incentives to states to offer state-supported child and maternal health services. Although the federal Children's Bureau, which administered the law, initially resisted including birth control among the services, the logjam broke, for two reasons. Eleanor Roosevelt lent her support, and as it became clear that women were needed to work in World War II war industries, it also became clear that contraception would help them continue working. As late as 1963, however, only thirteen states offered tax-supported family-planning programs, and not until 1967 did the federal government provide money consistently earmarked for family planning (Kadushin & Martin, 1988; Lundberg, 1947).

After World War II, Sanger continued her interest in research on improved methods of contraception. Gregory Pincus, the inventor of the birth control pill, received significant support from Margaret Sanger and the Planned Parenthood Federation of America, as it was then called. Kennedy (1970), who wrote a fairly negative biography of Sanger, believed that she thrived on conflict and pointed out that after the 1940s, when birth control became acceptable she slipped from the leadership.[14] But even he acknowledged that the result of her work has been the freeing of both women and men from the burden of unwanted and unplanned children.

Sanger's legacy of family-planning services is readily available to those who can afford to use them. Although there are public clinics that provide family-planning services, they tend to be used by poor women and adolescents. People wait long hours in public clinics, which tend to be overcrowded, not well furnished, and lacking in additional medical services (Radecki & Bernstein, 1989). Even so, such services are not free, despite earlier legislation (Family Planning and Population Research Act, 1970) designed to provide free birth control services. During the Reagan presidency, birth control services were cut back by 43 percent, despite additional demands on these clinics to use screening programs to detect sexually transmitted diseases, including AIDS. Pointing to the continued high rate of unintended pregnancies, one concerned observer called on President George Bush "to promote a 'kinder, gentler' nation by ensuring that the family planning program is allowed to achieve its potential. Both common humanity and an attractive cost-benefit ratio are forceful arguments for immediate responsive action" (Dryfoos, 1989, p. 690). We have not yet reached the point, however, at which family-planning information and contraceptive devices are so freely available that abortion as an option has lost its importance.

Roe v. Wade and the Legalization of Abortion

The regulation of abortion is an area of legislation that has always been left to the states, and in that matter the states have been influenced by physicians. After 1870, when restrictive abortion legislation was passed, those who needed abortions obtained them illegally or found sympathetic physicians who helped them out. Things changed in the 1960s.

The 1950s were the years of the postwar population explosion, and its sequelae of suburbs and raising children dominated our thinking about families. This thinking, though, did not quite match the reality. Women who had worked during the war were encouraged to stay home, and although many did so, the numbers of women in the labor force never declined to the prewar figure and, in fact, increased steadily during the full employment years of the 1950s. Many women who found domestic life unfulfilling, and the many others who found themselves in the labor force, yet trying vainly to carry out the ideals of the "feminine mystique," responded to Betty Friedan's famous book of that name (Friedan, 1963). In 1966, Friedan helped establish the National Organization for Women (NOW), one of whose aims from the beginning was increased reproductive rights for women. Abortion also came up repeatedly in consciousness-raising groups, an important organizing tool for the feminist movement, as a part of the new discussion of female sexuality.

During the 1960s, many young lawyers, sponsored by legal services corporations for the poor and inspired by *Brown v. Board of Education,* wanted to use the law to promote social change. Pregnancy counseling services opened in many communities in the 1960s, in part as a response to the awareness of the many illegal abortions that were taking place (it was estimated that between a million and a million and a half illegal abortions were performed each year, according to Jaffe, Lindheim, & Lee, 1981). The counseling services referred women to adoption agencies or to safe sources of abortion, sometimes in foreign countries where they were legal.

Reform was in the air in those days. Before the *Roe v. Wade* decision, the U.S. Supreme Court had carved out a right to privacy in *Griswold v. Connecticut* (1965) and *Eisenstadt v. Baird* (1972). The *Griswold* case had been planned in 1961 by Estelle Griswold, executive director of a newly opened Planned Parenthood clinic in New Haven, Connecticut, and C. L. Buxton, a respected Yale University professor of gynecology and obstetrics. Similar to Margaret Sanger's experience almost fifty years earlier, they were arrested a few days after they opened their clinic to dispense contraceptive devices and information. When they lost at trial, they appealed through the Connecticut courts and then took their case to the U.S. Supreme Court where they eventually won. In *Griswold,* the Court recognized a right to privacy in intimate marital relationships. In *Eisenstadt v. Baird,* the justices extended the privacy concept to contraceptive use by unmarried persons as well. The U.S. Supreme Court was asserting that an individual has the right to be protected against unwarranted government intrusion into a fundamental personal decision to bear children.

It was against that background that *Roe v. Wade* worked its way up to the U.S. Supreme Court.[15] In the late 1960s a woman named Norma McCorvey had consulted two young lawyers, Linda Coffee and Sara Weddington, because she was pregnant, allegedly as a result of a rape, and she was unable to obtain an abortion. Committed feminists, aroused by the prospect of using the law to bring about social change, the two attorneys had been looking for a plaintiff who would bring an abortion suit. Norma McCorvey met their requirements: a pregnant woman who wanted an abortion but could not obtain one on any legal ground. Once McCorvey came to them, they did the arduous legal research to shape the case.

The two lawyers were working in uncertain and, at that time, relatively unexplored legal territory. Their problem was to frame the case in such a fashion that statutes making abortions illegal would be held to violate a fundamental right belonging to a woman. The defendant in the suit, Henry Wade, was the Dallas district attorney. When the suit was filed, the court was asked to issue an injunction ordering Wade not to prosecute abortion cases. At the time abortion reform was in the wind nationally, as well as in Texas. By then, seventeen states had already reformed or repealed their abortion laws. The case developed in the midst of complex community processes in which a number of groups interested in abortion reform, and others opposed to it, were beginning to interact. The plaintiffs won at a hearing before a federal court panel of judges, but they were not given the injunction that they initially sought to prevent enforcement of the Texas abortion statute. The state also appealed.

As word circulated about the abortion case moving up to the Supreme Court, opposition formed. The Catholic church found that its influence was insufficient to prevent some abortion reform laws from passing. Right-to-life activists began to organize as well, and a Right-to-Life party was formed in New York State in 1969. The case went forward within crosscurrents of activity in the abortion reform movement, feminist groups, the Catholic church, and other groups opposed to abortion.

The case moved forward within a complex legal social context as well, as many such cases do. Numerous groups had an interest in it. Procedural issues required technical legal expertise. Who would receive credit for arguing the case and which line of argument should be followed were potential sources of conflict for the legal professionals preparing the case for the Supreme Court.

After oral arguments, the justices took the case under advisement. Abortion was growing as a political issue. It became a part of the 1972 presidential campaign when Senator Edmund Muskie, running in primary elections, expressed some opposition to abortion, and Richard M. Nixon followed suit in his presidential campaign. Within the Court, the case was assigned to Justice Harry A. Blackmun, whose work as counsel to the Mayo Clinic in Rochester, Minnesota, had given him an excellent understanding of medical issues. After some delay and reargument, the decision was finally announced in January 1973.

The Supreme Court's decision did not end the matter, as any reader of a daily newspaper knows. The ruling took effect when it was announced: Restrictive abortion laws could not be enforced in all states that still had them.

Justice Blackmun's opinion for the majority was criticized by some legal scholars and others because it had read a right of privacy into the Constitution that was not stated explicitly in that document. Other legal scholars have supported the decision as entirely correct, pointing out that many rights are not specifically listed in the Constitution and that the *Roe* decision followed a series of Supreme Court decisions regarding privacy in reproductive matters (Tribe, 1990, chap. 5). Antiabortion forces became more active, working at the state level to test the limits of the abortion decision by successfully lobbying for laws that added restrictions to the right to abortion. For the most part, when tested in the federal courts, the restrictions were struck down. But by 1980 the opponents of abortion won some important victories, most notably when the Supreme Court upheld the Hyde amendment, a 1976 congressional ban on the use of Medicaid funding for abortion if the states so choose (*Harris v. McRae,* 1980; Tribe, 1990, pp. 151–59).

Rather than settling the abortion debate, the *Roe* decision caused abortion to become an important political issue. Tribe traces the emergence of single issue politics to 1976 (Tribe, 1990, pp. 147–50), and since 1980, Presidents Reagan and then Bush have courted right-wing support by speaking against abortion. In the mid-1980s, abortion opponents stepped up their demonstrations against abortion clinics with some extremists resorting to violence, entering clinics and damaging equipment (*Northeast Women's Center v. McMonagle,* 1989). They also applied economic pressures and opened counseling services to attract women away from the abortion centers.

At this writing, groups opposed to abortion are continuing the nationwide campaign to harass and shut down abortion centers and to prevent research on or to import new methods of contraception (Palca, 1989; Riding, 1990). They are not unopposed. Prochoice forces rallied as it became apparent that the right to abortion was being threatened, especially with the changed composition of the Supreme Court, owing to retirements and appointments by Presidents Reagan and Bush. On April 9, 1989, crowds—variously estimated at between 300,000 and 600,000—marched in Washington, D.C., in support of the prochoice position, a reflection of the organization of myriad grass-roots prochoice groups nationwide. The march may have marked the beginning of single-issue politics for those in favor of choice, for since that time, a number of state and local elections have been decided on the basis of the candidates' positions on abortion.

The efforts at restricting access to abortions through legislation came to fruition in cases decided by the Supreme Court in 1989 and 1990. In the case of *Webster v. Reproductive Health Services* (1989), the Court allowed the states to legislate restrictions that probably would not have been allowed under a stricter reading of *Roe v. Wade*. In the spring term of 1990, the Supreme Court enabled the states to restrict teenagers' access to abortions when it upheld a Minnesota statute that required a minor to notify both parents or seek a judge's permission if she wanted an abortion (*Hodgson v. Minnesota,* 1990) and an Ohio statute requiring the notification of one parent, with the judicial bypass procedure (*Ohio v. Akron,* 1990).

After the *Webster* decision, the battle lines shifted even more to the state houses. Several state legislatures passed laws making it more difficult or more expensive to obtain abortions. In the foreseeable future, it is likely that the U.S. Supreme Court will hear more cases with different restrictions. Unless a majority of the justices vote to overrule *Roe v. Wade,* they will probably continue to uphold laws that impose restrictions on obtaining an abortion. Even if they continue to strike down laws that make it a crime for a woman to have an abortion or for a licensed physician to perform one, access to the procedure will become increasingly confined to older and more affluent women.

This issue will be decided politically, and it will affect the more than one million women and their male partners who each year deal with unwanted pregnancies by resorting to abortions. It will also have an important effect on the hundreds of thousands of adolescents who have abortions each year (see Joyce & Mocan, 1990). If restrictive laws result in a higher rate of birth of unwanted babies, then we will be faced with a social problem affecting every agency that deals with teenage adolescent mothers and their children. The repercussions of the abortion battle will be felt throughout society and will be especially important to people who work in the helping professions.

Summary

The regulation, or lack thereof, of reproduction, has always served social, economic, and psychological interests. Limitations on birth control information stemmed from the United States' Puritan heritage. In the 1870s and later, Anthony Comstock reasserted what many thought of as traditional values, by suppressing anything that attacked those values. He was able to act because the elected legislators, presumably reflecting the popular will in this matter, gave him the legal tools to do battle. Margaret Sanger, a product of the pre–World War I reform era, fought for social change and in the process created a specialized service, the family-planning clinic. Sanger began her fight out of a concern for social justice and for women's rights, but she also was concerned about sexual liberation. Her work took hold in the 1920s when sexual manners and mores were undergoing dramatic changes.

The availability of sex education and information about contraception is important to any program to prevent unwanted pregnancies. Similarly, the availability of abortions is necessary to prevent unwanted births. Reasoning from one moral perspective, the only choice is whether or not to risk pregnancy; once pregnant, a person has no choice. Following from this premise, it is essential to have education to avoid unwanted pregnancy and to have prenatal and postnatal care, assistance in child rearing, and help with decisions about abortion. If one accepts the position that there is a choice, then the pregnant person also needs counseling about and access to abortions. Although the majority of women who choose to have an abortion do not appear to suffer adverse psychological reactions, a few do, and they may well be in need of assistance. We can see from history that birth control and abortion services are closely tied to the social structure. The differences between the modern-day Comstocks and the modern-day Sangers will continue to affect us all as we strive to define and implement fundamental values.

14

Social Change and Helping Forms

Thus far we have stated a simple correlation between the form of help and the general social ethos. In the very language in which we have expressed our views we have revealed a bias toward solutions emphasizing changes in the social order. We admit to such a bias because we are writing to call the attention of the mental health professions to continuing and renewing the consideration of new forms of help, just as many did with the advent of the community mental health movement of the 1960s. We do not intend to denigrate current forms of practice but, rather, to express our understanding of the contemporary social situation.

We began our study of history in 1963, out of an enthusiasm for those accounts of the past that showed a striking resemblance to the 1960s. But then we asked ourselves to suppose that both the attempt to change social institutions and the attempt to change individuals stemmed from accurate perceptions of the social world at a given time. What followed? How could we understand the correlation in broader theoretical terms?

Let us assert as a basic assumption that people organize their lives largely according to the ways in which they earn their livings. Most of the major social institutions—be they educational, legal, religious, recreational, or health—help maintain that way of life. In the periods under discussion in this book, the United States changed from a predominantly rural and agricultural society to a predominantly urban and industrial society and now to a postindustrial society. The cities and factories absorbed successive waves of peoples with foreign and rural folkways and mores, but the social institutions of the newly developing cities were based on traditions relevant to a vanishing social order. Because the way of life was changing, irrelevant social institutions were under pressure to change, for good and sufficient reasons, and new helping services appeared. When institutions are obstructive and irrelevant, we speak of the alienation of individuals from their society. The problem of alienation, whether or not that early Marxian term was used, preoccupied the attention of the Progressives and of the reformers of the pre–World War I period and, we think it is safe to say, of those who created new services in the earlier period of immigration and later in the 1930s and the 1960s.

Customs, laws, beliefs, and values all change slowly. Systems arising from one way of life do not readily adapt to another, and new social institutions are not readily created by a society that does not believe in interference with natural social evolution. When the socially concerned mental health workers of an earlier day argued that social institutions were causing problems in living, they were right. A way of life had changed, but it took a long time and much struggle before social institutions changed to reduce alienation by serving people in terms of how people actually lived. In a later time, when the institutions had changed to some extent,[1] the people's problems were not seen as clearly related to the existence of faulty institutions, and new forms of help focused on helping individuals adapt to a new life-style. Changes in the pre–World War I period centered on the schools, courts, social agencies, health and recreational facilities, housing, and working conditions. Changes in the postwar period centered on manners and morals and on how people lived with themselves and with one another in the more stabilized urban, industrial society. This pattern was repeated during the Great Depression of the 1930s when massive programs were initiated to provide income aid to the elderly, the dependent, and the disabled. In the 1950s we worried about overly conforming youth, the psychodynamics of delinquency, autism, and childhood schizophrenia. With the War on Poverty in the 1960s, change focused once again on society's institutions rather than on individuals' shortcomings.

Historians and political scientists refer to historical periods as either reform or conservative (Phillips, 1990). *Conservative* describes governments dominated by business interests or governments in which some tradition—or some conventional wisdom, to use Galbraith's (1958) helpful term—is supported. *Reform* describes the efforts of the socially conscious to achieve changes in the service of the less fortunate, or changes in the service of a different view of the social order. Our reading of history suggests that the term *acute social change* may be more descriptive than *reform* is. When there are major political reforms, we suspect that it can be demonstrated that the rest of the major institutions of society are also under pressure to change. The introduction of the factory system, for example, and inventions that captured new sources of energy led to profound changes in a whole way of life, and not only to municipal reform.

The massive social problems of the present day show that once again we are in a period of acute social change. We are now experiencing the consequences of a fairly drastic shift in the way in which people earn a living. We have mastered new, powerful sources of energy, highly efficient and flexible materials and machinery, extraordinarily sophisticated devices for communication, and speedy devices for transportation. Writing from the perspective of the 1960s, it appeared to a number of observers that the need no longer existed for as many people to engage in economically useful work, as that term had been defined in the past (AFL-CIO, 1959, 1963; Bazelon, 1962; Brightbill, 1960; Charlesworth, 1964; De Grazia, 1962; Galbraith, 1958; Sarnoff, 1964; Theobald, 1961). Given the advances in technology, it seemed that relatively few people could do most of the work. Unskilled and semiskilled production and maintenance jobs were rapidly disappearing, whereas service positions, professional and technical employment, and supporting staffs were rising in numbers. According to the then labor secretary, Willard Wirtz, of 4.3 million new jobs created in the six years before 1964, 63 percent were in state and local governments and in nonprofit organizations, and only 4.7 percent were in private, profit-making, non-war-related industries (Wirtz, 1964). Machlup (1962) showed that one-third of the labor force was employed in the knowledge industries. Clearly, the need for new workers to produce goods was not very strong.

In the late 1960s, when we wrote the conclusion to our earlier book, we thought that were it not for the problems of distribution of those economic goods and the deeply ingrained moral principle that each person must work, the number of employed people and the number of hours required of each employed person could be quite low. We did not foresee, however, that with the increased scope of international conglomerate corporations, real wages would decline in our service economy, that in order to make ends meet, families would need two earners, and that many people would be forced to hold second jobs (Bluestone & Harrison, 1982; Currie and Skolnick, 1988). We underestimated the impact of changing divorce laws and the sexual revolution, as well as the greater numbers of out-of-wedlock pregnancies and growing welfare responsibilities. We did not anticipate the destructiveness of the Vietnam War, not only for those who participated in it, but also for our economic prospects when it became clear that we could not have both guns and butter. We could not foresee that the end of the costly cold war tensions in 1989 would result not in a flow of benefits from a ''peace dividend'' but to new and costly global military alignments. Our optimistic emphasis in the first edition on the implications for the social order of a trend toward greater leisure time proved misplaced, and so we have decided to leave to others the fine art of forecasting specifics.

Nonetheless, it is still instructive to examine the implications of a changing economy for the psychology of a people. Freud argued that civilization, a condition of social order, requires a renunciation of instinctual expression and the development of internalized controls in the form of a conscience. Delays in gratification must be accepted and enforced. Freud also contended that for socialization to occur, instinctual expression must be restricted, and as evidence for his concept, he pointed to the educator's repressive attitude toward most forms of libidinal expression in the young. Society's need for order, with the resultant renunciation of instinctual expression, will always conflict with individual desire for unhampered libidinal satisfaction, and thus discontent is inherent in civilization (Freud, 1930, 1938, 1959).

Let us add to Freud's argument the proposition that the degree of instinctual renunciation required is directly related to the amount of human energy required by any social group to earn its living. A society that masters some new source of physical energy or develops substitutes for human labor needs less human energy for its maintenance. An appropriate expansion in the permissible forms of libidinal expression should follow.[2] At one level, the viewpoint is highly oversimplified, and the closed-energy model leaves much to be desired in detail. Still, the model is helpful when we consider the nature of future helping forms or preventive forms.

Our contemporary economy makes us want to shift our basic attitudes toward instinctual renunciation. Galbraith (1958) spoke of the creation of new consumer desires as vital to enable the 1950s-style economy to maintain full production and full employment. Henry (1965) wrote about the role of advertising in both creating such desires and promoting the irrational thinking that makes it easy to create them. Henry believed that our way of life is increasingly geared to the immediate gratification of consumer desires, with the consequence that there may be an erosion of our collective willingness to delay satisfactions. The press toward gratification as a consequence of available time and energy is supplemented, then, by a pull toward impulse expression through advertising and available credits. These trends have been exacerbated by the widespread availability of television showing everyone what the "good life" actually looks like in the midst of reduced possibilities for many of achieving that life, because of poorer prospects for good jobs at good wages.

Galbraith (1958) observed that attempting to satisfy ever-increasing consumer wants is neither a necessary aspect of a full economy nor an essential emphasis in the life of an individual. If there were national goals and means of deciding personal worth other than a contribution to production or a propensity to consume, we would be freer to consider other acceptable ways of being. What those newer ways of being might be remain to be developed. We have a great many social problems, many related to poverty and a lack of opportunity but also many related to other aspects of social change.

When the mental health professional functions as an agent of deviancy control, deviance is defined in terms of variation from "the comfortable fit of an individual into a particular set of social circumstances" (Group for the Advancement of Psychiatry, 1966). The emphasis is on "the comfortable fit," not on the "particular set of social circumstances." We rarely see a discussion of the nature of schools, neighborhoods, housing, recreational facilities, or other environmental factors in relation to healthy development.[3] Only the nuclear family unit is considered to be a socializing agent, with attention concentrated on the role of mother and father. There is complete silence concerning the future life circumstances toward which each person is developing. There is an implicit acceptance of the part of socializing agents in preparing individuals for future life roles. Because the definition of the mature personality includes an ability to work, we can safely conclude that any characteristic that seems to foreshadow difficulty in meeting the demands of the adult role, as that role is currently understood, will be viewed as deviant and therefore dangerous.

The lessons for mental health professionals seem clear. We cannot permit ourselves to function as agents of deviancy control, regarding each new development as an alarming manifestation of social pathology. In so doing we promote social problems by supporting outmoded norms of conduct, thus blocking solutions by retarding the reduction of cultural lag. Because of our own drive for professional status, mental health professionals tend to be narrowly focused on the problems of individuals. Because of our theories and clinical methods, we are less likely to consider the social, economic, and institutional factors affecting problems of living. But, social action and resort to efforts to change laws or institutions that are oppressive—in the sense of no longer fulfilling human need—should not be foreign to our traditions if we trace their ancestry to the settlement house movement and feel a kinship with social activists who had a vision of achieving large-scale change to improve conditions of living.

We cannot readily extrapolate from one historical period to another, but we can learn something from the experiments in nature that our intellectual ancestors have provided. Within the next ten years, we will see a more concerted effort to come to terms with many of the social problems that are associated with poverty. We will be calling on our schools and other social agencies to change, to do more, or to do something different. Some social scientists and mental health workers will be defining problems and trying to convince decision makers to take appropriate action. Although the federal judicial climate may be less receptive than it was in the 1960s and 1970s to litigation protecting the rights of those less able to care for themselves, that route is not blocked, by any means. Those in mental health fields interested in prevention will have to learn to collaborate with lawyers and other social activists to achieve changes through law.

Our society's analyses are not always precise, and our society's solutions have not always been fully effective. Sometimes the solutions of the past became part of the problem. Nonetheless, we can still look to historical examples to see that if we do not limit our horizons, we can help ameliorate the problems of children and youth in our own day.

We suggest that some mental health professionals need to direct their attention to techniques for reducing cultural lag and to diagnosing the institutional contributions to problems in living. If our society is now in a period of acute change, then history and theory demand that mental health professionals concentrate on understanding and facilitating concomitant changes. Today then is the time for action; the challenge we face is to turn the concepts we acquired yesterday to the problems we will face tomorrow.

NOTES

Chapter 4

1. Abell (1943) points out that the settlement house movement developed partly as a reaction to the Salvation Army, itself an outgrowth of a strong religious mission movement. In England and the United States, the Salvation Army, with its military trappings, was characterized at first as vulgar and ridiculous and even as a threat to established religious groups. Originally based on a far-reaching plan for rehabilitating the poor, the Salvation Army was not merely an evangelical group with uniforms, cornets, soup kitchens, and street-corner revival meetings. General William Booth, its founder, proposed establishing a series of colonies to retrain men to be self-sufficient. The first step was an urban colony in which the men were to be given food, shelter, and work, in order that they develop a willingness to work and freedom from the "more repulsive habits." Once they had acquired basic work skills and attitudes, they were to be placed in a cooperative industrial-farm village where they were taught habits of self-reliance and resourcefulness. When this degree of rehabilitation was achieved, they were to be sent to an overseas colony and given suitable land. This plan was actually put into effect in the United States, and by 1897 there were three "colonies" in California, Colorado, and Ohio, with some two hundred inhabitants. The Salvation Army also formed a widespread network of residences, schools, orphan asylums, employment bureaus, legal aid societies, life insurance companies, day nurseries, halfway houses for prisoners, a youth corps, and similar services for the urban poor.

Stanton Coit opens his book with a description of his Neighborhood Guild concept. "No one has yet accepted, in the full sense in which he meant it, General Booth's challenge to bring forward a better scheme than his for lifting the fallen classes of society into independence and prosperity" (Coit, 1891, p. 1). He explicitly presents the settlement concept, with its emphasis on neighborhood organization and self-help, as an alternative to the Salvation Army.

2. This paper by David Levine and Zachary Levine received a second place in the annual Hamden, Connecticut, Science Fair in April 1967.

Chapter 5

1. The phrase *country care or medical advice* probably refers to the care of patients with tuberculosis, although it may be a euphemism for the care of unwed mothers.

Chapter 8

1. Lindsey's court was undoubtedly an exception to the way in which many juvenile courts actually operated. Unfortunately, many adopted the informality of procedure but not

the substance of Lindsey's "human artistry." Serious criticism was directed at the juvenile court concept very early (Eliot, 1914; Ryerson, 1978). In the late 1960s the supreme court stated that the juvenile court does not treat but violates the constitutional rights of individuals who appear before it and recommended that it be modified (*In re Gault,* 1967). Polier's (1964) survey of the operation and effectiveness of New York's juvenile court system indicates why criticism of the juvenile court was warranted.

2. Chicago, the court with the poorest record, had civil service probation workers who complained even then they did not have adequate time to supervise their probationers. Philadelphia had a decentralized, privately supported probation officer system, with the workers, all women, selected by the New Century Club, a women's group. The Indianapolis court extensively used volunteers for probation work, as did Judge Hoffman in Cincinnati. Hoffman stated: "It has been tacitly, if not expressly, determined by the social agencies of Cincinnati, the public schools and all the civic organizations that no child manifesting symptoms of conduct disorders shall enter a criminal career" (Hoffman, 1925, p. 259). The early courts also extensively used the volunteer system and the Big Brother movement (Coulter, 1913), but not without problems (Eliot, 1914; International Prison Commission, 1904). Ryerson (1978) believes that the data regarding the limitations of the Chicago court demoralized those who supported the juvenile court experiment.

3. Many of these quotations read like movie scripts, but they are not accurate transcripts. O'Higgins, one of Lindsey's collaborators (Lindsey & O'Higgins, 1910), was a professional writer. He was on the staff of *Everybody's Magazine* and was assigned to prepare a series of articles with Lindsey. The material in their book was reworked from Lindsey's dictated notes. Borough, another of his coauthors (Lindsey & Borough, 1931), was a newspaper editor who also worked from notes dictated to him by Lindsey. Borough (personal communication) had the advantage of seeing Lindsey at work in the Los Angeles court. There also are firsthand journalistic accounts by men as distinguished as Lincoln Steffens (1909), Franklin P. Adams (1915), and William McLeod Raine (1907) of Lindsey's approach in his court, and these accounts differ qualitatively only slightly from Lindsey's autobiographical accounts of his techniques. Lindsey the man and his basic approach come through with such clarity that we want to quote extensively from these various accounts.

4. In a typical month, less than 10 percent of the reports were poor, and about 7 percent were excellent (Lindsey, 1904). Probation workers today often avoid contacts with the school because they fear prejudicing teachers against a child who has become involved with the court. Lindsey, on the other hand, consulted teachers and principals concerning the disposition of their children. There was another form of follow-up as well: Teachers knew which of their boys were on probation. If a boy was absent, the teacher would call the probation officer, who immediately went out to investigate his absence (Vollmer, 1906).

5. The judge had a reputation among the boys for being able to ferret out the truth. He would not accept lies, and he seemed to know instinctively when a boy was not leveling with him. From the boys' point of view, he was fair, firm, and anything but gullible or soft. Lindsey once said, "Children don't rebel at authority, only ignorant authority" (Steffens, 1909, p. 156).

6. His use of the group is critical. Eddie's being sent away to Golden was not an arbitrary act by the judge. Rather, it was clear to all that Eddie had been given numerous chances and had not been able to help himself.

7. The discussion of Lindsey's technique in sending boys to Golden without escort shows that Lindsey did visit the state industrial school with some regularity and that he did follow up as he said he would. His visits probably helped the institution maintain its standards.

8. Lindsey stood firmly with the law, although he encouraged fights against injustice. When he suggested that a boy kept in chains run away from a prison (certain that the guards would not shoot the boy), he was expressing his indignation at the maltreatment of children.

9. Lindsey is appealing to the boy's strength and to his desire to be accepted in the community.

10. Lindsey's understanding of the subculture with which he was dealing is beautifully expressed here. Lindsey was sometimes criticized for using boys' language in court or for not correcting their language. His appreciation of the need to communicate in appropriate terms is expressed in the following anecdote: Lindsey had worked out an arrangement with the leader of a gang of young Italian Catholics to stop vandalizing a Protestant church in their neighborhood. The gang had chosen the boy to supervise the protection of the church. The young gang leader tried to get the judge to give him the right to beat up any boys who violated the agreement. Lindsey refused to give him the right and instead encouraged the boy just to tell his gang.

> And following my advice he immediately proceeded to tell the gang, with the accompaniment of the most violent swearing, how he would break their damn necks if they didn't cut it out.
>
> The good old deaconess from the church who had watched these proceedings with undisguised satisfaction, now suddenly screamed in protest, "Judge, stop that boy, he is swearing."
>
> The astonished Tony hesitated as he turned to the lady with the explanation, "Why, don't you want me to tell 'em?"
>
> "Oh yes, but not to swear at them."
>
> "Well," said Tony, "if I didn't, they wouldn't understand me." (Lindsey & Borough, 1931, p. 136)

11. This section provides an excellent example of how Lindsey used the concept of role in his work with the children. The example also shows how he used the boys to do the work of the court in the community. We have already told how Mickey rounded up boys to testify at a hearing that Lindsey held to show inequities in the treatment of children. Mickey was also used, with the court's support, to organize children into constructive activities. Lindsey delegated "police authority" to a gang leader to help control vandalism against a Protestant church in a Catholic neighborhood. Lindsey used boys as unofficial probation officers, and in the case of Eddie, Lindsey used them in the state school as his assistants in Eddie's therapy. In Lindsey's numerous battles with corrupt police and politicians, the boys in the neighborhoods were his eyes and ears. A policeman taking graft, a dive supposedly closed and then reopened, saloon keepers who sold liquor to minors or who seduced or forced children into prostitution thus became known to Lindsey, and he used the information constructively. In their turn the children campaigned for Lindsey during his election fights, and newsboys shouted "extra" whenever Lindsey wrote a column or a letter to the editor. Lindsey succeeded in making the people part of the court, thus encouraging a view of the court as an institution that helps and does not only punish.

12. His ability to empathize with his children is shown clearly in this statement. Borough (personal communication) expressed the opinion that a part of Lindsey always remained a child. He felt that this was why Lindsey could gain such an immediate rapport with most of the children who came before him.

13. Lindsey's remarkable understanding of Jimmie's position in his group is brought out beautifully by this comment.

14. Note Lindsey's use of the term *citizen.* He includes these children as part of the greater society, privileged to enjoy the benefits of its institutions, not simply subject to the coercion of the court. As a citizen, the child is seen as part of the community.

15. At that time, the sheriffs received a fee for each prisoner delivered to the jails. If a sheriff escorted more than one prisoner on a trip, then he would receive a fee for each person he delivered. When Lindsey began this practice, the sheriffs protested, and several grand jury investigations of his work followed. One of these investigations revealed that Lindsey had never lost a prisoner he had sent alone, whereas the sheriffs, who took their prisoners to jail in chains, had had a fair number of successful escapes. Lindsey proposed giving the fee to the boys, but it is not clear that he ever did so (Lindsey, 1904; Lindsey & Borough, 1931).

We have been unable to find any record of an independent investigation confirming his claim. But the claim is repeated over and over in political documents without any indication that his political opponents challenged it. A publication of the Denver Christian Citizenship Union (1910) asserts that the Public Utility Corporation, a political enemy, hired a detective who could not find anyone to challenge Lindsey's claim. One undated political pamphlet asserts that Lindsey's court was investigated and exonerated six times, but no specifics are offered. An item in the *Rocky Mountain News* on December 23, 1909, carries statements from police and probation officers denying that they or the judge ever accompanied any boy to Golden.

Political enemies charged that Lindsey was too soft on offenders, that his methods encouraged boys to be delinquent because they wanted to be part of the judge's gang, that no one interfered with his delinquents because they knew Lindsey would not act against the boys, that his tactics undermined respect for parental and school authority, but never that boys ran away when on their honor. The same article claimed that teachers and police could not really speak out because they feared reprisals from the politically powerful women's clubs that supported Lindsey (Colorado Humane Society, 1910). On the other hand are newspaper reports of boys going off on their own, correspondence between reformatory wardens and the Denver commissioner of safety referring to such cases, and old-time residents of Denver also claiming that he sent boys off alone. The figure of five and then six runaways remains constant in documents from 1910 to 1931. Although we cannot verify the figures, there is little doubt that Lindsey sent hundreds of boys to reform school on their honor and that he lost few, if any of them.

16. The appeal to loyalty to the judge had apparently worked. Lindsey (1903) often emphasized his interdependence with his boys. He told them that he was chancing his career on them and that they were chancing an unknown and likely tougher judge if he were forced to leave the bench.

17. As part of the "game of correction," Lindsey sometimes appealed to the offenders' hostility to the law. Lindsey encouraged the youth to regard his taking himself to reform school as a form of "fooling the cop" (Lindsey & Borough, 1931, p. 194).

18. Lindsey would give a boy paper and an envelope and ask him to write to him if he weakened or when he arrived at the reform school. He stayed in contact with the boy.

19. Although Lindsey never established such an institution, in 1914, Cincinnati established a court of domestic relations that had exclusive jurisdiction over all cases involving children or the family. All divorce and alimony cases, all cases of juvenile delinquency, all cases of neglect, and all cases concerning the administration of mothers' pensions came to this court. The court established a central registry so that it had available all salient facts concerning individuals and families who had any contact with public or semipublic agencies in the city over the preceding few years. As part of the court organization, there were two psychologists and a psychiatrist, and there was a working relationship with the vocational bureau of the public school system and the Central Mental Hygiene Clinic. The director of the psychological clinic of the school's vocational bureau was a supervisor in the court clinic. A further interlocking relationship was that among the court and the Central Mental

Hygiene Clinic and the Cincinnati General Hospital. There also were close relationships with the Boarding Home Bureau, which provided foster homes for children needing special care, and with the Big Brothers organizations, which maintained paid directors and a staff of trained social workers to supervise children entrusted to their care. The Rotary Club and other civic and luncheon clubs helped care for groups of both delinquent and dependent children. "It has been tacitly, if not expressly, determined by the social agencies of Cincinnati, the public schools and all the civic organizations that no child manifesting symptoms of conduct disorders shall enter a criminal career" (Hoffman, 1925, p. 259).

It became the policy of the court to send to the clinic all delinquent children and others who needed special care and treatment, and it permitted the directors of the clinic and the school authorities to prescribe treatment without the children's having to appear in the court. The court supervised the work and used its powers of coercion when parents would not cooperate or when legal commitment was necessary. The court proposed and worked toward an ideal in which the public schools would take responsibility for caring for delinquent and predelinquent cases so that the stigma of criminality would not be imposed on them.

The Cincinnati court tried to convert itself into an agency that referred cases for diagnosis and treatment to the organization, institution, or individual best qualified for the purpose. It saw itself as eliminating from the law hostility toward the lawbreaker and as substituting a social objective. It operated under the theory that the principle of retribution was unworkable and that a "kindly, sympathetic and helpful attitude toward those who fall is the only one that can possibly succeed, either in rehabilitating the individual or in protecting personal and property rights" (Hoffman, 1925, p. 263).

Judge Hoffman claimed that no girls had been committed to the State Industrial School for Girls from Cincinnati for the previous three years and that out of a population of twelve hundred boys in the state industrial school, only six were from Cincinnati. Official juvenile court hearings had practically ceased in Cincinnati, and there is evidence that the Cincinnati system was at least as successful as were other juvenile courts of its day in preventing further delinquency. Thomas and Thomas (1928), however, had a less positive view of the Cincinnati system, believing that its casework standards were poor and that only a few cases were actually served by the court's clinical facilities.

Chapter 9

1. E. K. Wickman, a psychologist, and Mildred Scoville, a social worker, both members of the earliest demonstration clinic teams, coined the name *child guidance*. Wickman (personal communication) recalls that he and Scoville were looking for a name that would avoid the stigma associated with the label *psychiatry*, a label that then meant a profession dealing with crazy people who were locked up. They settled on child guidance clinic, a variation of the name of the earlier established Bureau of Children's Guidance.

2. Copyright, American Orthopsychiatric Association, Inc., and reproduced by permission.

3. The selection of Commonwealth Fund psychiatric fellows and clinic directors was influenced by the anti-Semitism prevalent in the 1920s and also by the prejudice against women (Horn, 1989).

4. Why the almost exclusive selection of women for the first program in psychiatric social work? One factor was World War I. Because men were the shell-shocked veterans, the Smith College program was conceived as a contribution by a women's group to the war effort. Moreover, Southard, an influential psychiatrist, felt that "women passed through

more changes of an emotional sort in one year than men do in five, and yet are more rational than men'' (Cohen, 1958, p. 133). A third factor is that marginal people, women, and minority group members tend to flow into new professions rather than into established ones (Klein, 1946). The domination of the field by women at the beginning gave it a direction that only recently has begun to show signs of change.

5. Among the people who taught in the original Smith College summer session were J. J. Putnam, William Healy, H. W. Frink, and A. A. Brill, all well-known analysts. Jessie Taft, who became identified with the Rankian viewpoint, was also an instructor in the first summer session. Others with differing orientations included E. E. Southard, Lawson Lowrey, and Adolph Meyer, all of whom taught at that first session (Spaulding, 1918)

George S. Stevenson, who was a resident at the Phipps Psychiatric Clinic of Johns Hopkins University in 1918, remembered that some of the first psychiatric social work students were sent there for field experience. The women sat on the steps of the clinic in the evening and sang their college songs, many of which were Freudian parodies. We are indebted to Dr. Stevenson, whose prodigious memory retained the two following ditties. The first was sung to the tune of ''I'm Just Wild About Harry'':

> I'm wild, simply wild over Freud,
> With his psychoanalysis, we're overjoyed.
> Our libido knows no fright,
> We dissect our dreams each night,
> Upon repression, transference, we delight.
> No dismay we display over sex,
> It is we who are free from complex.
> But we know from what we dream,
> That you are not what you seem,
> We are hep, full of pep over Freud.

The second song was a parody of ''There's a Long, Long Trail A-Winding, into the Land of My Dreams.''

> There's a wrong, wrong trend that's trying
> To get you way off the track
> But until you're psychoanalyzed, you never will get back.

6. Although Healy, and Witmer before him, had used tests extensively, the 1920s was the decade in which mass testing became popular. Group tests of intelligence, developed during World War I, spread rapidly. Wickman, one of the first psychologists with the demonstration clinics, received his first training in individual testing as part of an army program to validate group tests, another effect of war on professional functioning (Wickman, personal communication). With the spread of interest in intelligence tests in the 1920s, it was inevitable that the clinics would use them.

7. Wickman's study, the forerunner of a great deal of later research and still widely cited, was not acceptable as a Ph.D. dissertation. He was told that he could take it to a department of education but that it was not psychology! (Wickman, personal communication).

8. Healy instituted the practice of the psychiatrist's making physical examinations in child guidance work. Later, when psychoanalytic concepts guided psychotherapeutic practices, many psychiatrists stopped conducting physicals on the grounds that the examination represented either a castration threat or a sexual assault and therefore was antitherapeutic (Senn & Stevenson, personal communications). Other considerations led the clinics to minimize physicals. Allen (1948), for example, felt parents would be confused if the clinic stressed physical examinations.

9. Such practices would likely be frowned on, if not actually prohibited, in most psychotherapeutically oriented clinics. In the 1950s, a worker who took a client to her own home would likely be accused of overidentifying or of having rescue fantasies. Part of the problem involves our concepts of a professional relationship. Likewise, schoolteachers express similar concerns about whether it is professionally correct to have close, personal relationships with their students (Sarason et al., 1966).

Chapter 11

1. Brace noted: "The colored people of the city seldom fall into such helpless poverty as the foreign whites; still there is a good deal of destitution and exposure to temptation among them" (1880, p. 219). Brace opened a "colored" school that evidently was segregated.

2. Statutes permitting the "binding out" of severely neglected or abandoned children were passed in the colonies as early as 1735. New York had such a law by 1824. Indenturing and apprenticing children ceased to be important after the Civil War, and it is possible that the Thirteenth Amendment, barring slavery, may have barred the indenturing of children as well (Folks, 1902).

3. The leaders of the Buffalo COS believed in accountability. Ansley Wilcox was first a director and then chairperson of the board of the Buffalo COS.

> For years, Wilcox had supported his mother, an aging widow in Augusta [Georgia]. Sending her weekly checks for over fifteen years, Wilcox insisted that she provide detailed accounts of how she spent his money. In her weekly thank-you note to him she documented her expenses: three dollars to repair the stove, thirty-six dollars to repair the roof, eight dollars for the dentist, two-fifty for books, and three dollars per day for food, "including milk."

In one note, thinking she had skipped a week, she wrote: "I believe I have failed to thank you for the check you mailed last week. I will try to use the money with discretion but there is always some most unexpected demand coming up. I am moderate in my outlays, yet the summing up is so big" (Goldman, 1983, p. 161).

4. A quaint method for receiving children in French foundling homes (and one that showed how frequently the service was needed) was the *tour*, or double cradle on a turntable. When the child had been placed on the cradle outside the building, a bell was rung, the roundtable was turned, the infant entered the institution, and an empty cradle was put into place to await the next baby (Warner, 1919).

5. Franklin D. Roosevelt was a victim of poliomyelitis. His interest in crippled children, reflected in his sponsorship of the March of Dimes, undoubtedly fostered public interest in this field. Seymour B. Sarason, who also had polio as a child, benefited from President Roosevelt's interest. When he wrote to the president asking for assistance because he needed an operation his family could not afford, the president's secretary answered the letter. Shortly afterward, Sarason learned from his physician that all costs would be covered by the New Jersey State Rehabilitation Commission (Sarason, 1988, p. 51).

6. The constitutional validity of the Social Security Act was upheld in *Steward Machine Co. v. Davis,* 301 U.S. 548 (1937). The U.S. Supreme Court held that the federal requirement that states pass legislation in conformity with federal law in order to receive federal funds was not coercive and did not violate state autonomy. In *Helvering v. Davis,* 301 U.S. 619 (1937), the Court also held that the Constitutional basis for the social security program was the power of Congress to spend money for the general welfare.

7. *King v. Smith,* 392 U.S. 309 (1968), struck down an Alabama regulation that

denied payments to mothers who cohabited either inside or outside the home with any able-bodied man. *Doe v. Shapiro,* 203 F. Supp. 7611 (D. Conn. 1969), appeal, dismissed 396 U.S. 488, rehearing denied, 397 U.S. 970 (1970), removed the requirement that a mother reveal the name of her illegitimate child's father before she could receive welfare. A requirement that welfare recipients permit home visits was upheld in *Wyman v. James,* 400 U.S. 309 (1971), but the Court's language made it clear that early-morning mass raids on the homes of welfare recipients, which had taken place in many communities, were not acceptable. The secretary of the Department of Health, Education and Welfare (now Health and Human Services) later changed the regulations to bar such midnight visits. Residence requirements for eligibility for welfare were found unconstitutional in *Shapiro v. Thompson,* 394 U.S. 618 (1969), because these unreasonably restricted the right to travel. In numerous other instances the courts did uphold restrictions on payments.

8. The nation initiated a large-scale food stamp program in 1964 (7 U.S.C. Sec. 2011–25), which some believe was one of the more revolutionary welfare measures of the 1960s. The program was designed to use the nation's agricultural surplus and to ensure the nutritional well-being of poor people. Eligible persons may buy food stamps at a discount and then spend them at regular food stores. Problems similar to those of establishing eligibility for welfare cropped up in this program as well. See *U.S. Department of Agriculture v. Moreno,* 413 U.S. 528 (1972); *U.S. Department of Agriculture v. Murry,* 413 U.S. 508 (1972); and *Dupler v. City of Portland,* 421 F. Supp. 1314 (D. Maine 1976).

9. *Goldberg v. Kelly,* 397 U.S. 254 (1970), gave due process rights to welfare recipients: They were entitled to a hearing before their welfare benefits could be terminated.

10. The material in this section is based largely on Bailis, 1974; Hertz, 1981; Piven & Cloward, 1971, 1977; and especially West, 1981.

11. In time, some local antipoverty agencies such as the MFY hired attorneys to pursue legal remedies. In the mid-1960s, the Office of Economic Opportunity, the parent agency for the antipoverty program, sponsored a neighborhood legal services program that also took up the cudgels in support of individual welfare recipients, resulting in some of the legal cases cited in the preceding footnotes.

12. The movement toward flat grants was not without its controversies as well, in terms of how the amounts should be established and what amounts should be allotted in order to take into account inflation: See *Johnson v. White*, 528 F. 2d 1228 (1st Cir. 1975).

13. In 1894, the COS in Buffalo, following the lead of other COSs, developed a "labor test" for welfare applicants:

> All able-bodied men who were applying to the COS for support [would] break six cubic yards of stone at a wage of seventy-five cents per yard. . . . [T]hose who stand this test, with the cold and exposure involved, are considered to have demonstrated sufficiently their willingness to work, and would therefore be deemed deserving of support. (Goldman, 1983, p. 162)

Chapter 12

1. The material in this section is based on stories that appeared in the *New York Times* between April 1874 and December 1875 and are reprinted in Bremner (1971) and in papers by Lazoritz (1990) and Stevens and Eide (1990).

2. The private societies to this day have a special right to assist in the prosecution of child abuse in family court. The New York Court of Appeals recognized this status when it allowed a nonlawyer social worker to represent the Westchester County Society for the Prevention of Cruelty to Children in presenting an abuse and neglect petition and in a fact-

finding hearing in family court, even though all other nonprofit corporations in New York must be represented by a member of the bar (*Matter of Sharon B.*, 1988).

3. Similar charges were levied against the Protestant charities that functioned in neighborhoods made up largely of Irish, German, and later Italian Catholic immigrants (Brace, 1880).

4. Contemporary research shows that a relatively high proportion of children return to their homes where they have been abused are reabused. Fifteen percent of those allowed to remain in their homes after a finding of abuse were removed within two years. Nearly half of those who remained in their own homes after a finding of abuse were neglected or reabused at least once more in the next two years (Wald, Carlsmith, & Leiderman, 1988).

5. This discussion is based on reports reprinted in the volume *The United States Children's Bureau 1912–1972* (1974). This book contains reprints of reports by Julia Lathrop, Grace Abbott, Dorothy Bradbury, Frederick Green, and others on the activities of the Children's Bureau. Nelson (1984) also discusses how the Children's Bureau, which at first had little interest in child abuse, came to play an important role beginning in the late 1950s.

6. It is an interesting sidelight that many of the SPCAs added work with children. In the late nineteenth century, there was a debate as to whether child protection and animal protection ought to be carried out by the same societies and even whether the SPCCs should be admitted to the American Humane Society (Lundberg, 1947; McCrae, 1910).

7. De Francis (1966) reviews the child abuse reporting legislation up to 1966. De Francis and Lucht (1974) offer a detailed analysis of the laws' provisions in all fifty states up to that date. Meriwether (1986) has a useful review of the more recent status of reporting laws and related legal issues.

8. The material in this section is based largely on Wheat and Lieber (1979).

Chapter 13

1. Hamburg (1986) notes, however, that early motherhood may present an alternative life course to entering the work force, especially for teenaged women from poverty groups who may find very limited employment opportunities. The pregnancies in those instances may be sought out rather than unwanted.

2. Right to Life and other conservative groups claim that school-based clinics refer teenagers to abortion services even when they are prohibited from doing so directly. They also insist that the clinics undercut parental authority by preserving confidentiality in matters of reproductive health. In most states, these clinics are obligated by law to preserve confidentiality. Conservative groups also question the cost effectiveness and the appropriateness of providing these services in the schools. Many clinics have been opened in predominantly black and other minority schools, leading some black groups to question the racist motives of their proponents. Opponents of school-based clinics also note, with some merit, that there is no evidence to support the hypothesis that school-based clinics result in fewer teenage pregnancies, although they agree that teen birthrates are falling. But they attribute the reduction in birthrate to abortions rather than to prevention (Glasow, 1988).

3. Today, the yellow pages of the telephone directory also list clinics or counseling centers that provide counseling against abortion. Some are openly antiabortion and advertise alternatives to abortions. Other ads are more ambiguous and are designed to attract the same clientele, namely, single women or couples who are concerned that they may be pregnant when they may not wish it or are ambivalent about having a child.

4. Unwanted infants are occasionally abandoned or murdered even today. Under the law in all states, women who have unwanted children may voluntarily give them up for

temporary foster care or may permanently relinquish parental rights and give up their infants for adoption through local social service agencies.

5. The increasing incidence of abortion stimulated a debate among feminists. Feminism as a movement and as an ideology began in the 1840s, in part as an outgrowth of the abolition movement, and it continued throughout the nineteenth century, with its adherents pressing for higher education, voting rights for women, and a redefinition of women's roles in society. Some feminists saw abortion as a matter of a woman's right to control her body, against men's ability to control women; they advocated family planning and distributed abortifacient information. Other prominent feminist leaders favored abstinence to avoid unwanted pregnancies and blamed men who ''forced themselves'' on their wives and then encouraged their wives to have abortions (Mohr, 1978). Some feminists linked birth control and abortion as part of a fight against prostitution, which they viewed as degrading to women (Reed, 1978).

6. Sec. 211, U.S. Statutes at Large, Containing the Laws and Concurrent Resolutions Enacted During the 1st Session of the 44th Congress.

7. The material in this section is based on Margaret Sanger's autobiography (Sanger, 1938), her own story of the fight for birth control (1931), a biography by Kennedy (1970), and biographical material contained in Reed's discussion of the history of contraception in the United States (Reed, 1978).

8. Margaret Sanger later remarried. Her second husband was a rich man who acceded to her request that they maintain separate residences and separate schedules, sometimes communicating through their secretaries. Her second husband was also a financial supporter of some of her projects.

9. Sanger approached a philanthropist for financial aid in publishing *Family Limitation*. The philanthropist was closely associated with A. A. Brill, who was then translating Freud's works into English. The following dialogue may illustrate why those devoted to psychoanalysis may not have been involved with social activism. Sanger described her conversation with the philanthropist:

> He asked whether I had been psychoanalyzed. ''What is psychoanalysis?''
>
> He looked at me critically as from a great height. ''You ought to be analyzed as to your motives. If, after six weeks, you still wish to publish this pamphlet, I'll pay for ten thousand copies.''
>
> ''Well, do you think I won't want to go on?''
>
> ''I don't only think so. I'm quite sure of it.''
>
> ''Then I won't be analyzed.'' (Sanger, 1938, p. 112)

10. Ellis, who also had an unusual living arrangement with his wife, apparently became Margaret Sanger's lover (Reed, 1978, p. 93). They later traveled together in Ireland.

11. It is always a problem for a lawyer to advise a client who has political aims. A lawyer views a case from a technical, legal viewpoint and advises a client in such a way as to maximize his or her freedom or gain or to minimize punishment or economic loss. Politicized clients are interested in getting their cause before the public and are less interested in what might happen to them. This problem is depicted vividly in Sanger's autobiography (Sanger, 1938, pp. 183–85; 225–28).

12. The attorney who represented her and her sister was a product of two New York City settlement houses.

13. Participation in research on a taboo subject can subject the research worker to sanctions. G. Stanley Hall refused to join the Massachusetts Birth Control League because he had been ''pilloried severely and ostracized'' by some of his friends when he wrote about adolescent sexuality and was involved in the social hygiene movement (Reed, 1978, p. 105).

14. Margaret Sanger's opponents are now resurrected in the form of an organization called STOPP, which claims to have a number of affiliates all over the country. STOPP is an acronym for Stop Planned Parenthood, and it is adamantly opposed to the sex education programs sponsored by Planned Parenthood, on the grounds that the programs promote teenage sexual expression and give short shrift to the concept of sin and to religiously based values of sexual abstinence outside marriage. The programs are said to interfere with parental prerogatives and to promote abortion. STOPP claims to have been successful in influencing some local United Way organizations to stop funding Planned Parenthood centers and in blocking attempts in some school districts to adopt a sex education curriculum sponsored by Planned Parenthood (See STOPP NEWS, January 1989, March 10, 1989).

15. The information in this section relies heavily on Faux (1988).

Chapter 14

1. We do not intend to imply that the changes were caused exclusively or even largely by the efforts of social reformers. Surely they helped, but the extent of their influence should not be overestimated.

2. Fuller development of the social implications of these principles may be found in M. Levine and A. Levine (1970).

3. The late 1970s and the 1980s were a conservative era. Responses to two major problems illustrate our thesis: During the years of the Reagan presidency, the major official response to the burgeoning traffic in drugs was "Just Say No," a slogan targeted to individual users of drugs. Homelessness—created by high unemployment, inflation, skyrocketing real estate costs, and drastic reductions in government support for affordable housing—was blamed on the mental illness, addiction, laziness, and other shortcomings of the homeless people (many of them children) themselves.

REFERENCES

Abbott, E., & Breckinridge, S. P. (1917). *Truancy and nonattendance in the Chicago schools*. Chicago: University of Chicago Press.

Abbott, G. (1925). History of the juvenile court movement throughout the world. In J. Addams (Ed.), *The child, the clinic and the court*, pp. 267–73. New York: New Republic.

Abell, A. I. (1943). *The urban impact on American Protestantism: 1865–1890*. Cambridge, MA: Harvard University Press.

Abelson, W. D. (1966). *A clinic in the community*. New Haven, CT: Clifford W. Beers Guidance Clinic.

Adams, F. P. (1915). Personal glimpses: A glimpse of Ben Lindsey justice. *Literary Digest* 50: 762–66.

Addams, J. (1902). *Democracy and social ethics*. New York: Macmillan.

———. (1909). *The spirit of youth and the city streets*. New York: Macmillan.

———. (1910). *Twenty years at Hull House*. New York: Macmillan.

———. (1912). *A new conscience and an ancient evil*. New York: Macmillan.

———. (1929). A toast to John Dewey. *Survey* 63: 203–4.

Additon, H. (1928). City planning for girls. *Social Service Monograph* 5. Chicago: University of Chicago Press.

Adler, H. M. (1922). A behavioristic study of delinquency. *Proceedings of forty-sixth annual session of the American Association for the Study of the Feebleminded*. St. Louis.

Adler, N. E., David, H. P., Major, B. N., Roth, S. H., Russo, N. F., & Wyatt, G. E. (1990). Psychological responses after abortion. *Science* 248: 41–44.

AFL–CIO Community Services Activities. (1963). The shorter work week and the constructive use of free time. *Proceedings of the Eighth Annual AFL–CIO National Conference on Community Services*. New York.

AFL–CIO Department of Research. (1959). *Labor looks at automation*. New York: AFL–CIO.

Alinsky, S. (1971). *Rules for radicals*. New York: Vantage Press.

Allen, E. (1929). A mental hygiene program in grade schools. *Mental Hygiene* 13: 289–97.

Allen, F. H. (1948). The Philadelphia Child Guidance Clinic. In L. G. Lowrey and V. Sloane (Eds.), *Orthopsychiatry, 1923–1948: Retrospect and prospect*, pp. 394–413. New York: American Orthopsychiatric Association.

Allen, F. L. (1959). *Only yesterday*. New York: Bantam.

Altmeyer, A. J. (1968). *The formative years of social security*. Madison: University of Wisconsin Press.

Atkinson, C., & Maleska, E. T. (1964). *The story of education*. New York: Bantam.

Bailis, L. N. (1974). *Bread or justice*. Lexington, MA: Lexington Books.

Bailyn, B. (1960). *Education in the forming of American society*. New York: Vintage Books.

Barden, J. C. (1991). Washington to give up direction of foster care. *New York Times*. Sunday, July 14: A12.

Barker, L. F. (1918). The first ten years of the National Committee for Mental Hygiene, with some comments on its future. *Mental Hygiene* 2: 557–81.

Barnett, H. (1950). The beginning of Toynbee Hall. In L. M. Pacey (Ed.), *Readings in the development of settlement work,* pp. 9–20. New York: Association Press.

Bazelon, D. T. (1962). The paper economy. *Commentary* 33: 185–97.

Beard, C. A., & Beard, M. R. (1927). *The rise of American civilization*. Vol. 2, *The industrial era*. New York: Macmillan.

Beers, C. W. (1933). *A mind that found itself* (Originally published 1908). New York: Doubleday Doran.

Beilen, H. (1962). Teachers' and clinicians' attitudes toward the behavior problems of children, a reappraisal. In V. Noll & R. Noll (Eds.), *Readings in educational psychology,* pp. 361–80. New York: Macmillan.

Bender, L., & Blau, A. (1937). The reaction of children to sexual relations with adults. *American Journal of Orthopsychiatry* 7: 500–18.

Bender, L., & Grugett, A. E., Jr. (1952). A follow-up report on children who had atypical sexual experience. *American Journal of Orthopsychiatry* 22: 825–37.

Bennett, D. M. (1878). *Anthony Comstock: His career of cruelty and crime*. New York: Liberal and Scientific Publishing House. Reprint, New York: Da Capo Press.

Bernstein, P. (1966). *The lean years: A history of the American worker, 1920–1933*. Baltimore: Penguin Books.

Bernstein, R. J. (1967). *John Dewey*. New York: Washington Square Press.

Blanton, S. (1925). The function of the mental hygiene clinic in the schools and colleges. In J. Addams (Ed.), *The child, the clinic and the court,* pp. 93–101. New York: New Republic.

Bluestone, B., & Harrison, B. (1982). *The deindustrialization of America*. New York: Basic Books.

Bordin, R. (1964). Emma Hall and the reformatory principle. *Michigan History* 48: 315–32.

Boring, E. G. (1957). *A history of experimental psychology*. 2nd ed. New York: Appleton–Century–Crofts.

Bourne, R. S. (1916). *The Gary schools*. Boston: Houghton Mifflin.

Bowen, J. T. (1925). The early days of the juvenile court. In J. Addams (Ed.), *The child, the clinic and the court,* pp. 298–309. New York: New Republic.

Brace, C. L. (1880). *The dangerous classes of New York and twenty years' work among them*. 3rd ed. New York: Wynkoop & Hallenbeck.

Bremner, R. H. (1971). *Children and youth in America. A documentary history*. Vol. 2: *1866–1932*. Cambridge, MA: Harvard University Press.

Brightbill, C. K. (1960). *The challenge of leisure*. Englewood Cliffs, NJ.: Prentice-Hall.

Brill, A. A. (1939). The introduction and development of Freud's work in the United States. *American Journal of Sociology* 45: 318–25.

Brotemarkle, R. A. (Ed.). (1931). *Clinical psychology: Studies in honor of Lightner Witmer*. Philadelphia: University of Pennsylvania Press.

Broun, H., & Leech, M. (1927). *Anthony Comstock, Roundsman of the Lord*. New York: A. & C. Boni.

Bryant, J. E. (1906–7). A method for determining the extent and causes of retardation in a city school system. *The Psychological Clinic* 1: 41–52.

Bulkley, J. A. (1989). The impact of new child witness research on sexual abuse prosecutions. In S. J. Ceci, D. F. Ross, & M. P. Toglia (Eds.), *Perspectives on children's testimony,* pp. 208–29. New York: Springer-Verlag.

Bureau of Education (1880). *Legal rights of children.* Washington, DC: U.S. Government Printing Office.

Burgess, E. W. (1939). The influence of Sigmund Freud on sociology in the United States. *American Journal of Sociology* 45: 356–74.

Callahan, D. (1967). The quest for social relevance. *Daedalus* 96: 151–79.

Campbell, C. M. (1918). A city school district and its subnormal children; with a discussion of some social problems involved and suggestions for constructive work. *Mental Hygiene* 2: 232–44.

———. (1919). Education and mental hygiene. *Mental Hygiene* 3: 398–408.

Caplow, T. (1964). *The sociology of work.* New York: McGraw-Hill.

Carmichael, S., & Hamilton, C. V. (1967). *Black power.* New York: Vintage Books.

Chambers, C. A. (1967). *Seedtime of reform: American social service and social action, 1918–1933.* Ann Arbor: University of Michigan Press.

Charlesworth, L. C. (Ed.). (1964). Leisure in America: Blessing or curse. *Monograph* 4. American Academy of Political and Social Science.

Children's Bureau (1963a). *The abused child: Principles and suggested language for legislation on reporting of the physically abused child.* Children's Bureau, Welfare Administration, U.S. Department of Health, Education and Welfare. Washington, DC: U.S. Government Printing Office.

———. (1963b). Juvenile court statistics. *Children's Bureau Statistical Series* 79. Washington, DC: U.S. Department of Health, Education and Welfare.

Chilman, C. (1986). Some psychosocial aspects of adolescent sexual and contraceptive behaviors in a changing American society. In J. B. Lancaster & B. A. Hamburg (Eds.), *School-age pregnancy and parenthood: Biosocial dimensions,* pp. 191–217. Hawthorne, NY: Aldine.

Clark, B. R. (1964). Sociology of education. In R. L. Faris (Ed.), *Handbook of modern sociology,* pp. 734–69. Chicago: Rand McNally.

Cohen, N. E. (1958). *Social work in the American tradition.* New York: Holt, Rinehart and Winston.

Coit, S. (1891). *Neighborhood guilds: An instrument of social reform.* London: Swan Sonnenschein.

Coll, B. (1969). *Perspective in public welfare: A history.* Washington, DC: U.S. Government Printing Office.

Colorado Humane Society. (1910). *Child and animal protection* 3.

Committee of Syracuse Board of Education to investigate the school system of Gary, Indiana. (1915). *The Gary system.* Syracuse, NY: C. W. Bardeen.

Commonwealth Fund. (1922). *Fourth annual report, 1921–1922.* New York: Commonwealth Fund.

———. (1963). *Historical sketch, 1918–1962.* New York: Harkness House.

Comstock, A. (1967). *Traps for the young.* Cambridge, MA: Harvard University Press (Originally published 1883).

Cornman, O. P. (1906–7). The retardation of the pupils of five city school systems. *The Psychological Clinic* 1: 245–57.

Corwin, D. L., Berliner, L., Goodman, G., Goodwin, J., & White, S. (1987). Child sexual abuse and custody disputes: No easy answers. *Journal of Interpersonal Violence* 2: 91–105.

Coulter, E. K. (1913). *The children in the shadow.* New York: McBride, Nast.

Courtis, S. A. (1919). *Measurement of classroom products.* New York: General Education Board.

Cox, H. G. (1967). The "new breed" in American churches: Sources of social activism in American religion. *Daedalus* 96: 135–50.

Cremin, L. A. (1964). *The transformation of the school.* New York: Vintage Books.

Cross, W. L. (Ed.). (1934). *Twenty-five years after: Sidelights on the mental hygiene movement and its founder.* New York: Doubleday Doran.

Currie, E., & Skolnick, J. (1988). *America's problems.* 2nd ed. Glenview, IL: Scott-Foresman.

Curti, M. (1951). *The growth of American thought.* New York: Harper.

D'Augelli, A. R. (1982). Historical synthesis of consultation and education. In D. R. Ritter (Ed.), *Consultation, education and prevention in community mental health,* pp. 3–50. Springfield, IL: Thomas.

Davie, M. (1940). *Sumner today: Selected essays of William Graham Sumner.* New Haven, CT: Yale University Press.

Davis, A. F. (1959). Spearheads for reform: The social settlements and the progressive movement, 1890–1914. Ph.D. diss., University of Wisconsin (University Microfilms 59.5758).

De Francis, V. (1966). *Child abuse legislation: Analysis of reporting laws in the United States.* Denver: Children's Division, American Humane Association.

De Francis, V., & Lucht, C. L. (1974). *Child abuse legislation in the 1970's.* Rev. ed. Denver: Children's Division, American Humane Association.

Degler, C. N. (1959). *Out of our past.* New York: Harper & Row.

De Grazia, S. (1962). *Of time, work and leisure.* New York: Twentieth Century Fund.

Denver Christian Citizenship Union. (1910). The civic review. *Quarterly Bulletin* 3: 3–9.

Devereux, H. T. (1909–10). Report of a year's work on defectives in a public school. *The Psychological Clinic* 3: 45–48.

Dewey, J. (1897). *My pedagogic creed.* New York: G. L. Kellogg. Reprinted (1964) in R. D. Archambault (Ed.), *John Dewey on Education.* New York: The Modern Library.

Dewey, J., & Dewey, E. (1915). *Schools of tomorrow.* New York: Dutton.

DHEW (1979). *Preventing disease/promoting health: Objectives for the nation.* Washington, DC: U.S. Department of Health, Education and Welfare.

Dryfoos, J. G. (1989). What President Bush can do about family planning. *American Journal of Public Health* 79: 689–90.

Duffus, R. L. (1938). *Lillian Wald: Neighbor and crusader.* New York: Macmillan.

Dummer, E. S. (1948). Life in relation to time. In L. G. Lowrey & V. Sloane (Eds.), *Orthopsychiatry, 1923–1948: Retrospect and prospect,* pp. 3–13. New York: American Orthopsychiatric Association.

Eckenrode, J., Powers, J., Doris, J., Mansch, J., & Balgi, N. (1988). Substantiation of child abuse and neglect reports. *Journal of Consulting and Clinical Psychology* 56: 9–16.

Eels, K., Davis, A., Havighurst, R. J., Herrick, V. E., & Tyler, R. M. (1951). *Intelligence and cultural differences.* Chicago: University of Chicago Press.

Eisenberg, L., & De Maso, D. R. (1985). 50 years of the American Journal of Orthopsychiatry: An overview and introduction. In E. Flaxman & E. Herman (Comps.), *American Journal of Orthopsychiatry: Annotated index.* Vols. 1–50, *1930–1980,* pp. viii–xxxviii. Greenwich, CT: JAI Press.

Eliot, T. D. (1914). *The juvenile court and the community.* New York: Macmillan.

Everson, M. D., Hunter, W. M., Runyon, D. K., Edelsohn, G. A., & Coulter, M. L.

(1989). Maternal support following disclosure of incest. *American Journal of Orthopsychiatry* 59: 197–207.

Farber, M. A. (1990). The tormenting of Hilary. *Vanity Fair*. June: 120–27, 193–204.

Faux, M. (1988). *Roe v. Wade*. New York: Macmillan.

Flaxman, E., & Herman, E. (Comps.). (1985). *American Journal of Orthopsychiatry: Annotated index*. Vols. 1–50, *1930–1980*. Greenwich, CT: JAI Press.

Flexner, A. (1910). *Medical education in the United States and Canada: A report to the Carnegie Foundation for the Advancement of Teaching*. Washington, DC: Carnegie Foundation for the Advancement of Teaching (Reprinted 1960).

———. (1940). *I remember*. New York: Simon & Schuster.

———. (1961). Is social work a profession? In R. E. Pumphrey & M. W. Pumphrey (Eds.), *The heritage of American social work,* pp. 301–7. New York: Columbia University Press.

Flexner, A., & Bachman, F. P. (1918). *The Gary schools: A general account*. New York: General Education Board.

Flexner, E. (1966). *Century of struggle*. Cambridge, MA: Harvard University Press.

Folks, H. (1902). *The care of destitute, neglected and delinquent children*. New York: Macmillan.

Freeman, H. F. (1961). University settlement. In R. E. Pumphrey & M. W. Pumphrey (Eds.), *The heritage of American social work,* pp. 197–201. New York: Columbia University Press.

French, L. M. (1940). *Psychiatric social work*. New York: Commonwealth Fund.

Freud, S. (1930). *Civilization and its discontents*. London: Hogarth Press.

———. (1938). Three contributions to the theory of sex. *Basic writings of Sigmund Freud,* pp. 553–629. New York: Modern Library.

———. (1959). Instincts and their vicissitudes. *Collected Papers,* vol. 4, pp. 60–83. New York: Basic Books.

Friedan, B. (1963). *The feminine mystique*. New York: Norton.

Funicello, T., & Schram, S. F. (1989). Welfare in the Cuomo years: Less is less. In P. W. Colby & J. W. White (Eds.), *New York state today: Politics, government, public policy*. 2nd ed., pp. 281–91. Albany: State University of New York Press.

Furman, S. S. (1965). Suggestions for refocusing child guidance clinics. *Children* 12: 140–44.

Furman, S. S., Sweat, L. G., & Crocetti, G. M. (1965). Social class factors in the flow of children to outpatient psychiatric facilities. *American Journal of Public Health* 55: 385–92.

Furstenberg, F. F., Jr., Brooks-Gunn, J., & Morgan, S. P. (1987). *Adolescent mothers in later life*. Cambridge: Cambridge University Press.

Galbraith, J. K. (1958). *The affluent society*. New York: Mentor Books.

Garbarino, J., Guttman, E., & Seeley, J. W. (1986). *The psychologically battered child*. San Francisco: Jossey-Bass.

Garn, S. M., Pesick, S. D., & Petzold, A. S. (1986). The biology of teenage pregnancy: The mother and the child. In J. B. Lancaster & B. A. Hamburg (Eds.), *School-age pregnancy and parenthood: Biosocial dimensions,* pp. 77–93. Hawthorne, NY: Aldine.

Gelles, R. J. (1986). School-age parents and child abuse. In J. B. Lancaster & B. A. Hamburg (Eds.), *School-age pregnancy and parenthood: Biosocial dimensions,* pp. 347–59. Hawthorne, NY: Aldine.

George, H. (1879). *Progress and poverty*. New York: Doubleday & McClure.

Gesell, A. (1928). *Infancy and human growth*. New York: Macmillan.

Glasow, R. D. (1988). *School-based clinics: The abortion connection.* Washington, DC: National Right to Life Educational Trust Fund.

Glazer, N., & Moynihan, D. P. (1963). *Beyond the melting pot.* Cambridge, MA: MIT Press.

Glueck, B. (1919). Special preparation of the psychiatric social worker. *Mental Health* 3: 409–19.

Gold, M. (1930). *Jews without money.* New York: Liveright.

Golding, W. (1954). *Lord of the flies: A novel.* New York: Coward-McCann.

Goldman, E. F. (1956). *Rendezvous with destiny.* New York: Vintage Books.

Goldman, M. (1983). *High hopes: The rise and decline of Buffalo, New York.* Albany: State University of New York Press.

Goldstein, J., Freud, A., & Solnit, A. J. (1973). *Beyond the best interests of the child.* New York: Free Press.

———. (1979). *Before the best interests of the child.* New York: Free Press.

Goodman, G. S. (Ed.). (1984). The child witness. *Journal of Social Issues* 40: 1–7.

Gordon, L. (1988). *Heroes of their own lives: The politics and history of family violence, Boston 1880–1960.* New York: Viking.

Goren, A. A. (1970). *New York Jews and the quest for community.* New York: Columbia University Press.

Gouldner, A. W. (1968). The sociologist as partisan: Sociology and the welfare state. *American Sociologist* 3: 103–16.

Grant, Q. A. R., & Stringer, L. A. (1964). Design for a new orthopsychiatric discipline. *American Journal of Orthopsychiatry* 34: 722–29.

Grob, G. N. (1966). *The state and the mentally ill.* Durham: University of North Carolina Press.

Group for the Advancement of Psychiatry. (1966). *Psychopathological disorders in childhood: Theoretical considerations and a proposed classification.* New York: Group for the Advancement of Psychiatry.

Hall, G. S. (1923). *Life and confessions of a psychologist.* New York: Appleton.

Hamburg, B. A. (1986). Subsets of adolescent mothers: Developmental, biomedical, and psychosocial issues. In J. B. Lancaster & B. A. Hamburg (Eds.), *School-age pregnancy and parenthood: Biosocial dimensions*, pp. 115–45. Hawthorne, NY: Aldine.

Handler, J., & Hollingworth, E. J. (1971). *The "deserving poor": A study of welfare administration.* Chicago: Markham.

Handlin, O. (1951). *The uprooted.* New York: Grosset & Dunlap.

———. (1959). *Immigration as a factor in American history.* Englewood Cliffs, NJ: Prentice-Hall.

———. (1970). Boston's immigrants: The economic adjustment. In P. Kramer and F. L. Holborn (Eds.), *The city in American life.* New York: Capricorn Books.

Hapgood, H. (1967). *The spirit of the ghetto.* Cambridge, MA: Harvard University Press.

Haring, N. G., & Phillips, E. L. (1962). *Educating emotionally disturbed children.* New York: McGraw-Hill.

Harrington, M. (1984). *The new American poverty.* New York: Holt, Rinehart and Winston.

Haslam, M. T. (1989, June). Child sexual abuse—The Cleveland experience. Paper presented at the International Congress of Law & Psychiatry, Jerusalem.

Haugaard, J. J., & Reppucci, N. D. (1988). *The sexual abuse of children.* San Francisco: Jossey-Bass.

Hays, S. P. (1957). *The response to industrialism.* Chicago: University of Chicago Press.

Healy, W. (1915). *The individual delinquent.* Boston: Little, Brown.

Healy, W. (1925). The psychology of the situation: A fundamental for understanding and

treatment of delinquency and crime. In J. Addams (Ed.), *The child, the clinic and the court,* pp. 37–52. New York: New Republic.
Healy, W., & Bronner, A. F. (1915). An outline for institutional education and treatment of young offenders. *Journal of Educational Psychology* 6: 301–16.
———. (1926). *Delinquents and criminals: Their making and unmaking*. New York: Macmillan.
———. (1948). The child guidance clinic: Birth and growth of an idea. In L. G. Lowrey & V. Sloane (Eds.), *Orthopsychiatry, 1923–1948: Retrospect and prospect,* pp. 14–49. New York: American Orthopsychiatric Association.
Healy, W., Bronner, A. F., & Bowers, A. M. (1931). *The structure and meaning of psychoanalysis*. New York: Knopf.
Healy, W., & Healy, M. T. (1926). Pathological lying, accusation, and swindling. *Criminal Science Monograph* 1. Boston: Little, Brown.
Hechler, D. (1988). *The battle and the backlash: The child sex abuse war*. Lexington, MA: Lexington Books.
Heilman, J. D. (1906–7). The need for special classes in the public schools. *Psychological Clinic* 1: 104–14.
Henderson, N. (1910–11). *Visiting teachers: The work of visiting teachers employed by the Public Education Society, 1910–1911*. New York: Public Education Society.
Henry, J. (1965). *Culture against man*. New York: Random House.
Henshaw, S. K., & Silverman, J. (1988, July/August). The characteristics and prior contraceptive use of U.S. abortion patients. *Family Planning Perspectives* 20: 158–68.
Hertz, S. H. (1981). *The welfare mothers movement: A decade of change for poor women?* Washington, DC: University Press of America.
Hoffman, C. W. (1925). Organization of family courts with special reference to the juvenile court. In J. Addams (Ed.), *The child, the clinic and the court,* pp. 255–66. New York: New Republic.
Hofstadter, R. (1955). *Social Darwinism in American thought*. Rev. ed. Boston: Beacon Press.
Holden, A. C. (1922). *The settlement idea: A vision of social justice*. New York: Macmillan.
Hollingshead, A. B. (1949). *Elmtown's youth*. New York: Wiley.
Hollingshead, A. B., & Redlich, F. C. (1958). *Social class and mental illness*. New York: Wiley.
Hopkins, C. K. (1940). *The rise of the social gospel in American Protestantism, 1865–1915*. New Haven, CT: Yale University Press.
Horn, M. (1989). *Before it's too late: The child guidance movement in the United States, 1922–1945*. Philadelphia: Temple University Press.
Horwitz, M. J. (1977). *The transformation of American law, 1780–1860*. Cambridge, MA: Harvard University Press.
Hunter, W. J. (1900). *Manhood wrecked and rescued: How strength, or vigor, is lost, and how it may be restored by self-treatment*. Passaic, NJ: Health-Culture Co.
Interdepartmental Committee on Children and Youth. (1951). *Programs of the federal government affecting children and youth*. Washington, DC: U.S. Government Printing Office.
International Prison Commission. (1904). *Children's courts in the United States: Their origin, development, and results*. Washington, DC: U.S. Government Printing Office.
Jaffe, F. S., Lindheim, B. L., & Lee, P. R. (1981). *Abortion politics: Private morality and public policy*. New York: McGraw-Hill.
James, H. (Ed.). (1926). *The letters of William James*. Vols 1–2. Boston: Little, Brown.

James, W. (1899). *Talks to teachers*. New York: Holt.

Jansson, B. S. (1988). *The reluctant welfare state: A history of American social welfare policies*. Belmont, CA: Wadsworth.

Jarrett, M. C. (1918). Psychiatric social work. *Mental Hygiene* 2: 283–90.

Joffe, C. (1986). *The regulation of sexuality: Experiences of family planning workers*. Philadelphia: Temple University Press.

Johnson, G. J. (1984). Family, law and culture: An overview of parental rights. In E. J. Hunter & D. B. Hunter (Eds.), *Professional ethics and law in the health sciences*, pp. 1–30. Malibar, FL: Robert E. Krieger.

Johnson, H. M. (1916). *The visiting teacher in New York City*. New York: Public Education Association of the City of New York.

Johnson, P. E. (1978). *A shopkeeper's millennium*. New York: Hill & Wang.

Joyce, T. J., & Mocan, N. H. (1990). The impact of legalized abortion on adolescent childbearing in New York City. *American Journal of Public Health* 80: 273–78.

Kadushin, A., & Martin, J. A. (1988). *Child welfare services*. 4th ed. New York: Macmillan.

Kalisch, B. J. (1978). *Child abuse and neglect: An annotated bibliography*. Westport, CT: Greenwood Press.

Kanner, L. (1962). Emotionally disturbed children: A historical review. *Child Development* 33: 97–102.

———. (1964). *A history of the care and study of the mentally retarded*. Springfield, IL: Thomas.

Katz, M. B. (1986). *In the shadow of the poorhouse: A History of welfare in America*. New York: Basic Books.

Kelley, F. (1882). On some changes in the legal status of the child since Blackstone. *The International Review*, August, pp. 83–98. Reprinted in S. N. Katz (Ed.), (1974), *The legal rights of children*, pp. 83–98. New York: Arno Press.

Kempe, C. H. (1978). Sexual abuse, another hidden pediatric problem: The 1977 C. Anderson Aldrich Lecture. *Pediatrics* 62: 382–89.

Kempe, C. H., & Helfer, R. E. (Eds.). (1972). *Helping the battered child and his family*. Philadelphia: Lippincott.

Kempc, C. H., Silverman, F., Steele, B., Droegmueller, W., & Silver, H. (1962). The battered child syndrome. *Journal of the American Medical Association* 181: 17–24.

Kennedy, D. M. (1970). *Birth control in America: The career of Margaret Sanger*. New Haven, CT: Yale University Press.

Kessen, W. (1965). *The child*. New York: Wiley.

Key, E. (1909). *Century of the child*. New York: Putnam.

Klein, V. (1946). *The feminine character*. New York: International Universities Press.

Klerman, L. V. (1986). The economic impact of school-age child rearing. In J. B. Lancaster & B. A. Hamburg (Eds.), *School-age pregnancy and parenthood: Biosocial dimensions*, pp. 361–77. Hawthorne, NY: Aldine.

Konner, M., & Shostak, M. (1986). Adolescent pregnancy and childbearing: An anthropological perspective. In J. B. Lancaster & B. A. Hamburg (Eds.), *School-age pregnancy and parenthood: Biosocial dimensions*, pp. 325–45. Hawthorne, NY: Aldine.

Lancaster, J. B., & Hamburg, B. A. (1986). The biosocial dimensions of school-age pregnancy and parenthood: An introduction. In J. B. Lancaster & B. A. Hamburg (Eds.), *School-age pregnancy and parenthood: Biosocial dimensions*, pp. 3–13. Hawthorne, NY: Aldine.

Lasch, C. (Ed.). (1965). *The social thought of Jane Addams*. Indianapolis: Bobbs-Merrill.

———. (1967). *The new radicalism in American, 1889–1963: The intellectual as a social type*. New York: Vintage Books.

Lathrop, J. C. (1925). The background of the juvenile court in Illinois. In J. Addams (Ed.), *The child, the clinic and the court*, pp. 290–97. New York: New Republic.

Lazoritz, S. (1990). Whatever happened to Mary Ellen? *Child Abuse and Neglect* 14: 143–49.

Lee, P. R., & Kenworthy, M. E. (1929). *Mental hygiene and social work*. New York: Commonwealth Fund.

Levine, A. (1977). Women and work in America. In H. R. Kaplan (Ed.), *American minorities and economic opportunity*, pp. 198–256. Itasca, IL: Peacock.

Levine, A., & Levine, M. (Eds.). (1970). *The Gary Schools* by R. Bourne, and *The Gary Schools* by A. Flexner and F. Bachman. Cambridge, MA: MIT Press.

Levine, D., & Levine, Z. (1967, April). Political activity in the social settlements. Unpublished Science Fair Study report. Hamden, CT.

Levine, M. (1981). *The history and politics of community mental health*. New York: Oxford University Press.

———. (1986). The role of the special master in institutional reform litigation. *Law & Policy* 8: 275–321.

Levine, M., & Battistoni, L. (1991). The corroboration requirement in child sex abuse cases. *Behavioral Sciences and the Law* 9: 3–20.

Levine, M., & Doherty, E. (1991). The Fifth Amendment and therapeutic requirements to admit abuse. *Criminal Justice and Behavior* 18: 98–112.

Levine, M., & Levine, A. (1970). Social change and psychopathology: Some derivations from *Civilization and its discontents*. In G. D. Goldman and D. N. Milman (Eds.), *The contributions of psychoanalysis to community psychology*, pp. 25–45. Springfield, IL: Thomas.

———. (1977). The social context of evaluative research. *Evaluation Quarterly* 4: 515–42.

Levine, M., & Perkins, D. V. (1987). *Principles of community psychology*. New York: Oxford University Press.

Levine, M., & Wishner, J. (1977). The case records of the psychological clinic at the University of Pennsylvania (1896–1961). *Journal of the History of the Behavioral Sciences* 13: 59–66.

Levy, D. M. (1952). Critical evaluation of the present state of child psychiatry. *American Journal of Psychiatry* 108: 481–94.

Lindsey, B. B. (1903). The reformation of juvenile delinquents through the juvenile court. Paper read at the National Conference of Charities and Corrections, 30th annual meeting, Atlanta.

Lindsey, B. B. (1904). *Juvenile court of Denver: The problem of the children and how the state of Colorado cares for them*. Denver: Merchants Publishing.

———. (1925a). Colorado's contribution to the juvenile court. In J. Addams (Ed.), *The child, the clinic and the court*, pp. 274–89. New York: New Republic.

———. (1925b). The juvenile court of the future. *Annual report and proceedings of the National Probation Association*.

Lindsey, B. B., & Borough, R. (1931). *The dangerous life*. New York: Liveright.

Lindsey, B. B., & Evans, W. (1925). *The revolt of modern youth*. New York: Boni & Liveright.

Lindsey, B. B., & O'Higgins, H. J. (1910). *The beast*. New York: Doubleday, Page.

Linn, J. W. (1935). *Jane Addams: A biography*. New York: Appleton–Century.

Loring, E. R. (1920). Some adaptive difficulties found in school children. *Mental Hygiene* 4: 330–63.

Lowrey, L. G. (1948). Orthopsychiatric treatment. In L. G. Lowrey & V. Sloane (Eds.), *Orthopsychiatry, 1923–1948: Retrospect and prospect,* pp. 524–49. New York: American Orthopsychiatric Association.

Lowrey, L. G., & Smith, G. (1933). *The institute for child guidance, 1927–1933*. New York: Commonwealth Fund.

Lubove, R. (1965). *The professional altruist: The emergence of social work as a career, 1880–1930*. Cambridge, MA: Harvard University Press.

———. (1968). *The struggle for social security*. Cambridge, MA: Harvard University Press.

Lukas, A. (1967, September 28). Jewish federation: 50 rewarding years of diverse philanthropy. *New York Times,* pp. 49, 94.

Lundberg, E. C. (1947). *Unto the least of these: Social services for children*. New York: Appleton–Century–Crofts.

Lynd, R. S., & Lynd, H. M. (1929). *Middletown*. New York: Harcourt Brace.

Lynd, S. (1961). Jane Addams and the radical impulse. *Commentary* 32: 54–59.

Machlup, F. (1962). *The production and distribution of knowledge in the United States*. Princeton, NJ: Princeton University Press.

Mack, J. W. (1925). The chancery procedure in the juvenile court. In J. Addams (Ed.), *The child, the clinic and the court,* pp. 310–19. New York: New Republic.

Mason, P. T., Jr. (1972). Child abuse and neglect. Part 1, Historical overview: Legal matrix and social perspectives. *North Carolina Law Review* 50: 293–349.

Massimo, J., & Shore, M. (1963). The effectiveness of a comprehensive vocationally oriented psychotherapy program for adolescent delinquent boys. *American Journal of Orthopsychiatry* 33: 634–43.

May, E. (1964). *The wasted Americans*. New York: Signet Books.

McCrea, R. C. (1910). *The humane movement: A descriptive survey*. New York: Columbia University press.

Mensh, I. N. (1966). *Clinical psychology: Science and profession*. New York: Macmillan.

Meriwether, M. H. (1986). Child abuse reporting laws: Time for a change. *Family Law Quarterly* 20: 141–71.

Miller, K. D., & Miller, E. P. (1962). *The people are the city*. New York: Macmillan.

Miner, M. E. (1916). *Slavery of prostitution*. New York: Macmillan.

Mohr, J. C. (1978). *Abortion in America*. New York: Oxford University Press.

Monroe, P. (Ed.). (1911). *A cyclopedia of education*. New York: Macmillan.

Montessori, M. (1964). *The Montessori method*. New York: Schocken Books.

Moynihan, D. P. 1969). *Maximum feasible misunderstanding: Community action in the war on poverty*. New York: Free Press.

National Association of Visiting Teachers and Home and School Visitors. (1921). *The visiting teacher in the U.S.* New York: Public Education Association of New York.

Needham, G. C. (1887). *Street arabs and gutter snipes*. Boston: D. L. Guernsey.

Neilson, W. A. (1919). The Smith College experiment in training for psychiatric social work. *Mental Hygiene* 3: 59–64.

Nelson, B. J. (1984). *Making an issue of child abuse*. Chicago: University of Chicago Press.

Newman, R. G. (1967). *Psychological consultation in the schools: A catalyst for learning*. New York: Basic Books.

Nicholson, E. B., & Bulkley, J. (Eds.). (1988). *Sexual abuse allegations in custody and visitation cases*. Washington, DC: American Bar Association National Legal Resource Center for Child Advocacy and Protection.

Oberndorf, C. P. (1953). *A history of psychoanalysis in America*. New York: Grune & Stratton.

Oppenheimer, J. J. (1925). *The visiting teacher movement, with special reference to administrative relationships*. New York: Joint Committee on Preventing Delinquency.

Palca, J. (1989). The pill of choice? *Science* 245: 1319–23.

Parker, S. W. (1917–18). Orthogenic cases. No. 12: A study of the interplay of personality. *The Psychological Clinic* 11: 97–111, 129–41, 157–78.

Pastore, N. (1949). *The nature–nurture controversy*. New York: Kings Crown Press.

Paulsen, M. G. (1966). The legal framework for child protection. *Columbia Law Review* 66: 679–717.

———. (1967). Child abuse reporting laws: The shape of the legislation. *Columbia Law Review* 67: 1–49.

Personal Glimpses. (1915, December 25). Why children trust Ben Lindsey. *Literary Digest* 50: 1506–8.

Petchesky, R. P. (1984). *Abortion and woman's choice: The state, sexuality, and reproductive freedom*. New York: Longman.

Peterson, J. (1926). *Early conceptions and tests of intelligence*. Yonkers, NY: World Book.

Pfohl, S. J. (1977). The "discovery" of child abuse. *Social Problems* 24: 310–23.

Phillips, K. (1990). *The politics of rich and poor*. New York: Random House.

Piven, F. F., & Cloward, R. A. (1971). *Regulating the poor: The functions of public welfare*. New York: Pantheon.

———. (1977). *Poor people's movements: Why they succeed, how they fail*. New York: Pantheon.

Platt, A. M. (1969). *The child savers: The invention of delinquency*. Chicago: University of Chicago Press.

Polier, J. W. (1964). *A view from the bench: The juvenile court*. New York: National Council on Crime and Delinquency.

Powdermaker, H. (1966). *Stranger and friend*. New York: Norton.

President's Commission on Mental Health Task Panel on Prevention (1978). Prevention. In *Task Panel Reports Submitted to the President's Commission on Mental Health*. Vol. 4, *Appendix*, pp. 1822–63. Washington, DC: Superintendent of Documents, U.S. Government Printing Office.

Pumphrey, R. E., & Pumphrey, M. W. (Eds.). (1961). *The heritage of American social work*. New York: Columbia University Press.

Radecki, S. E., & Bernstein, G. S. (1989). Use of clinic versus private family planning care by low-income women: Access, cost, and patient satisfaction. *American Journal of Public Health* 79: 692–97.

Raine, W. M. (1907, May). How Judge Lindsey handles his boys. *Ladies Home Journal*.

Redl, F. (1966). *When we deal with children*. New York: Free Press.

Reed, J. (1978). *From private vice to public virtue*. New York: Basic Books.

Report of the City Superintendent of Schools. (1913–14). New York.

Report of the Hartford Vice Commission. (1913). Hartford, CT.

Report of the International Society for the Rescue of Jewish Women and Children. (1927). New York.

Report of the Juvenile Court: City and County of Denver, November 1, 1908–November 1, 1909.

Report of the Juvenile Court: City and County of Denver. November 1, 1909–October 31, 1910.

Report of Juvenile Division of County Court: Arapahoe County, CO. January 1901–July 1902.

Report of the Moral Survey Committee on the Social Evil. (1913). Syracuse, NY.

Report of the U.S. Immigration Commission on the Importation and Harboring of Women for Immoral Purposes. (1909). S. Doc. 196, 23.

Richmond, M. E. (1961). Good spirit and earnestness. In R. E. Pumphrey & M. W. Pumphrey (Eds.), *The heritage of American social work,* pp. 259–68. New York: Columbia University Press.

———. (1965). *Social diagnosis.* New York: Free Press (Originally published 1917).

Ridenour, N. (1948). Mental health education. In L. G. Lowrey & V. Sloane (Eds.), *Orthopsychiatry, 1923–1948: Retrospect and prospect,* pp. 550–71. New York: American Orthopsychiatric Association.

Riding, A. (1990, July 29). Abortion politics are said to hinder use of French pill. *New York Times,* pp. 1, 15.

Riis, J. (1902). *The battle with the slums.* New York: Macmillan.

———. (1917). *How the other half lives.* New York: Scribner.

Riordan, W. L. (1905). *Plunkitt of Tammany Hall.* New York: McClure Phillips.

Roback, A. A. (1961). *History of psychology and psychiatry.* New York: Citadel.

———. (1964). *A history of American psychology.* New and rev. ed. New York: Collier.

Robison, S. M. (1958). A study of delinquency among Jewish children in New York City. In M. Sklare (Ed.), *The Jews: Social patterns of an American group,* pp. 535–41. New York: Free Press.

Rothman, D. J. (1971). *The discovery of the asylum: Social order and disorder in the new republic.* Boston: Little Brown.

Ryerson, E. (1978). *The best laid plans: America's juvenile court experiment.* New York: Hill & Wang.

Sanger, M. (1931). *My fight for birth control.* New York: Holt & Rinehart.

———. (1938). *An autobiography.* New York: Norton.

Sarason, S. B. (1958). *Psychological problems in mental deficiency.* 3rd ed. New York: Harper & Row.

———. (1972). *The creation of settings and the future societies.* San Francisco: Jossey-Bass.

———. (1978). The nature of problem solving in social action. *American Psychologist* 33: 370–80.

———. (1988). *The making of an American psychologist.* San Francisco: Jossey-Bass.

Sarason, S. B., & Doris, J. (1969). *Psychological problems in mental deficiency.* 4th ed. New York: Harper & Row.

———. (1979). *Educational handicap, public policy and social history.* New York: Free Press.

Sarason, S. B., Levine, M., Goldenberg, I. I., Cherlin, D. L., & Bennett, E. M. (1966). *Psychology in community settings: Clinical, educational, vocational, social aspects.* New York: Wiley.

Sarnoff, D. (1964, July). The social impact of computers. Address to the American Bankers Association, National Automation Conference, New York World's Fair.

Sayles, M. B. (1926). *Three problem children: Narratives from the case records of a child guidance clinic.* New York: Joint Committee on Methods of Preventing Delinquency.

Scheff, T. S. (1966). *Being mentally ill: A sociological theory.* New York: Aldine.

Schlesinger, A. M. (1957). *The Age of Roosevelt: The crisis of the old order, 1919–1933.* Boston: Houghton Mifflin.

Schumacher, H. C. (1948). The Cleveland Guidance Center. In L. G. Lowrey & V. Sloane (Eds.), *Orthopsychiatry 1923–1948: Retrospect and prospect,* pp. 377–93. New York: American Orthopsychiatric Association.

Seidl, F. W. (1989). Is Buffalo the birthplace of American social work? *Reporter* 20: 12–13.

Seller, M. (1977). *To seek America: A history of ethnic life in the United States.* Englewood, NJ: Jerome S. Ozer.

Shakow, D. (1948). Clinical psychology: An evaluation. In L. G. Lowrey & V. Sloane (Eds.), *Orthopsychiatry 1923–1948: Retrospect and prospect,* pp. 231–47 New York: American Orthopsychiatric Association.

Smith, T. L. (1914). The development of psychological clinics in the United States. *Pedagogical Seminary* 21: 143–53.

Smuts, R. (1959). *Women and work in America.* New York: Columbia University Press.

Southard, E. E. (1914). Notes on public institutional work in mental prophylaxis. *Journal of the American Medical Association* 63: 1898–1903.

———. (1918). Mental hygiene and social work: Notes on a course in social psychiatry for social workers. *Mental hygiene* 2: 395–406.

Southard, E. E., & Jarrett, M. C. (1922). *The kingdom of evils.* New York: Macmillan.

Spaulding, E., III. (1918). The course in social psychiatry. *Mental Hygiene* 2: 586–89.

Spencer, H. (1873). *The study of sociology.* New York: Appleton.

Steffens, L. (1904). *The shame of the cities.* New York: McClure Phillips.

———. (1909). *Upbuilders.* New York: Doubleday, Page.

———. (1931). *The autobiography of Lincoln Steffens.* New York: Harcourt Brace.

Stein, H. D. (1958). Jewish social work in the United States: 1920–1955. In M. Sklare (Ed.), *The Jews: Social patterns of an American group.* New York: Free Press.

Stevens, P., & Eide, M. (1990, July/August). The first chapter of children's rights. *American Heritage,* pp. 84–91.

Stevenson, G. S. (1944). The development of extramural psychiatry in the United States. *American Journal of Psychiatry* 100: 147–50.

———. (1948). Child guidance and the National Committee for Mental Hygiene. In L. G. Lowrey & V. Sloane (Eds.), *Orthopsychiatry 1923–1948: Retrospect and prospect,* pp. 50–82. New York: American Orthopsychiatric Association.

Stevenson, G. S., & Smith, G. (1934). *Child guidance clinics: A quarter century of development.* New York: Commonwealth Fund.

STOPP NEWS. (1989, January), 5 (1).

STOPP NEWS. (1989, March 10), 5 (3).

Sullivan, H. S. (1964). *The fusion of psychiatry and social science* (Introduction by Helen Swick Perry). New York: Norton.

Summit, R. C. (1988). Hidden victims, hidden pain: Societal avoidance of child sexual abuse. In G. E. Wyatt & G. J. Johnson (Eds.), *Lasting effects of child sexual abuse,* pp. 39–60. Newbury Park, CA: Sage.

Sumner, W. G. (1883). *What social classes owe to each other.* New York: Harper's.

Sussman, A., & Cohen, S. J. (1975). *Reporting child abuse and neglect: Guidelines for legislation.* Cambridge, MA: Ballinger.

Sutherland, E. H., & Cressey, D. R. (1960). *Principles of criminology.* 6th ed. Philadelphia: Lippincott.

Taylor, J. S. (1916). A report on the Gary experiment in New York. *Educational Review,* pp. 8–28.

Theobald, R. (1961). *The challenge of abundance*. New York: Mentor Books.

Thomas, W. I., & Thomas, D. S. (1928). *The child in America*. New York: Knopf.

Tiffin, S. (1982). *In whose best interest? Child welfare reform in the progressive era*. Westport, CT: Greenwood Press.

Tribe, L. H. (1990). *Abortion: The clash of absolutes*. New York: Norton.

Tucker, G. (1855). The progress of the United States in population and wealth. In E. C. Rozwenc (Ed.), *Ideology and power in the age of Jackson*, pp. 3–14. New York: Anchor Books.

United States Children's Bureau, 1912–1972. (1974). Reprinted in *Children and youth: Social problems and social policy*. New York: Arno Press.

Van de Kamp, J. K. (1986). *Report on the Kern County child abuse investigation*. Sacramento: Office of the Attorney General, State of California.

Vandenpol, A. (1982). Dependent children, child custody and the mothers' pensions: The transformation of state–family relations in the early 20th century. *Social Problems* 29: 221–35.

Van Sickle, J. H., Witmer, L., & Ayres, L. P. (1911). Provision for exceptional children in public schools. *U.S. Bureau of Education Bulletin* 14: Whole No. 461.

Vollmer, H. (1906). The juvenile court. Paper presented at a meeting of the Contemporary Club, Davenport, IA.

Wadlington, W. (1984). *Domestic relations*. Mineola, NY: Foundation Press.

Wadlington, W., Whitebread, C. H., & Davis, S. M. (1983). *Cases and materials on children in the legal system*. Mineola, NY: Foundation Press.

Wald, L. (1915). *The home on Henry Street*. New York: Holt.

Wald, M. S., Carlsmith, J. M., & Leiderman, P. H. (1988). *Protecting abused and neglected children*. Stanford, CA: Stanford University Press.

Waller, W. (1932). *The sociology of teaching*. New York: Wiley.

Walsh, J. L., & Elling, R. H. (1968). Professionalism and the poor: Structural effects and professional behavior. *Journal of Health & Social Behavior* 9: 16–28.

Warner, A. G. (1919). *American charities*. 3rd ed. New York: Crowell.

Watson, H., & Levine, M. (1989). Psychotherapy and mandated reporting of child abuse. *American Journal of Orthopsychiatry* 59: 246–56.

Watson, J. B. (1928). *Psychological care of infant and child*. New York: Norton.

Watson, R. I. (1953). A brief history of clinical psychology. *Psychological Bulletin* 50: 321–46.

Weber, M. (1964). *The theory of social and economic organization*. Glencoe, IL: Free Press.

West, G. (1981). *The national welfare rights movement: The social protest of poor women*. New York: Praeger.

Wexler, R. (1990). *Wounded innocents. The real victims of the war against child abuse*. Buffalo, New York: Prometheus Books.

Wheat, P., & Lieber, L. L. (1979). *Hope for the children*. Minneapolis: Winston Press.

White House Conference (1933). Dependent and neglected children. White House Conference on Child Health and Protection. Reprinted in *Children and youth: Social problems and social policy*, pp. 353–89. New York: Arno Press, 1974.

Whittaker, J. K., Kinney, J., Tracy, E. M., & Booth, C. (Eds.). (1990). *Reaching high-risk families*. Hawthorne, NY: Aldine.

Wickman, L. K. (1928). *Children's behavior and teachers' attitudes*. New York: Commonwealth Fund.

Wiens, H. (1984). I was on the orphan train. *The Plain Truth* 49: 31–32, 43.

Wilensky, H. L., & Lebeaux, C. N. (1965). *Industrial society and social welfare*. New York: Free Press.

Wirt, W. A. (1912). *Newer ideals in education: The complete use of the school plant.* Philadelphia: Public Education Association.
Wirtz, W. (1964, February). Address to the annual Citizens Action Commission Meeting, New Haven, CT. *Community Progress* 1.
Witmer, H. L. (1940). *Psychiatric clinics for children.* New York: Commonwealth Fund.
Witmer, H. L., & students (1933). The outcome of treatment in a child guidance clinic. *Smith College Studies in Social Work* 4.
Witmer, L. (1897). The organization of practical work in psychology. *Psychological Review* 4: 116–17.
———. (1902). *Analytical psychology.* Boston: Ginn.
———. (1906–7). Clinical psychology. *The Psychological Clinic* 1: 1–9.
———. (1908–9a). The hospital school. *The Psychological Clinic* 2: 138–46.
———. (1908–9b). Retardation through neglect in children of the rich. *The Psychological Clinic* 2: 157–74.
———. (1908–9c). Retrospect and prospect: An editorial. *The Psychological Clinic* 2: 1–4.
———. (1908–9d). The treatment and cure of a case of mental and moral deficiency. *The Psychological Clinic* 2: 153–79.
———. (1911). *The special class for backward children.* Philadelphia: Psychological Clinic Press.
———. (1915). The exceptional child: At home and in school. In *University lectures delivered by members of the faculty in the free public lecture course, 1913–1914,* pp. 534–55. Philadelphia: University of Pennsylvania Press.
———. (1917–18). Diagnostic education: An education for the fortunate few. *The Psychological Clinic* 11: 69–78.
———. (1919–20). Orthogenic Cases, No. 14. Don: A curable case of arrested development due to a fear psychosis, the result of shock in a three-year-old infant. *The Psychological Clinic* 13: 97–111.
———. (1931). Psychological diagnosis and the psychonomic orientation of analytic science. An epitome. In R. A. Brotemarkle (Ed.), *Clinical psychology: Studies in honor of Lightner Witmer,* pp. 388–409. Philadelphia: University of Pennsylvania Press.
Witte, E. E. (1963). *The development of the Social Security Act.* Madison: University of Wisconsin Press.
Witty, P. A., & Theman, V. (1934). The psycho-educational clinic. *Journal of Applied Psychology* 18: 369–92.
Wolfe, D. A. (1987). *Child abuse: Implications for child development and psychopathology.* Beverly Hills, CA: Stage.
Woods, R. A. (1891). *English social movements.* New York: Scribner.
———. (1898). *The city wilderness: A settlement study.* Boston: Houghton Mifflin.
———. (1906). Social work: A new profession. *Charities* 15: 469–76.
Woods, R. A., & Kennedy, A. J. (1911). *Handbook of settlements.* New York: Russell Sage.
———. (1922). *The settlement horizon: A national estimate.* New York: Russell Sage.

Legal References

Brown v. Board of Education, 347 U.S. 483 (1954).
Carey v. Population Services, 431 U.S. 678 (1977).
Coy v. Iowa, 108 S. Ct. 2798 (1988).

Doe v. Shapiro, 203 F. Supp. 7611 (D. Conn. 1969), appeal dismissed 396 U.S. 488, rehearing denied, 397 U.S. 970 (1970).
Dupler v. City of Portland, 421 F. Supp. 1314 (D. Maine 1976).
Eisenstadt v. Baird, 405 U.S. 438 (1972).
Family Planning and Population Research Act, Title X, Public Health Service Law (1970).
Goldberg v. Kelly, 397 U.S. 254 (1970).
Griswold v. Connecticut, 381 U.S. 479 (1965).
Harris v. McRae, 448 U.S. 297 (1980).
Helvering v. Davis, 301 U.S. 619 (1937).
Hodgson v. Minnesota, 110 S. Ct. 2926 (1990).
In re Gault, 387 U.S. 1 (1967).
Johnson v. Johnson, 564 P. 2d 71, cert. denied 434 U.S. 1048 (1977).
Johnson v. White, 528 F. 2d 1228 (1st Cir. 1975).
King v. Smith, 392 U.S. 309 (1968).
Laws of New York, Ch. 676, Sec. II (1881).
Laws of the State of New York, ch. 205 (1875).
Maryland v. Craig, 110 S. Ct. 3157 (1990).
Matter of Knowack, 158 N.Y. 483 (1899).
Matter of Sharon B., 72 N.Y. 2d 394 (1988).
New York Social Services Law, Sec. 384.
Northeast Women's Center v. McMonagle, 868 F. 2d 1342 (3rd Cir. 1989).
Ohio v. Akron Center for Reproductive Health, 58 LW 4979 (1990).
People of the State of New York ex rel the State Board of Charities v. the New York Society for the Prevention of Cruelty to Children, 161 N.Y. 233 (1900).
Roe v. Wade, 410 U.S. 113 (1973).
Rust v. Sullivan, 59 LW 4451 (May 23, 1991).
Shapiro v. Thompson, 394 U.S. 618 (1969).
Steward Machine Co. v. Davis, 301 U.S. 548 (1937).
United States v. One Package, 86 F. 2d 737 (2d Cir. 1936).
U.S. Department of Agriculture v. Moreno, 413 U.S. 528 (1972).
U.S. Department of Agriculture v. Murry, 413 U.S. 508 (1972).
Watts v. Watts, 77 Misc. 2d 178 (N.Y. Family Court 1973).
Webster v. Reproductive Health Services, 109 S. Ct. 3040 (1989).
Wyman v. James, 400 U.S. 309 (1971).

NAME INDEX

SUBJECT INDEX